Integrated Analytical Systems

Series Editor
Radislav A. Potyrailo
Niskayuna, USA

This comprehensive and interdisciplinary series offers the most recent advances in all key aspects of development and applications of modern instrumentation for chemical and biological analysis on the microscale.

These key aspects will include (1) innovations in sample introduction through micro- and nano-fluidic designs, (2) new types and methods of fabrication of physical transducers and ion detectors, (3) materials for sensors that became available due to the breakthroughs in combinatorial materials science and nanotechnology, and (4) innovative data processing and mining methodologies that provide dramatically reduced rates of false alarms.

Clearly, a true multidisciplinary effort is required to meet objectives for a system with previously unavailable capabilities. This cross-discipline fertilization is driven by the expanding need for chemical and biological detection and monitoring and leads to the creation of instruments with new capabilities for new demanding applications. Indeed, instruments with more sensitivity are required today to analyze ultra-trace levels of environmental pollutants, pathogens in water, and low vapor pressure energetic materials in air. Sensor devices with faster response times are desired to monitor transient in-vivo events and bedside patients. More selective instruments are wanted to analyze specific proteins in vitro and analyze ambient urban or battlefield air. For these and many other applications, new features of modern microanalytical instrumentation are urgently needed. This book series is a primary source of both fundamental and practical information on both the current state of the art and future directions for microanalytical instrumentation technologies. This book series is addressed to the rapidly growing number of active practitioners and developers and those who are interested in starting research in this direction, directors of industrial and government research centers, laboratory supervisors and managers, students and lecturers.

More information about this series at http://www.springer.com/series/7427

Sang-Hyun Oh · Carlos Escobedo
Alexandre G. Brolo
Editors

Miniature Fluidic Devices for Rapid Biological Detection

Springer

Editors
Sang-Hyun Oh
Department of Electrical and Computer Engineering
University of Minnesota
Twin Cities, MN
USA

Alexandre G. Brolo
Department of Chemistry
University of Victoria
Victoria, BC
Canada

Carlos Escobedo
Department of Chemical Engineering
Queens University
Kingston, ON
Canada

ISSN 2196-4475 ISSN 2196-4483 (electronic)
Integrated Analytical Systems
ISBN 978-3-319-64745-6 ISBN 978-3-319-64747-0 (eBook)
https://doi.org/10.1007/978-3-319-64747-0

Library of Congress Control Number: 2017949153

Printed on acid-free paper

This Springer imprint is published by Springer Nature
The registered company is Springer International Publishing AG
The registered company address is: Gewerbestrasse 11, 6330 Cham, Switzerland

Preface

Rapid and sensitive detection of biomolecules and biological particles is vital to gaining insight into the fundamental life processes they enable, as well as discovering rare species linked to the pathogenesis of various disease states. While advances in nanotechnology have helped enable a new generation of miniaturized biosensors, this goal has proven difficult due to fundamental limitations of molecular capture and detection. For surface-based biosensors, detection of analyte is often limited by mass transport. Diffusion into the capture volume is typically a slow and low-yield process, with the problem being compounded for nanoscale sensors. Various schemes have been attempted to accelerate mass transport, such as electrokinetic pre-concentration of charged molecules, dielectrophoresis of neutral polarizable particles and molecules, or evaporation-driven concentration of sample droplets; while other schemes have focused on exploiting the nanometer-scale dimensions of nanostructured biosensors to reduce the transport time, forcing a sample solution through nanopores to shrink the diffusion distance.

In this book, the authors present a variety of strategies to beat the diffusion limit in miniaturized devices by integration of various concentrating or trapping methods to improve throughput and sensitivity. Chapter 1 begins with an overview of nanopore membranes and discusses their applications for separation, sensing, and fluidic control. Chapter 2 further develops this theme and presents nanoporous gold particles for label-free optical biosensing. Chapter 3 presents principles of microfluidic mixers, highlighting the importance of effective analyte delivery to the active sensing surface, which improves the performance of biosensors. Chapter 4 then focuses on transport of fluids, molecules, and nanoparticles through nanopore devices. Chapter 5 discusses active sample manipulation mechanism via dielectrophoresis. While this technique has been widely used for trapping and concentrating biological particles and molecules, recent advances in nanofabrication techniques have enabled dielectrophoresis in extremely scaled nanoelectrodes. This, in turn, allows efficient trapping and manipulation with very low voltages. Chapter 6 focuses on detection and imaging aspects based on lensless holography, which may find novel capabilities for miniaturized and rapid biological detection and imaging. Chapter 7 overviews the unique potential of digital microfluidics for sample preparation and rapid biological detection. Finally, Chapter 8 shows how to put together these principles into a microfluidic cell culture system with multi-tissue interactions that own a high versatility and can be integrated with multi-analyte biosensors. The field of miniaturized biosensors promises exciting further development and we hope that the readers will enjoy the selected topics presented herein.

Twin Cities, MN, USA Sang-Hyun Oh
Kingston, ON, Canada Carlos Escobedo
Victoria, BC, Canada Alexandre G. Brolo

Contents

Nanopore Membranes for Separation and Sensing 1
Gustav Emilsson and Andreas B. Dahlin

Nanoporous Gold Nanoparticles and Arrays for Label-Free Nanoplasmonic Biosensing 25
Camille G. Artur and Wei-Chuan Shih

Microfluidic Mixing for Biosensors 69
N. Scott Lynn Jr.

Self-directed Transport on Nanostructured Plasmonic Sensors 105
Shailabh Kumar

Enhanced Plasmonic Detection with Dielectrophoretic Concentration 123
Avijit Barik and Sang-Hyun Oh

An Introduction to Lensless Digital Holographic Microscopy 147
Nathan C. Lindquist

Biosensing on Digital Microfluidics: From Sample Preparation to Detection 171
Ehsan Samiei and Mina Hoorfar

Miniature Fluidic Microtissue Culturing Device for Rapid Biological Detection 207
Patrick M. Misun, Andreas Hierlemann and Olivier Frey

Index 227

Nanopore Membranes for Separation and Sensing

A "Prosporous" Future

Gustav Emilsson and Andreas B. Dahlin

Abstract During the last 20 years, the use of nanopore membranes to separate molecules depending on their size, charge or other characteristics, have increased in interest. These more ordered and defined nanopores have several advantages compared to traditional ultrafiltration membranes and provide possibilities to combine with, e.g., both electrical and optical sensing schemes. In this chapter, we discuss some of the more common nanopore membranes and how they can be used both for separation and sensing of analytes.

1 Introduction

Filtration and sieving of molecules is a research area which for a long time has attracted attention. The concept of separation is essentially the pursuit for ways to allow certain molecules, with some desired characteristics, to pass through an opening while at the same time preventing molecules, without the desired characteristics, from passing through. These desired characteristics could, e.g., be the size and/or charge of the molecule but also specific interactions between the molecule and a receptor.

Traditional membranes used in, e.g., dialysis and ultrafiltration can be used to separate molecules with a significant difference in size, such as ions or small ligands from macromolecules. Many of the limits with traditional membranes are related to their broad size distribution and high thickness. With the development of new nanopore membrane structures, which are more ordered and defined compared to traditional membranes, it is possible to achieve high ratiometric separation even for molecules with similar molecular mass. During the last 20 years there have been

G. Emilsson · A.B. Dahlin (✉)
Department of Chemistry and Chemical Engineering,
Chalmers University of Technology, 41296 Göteborg, Sweden
e-mail: andreas.dahlin@chalmers.se

G. Emilsson
e-mail: gusemi@chalmers.se

S.-H. Oh et al. (eds.), *Miniature Fluidic Devices for Rapid Biological Detection*,
Integrated Analytical Systems, https://doi.org/10.1007/978-3-319-64747-0_1

several examples of selective separation of proteins [1], enantiomers [2] and other types of molecular sieving [3, 4] using such kind of nanopores. Some of these types of membranes also show material compatibility for combinations with microfluidic techniques for lab-on-a-chip applications [5, 6].

Parallel to the development of advanced membranes for separation of molecules there has also been an increased interest in membrane sensing devices. The future prospects of combining the separation features of membranes with different types of sensors are promising. If the membrane itself acts as the sensing element it provides an efficient platform for delivery of molecules to the sensor area, compared to traditional surface-based sensors, due to the possibility of flow-through delivery [7]. It is also possible to further modify membranes using surface functionalization to realize different types of so called "smart" membranes with tunable gating of molecules. Together, these characteristics provide new and exciting possibilities to create new types of membranes with applications in separation, sensing, and fluidic control.

In this chapter, we discuss the development of different types of nanopore membranes used to separate molecules, how they can be functionalized to create "smart" filters, and also the implementation and use of nanopore membranes in sensing devices. We try to limit ourselves to membranes with relatively defined channels, so that for every entry point there is also a single exit.

2 Hindered Transport

To understand the advantages between different types of membranes we first need to discuss some of the different factors that govern the transport of molecules across membranes. The following geometrical characteristics affect the molecular selectivity and flux through nanopore membranes:

- The size and shape of the pores
- The thickness of the membrane
- The porosity of the membrane.

The diffusion of particles in solution with a size much greater than the solvent, as is the case for, e.g., biomolecules, can be explained using the Stokes–Einstein equation. For a spherical object this is given as

$$D_0 = \frac{k_B T}{6\pi\eta r_m} \tag{1}$$

Here η is the dynamic viscosity of the solution and r_m the radius of the corresponding hard sphere, T is the absolute temperature, and k_B is Boltzmann's constant. However, for a molecule to translocate across a constrained space, like the opening in a nanopore membrane, there are significant hindrances which affect its motion. There are two main effects to take into account to explain the hindrance imposed by the constrained space [8].

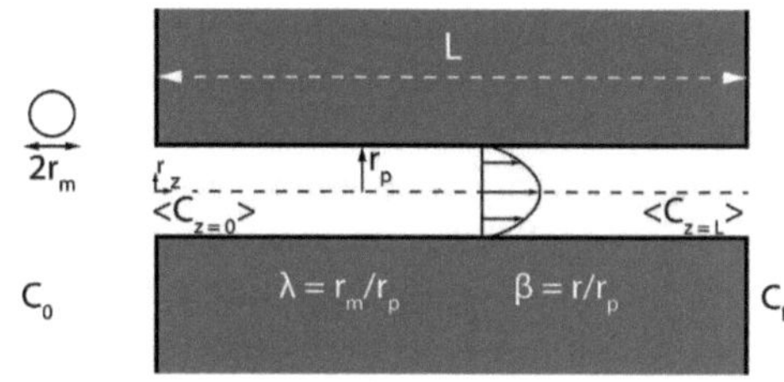

Fig. 1 Illustration of the transport of a spherical molecule across a cylindrical pore

The first one is the probability that the molecule will encounter the pore opening from the bulk solution. For a molecule to enter the constrained space there is a need to account for the fact that the molecule encounters both steric as well as long-range effects, such as electrostatic forces, from the pore which leads to radial variations in the molecule concentration. The partitioning of molecules from the bulk to the pore can be expressed as a partitioning coefficient at equilibrium

$$\phi = \frac{\langle C \rangle_{z=0}}{C_0} = \frac{\langle C \rangle_{z=L}}{C_L} = 2\int_0^{1-\lambda} e^{-E/k_B T} \beta d\beta \tag{2}$$

Describing the average concentration inside the pore opening, $\langle C \rangle_{z=0}$, and exit, $\langle C \rangle_{z=L}$, to that of the bulk solution on each side of the membrane. Where λ is the ratio of the molecule radius to pore radius, r_m/r_p, E is the potential of the long-range interactions, and β the dimensionless radial position, r/r_p (Fig. 1). As long as r_m is comparable to r_p the partition coefficient will not reach unity since the region $\beta > 1-\lambda$ is excluded for the molecules.

The second hindrance effect for molecules moving through the pore stems from the increased hydrodynamic hindrance, induced by the pore wall, which causes increased drag on the molecules within; and thereby slowing them down. The hydrodynamic hindrance depends on both λ and β. The average flux over the pore cross section, $\langle N \rangle$, taking into account both the steric and hydrodynamic hindrance, can be shown to follow [8]:

$$\langle N \rangle_{Diffusion} = \frac{H_D D_0}{L}(C_0 - C_L) \tag{3}$$

$$\langle N \rangle_{Convection} = H_C C_0 \langle V \rangle \tag{4}$$

The two expressions describe the flux when the transport is dominated by either diffusion (Eq. 3) or convection (Eq. 4) with an average fluid velocity $\langle V \rangle$.

The dimensionless coefficients describing the hindrance effects, H_D and H_C, has been shown to be, for most purposes, reasonably estimated using the so-called "centreline approximation" ($\beta = 0$) [8]. For a neutral molecule not influenced by the long-range interactions of the wall ($E = 0$) the values for the coefficients have been evaluated numerically [9]. There also exist analytical expressions, for the coefficients which show good agreement with experimental data [10–12]. All values range from zero, meaning no transport through the pore, to unity, meaning the transport is just like in a bulk environment. Care should be taken if the interactions between the molecule and the pore wall are attractive, as the centreline approximation is likely not going to be valid. More recent models average the cross-sectional diffusion instead of employing the centreline approximation [13]. These methods improve the fitting to experimental data compared to the centreline approximation, which for cylindrical pores has a root-mean-square error of 20% for diffusive hindrance and 6% for convective hindrance [13].

To express the flux of molecules across the membrane, as is commonly used as a benchmark of membrane performance, the two expressions for the average flux (Eqs. 3 and 4) is simply multiplied by the membrane porosity per surface area. The change in the hindrance factors, calculated using the analytical expression by Bungay and Brenner [10], for different values of λ is illustrated in Fig. 2. As can be seen from the figure, the diffusion dominated regime is to a higher degree dependant on the value of λ compared to the convection dominated case. From Eq. 3 we can also note that the rate of transport across the membrane is inversely proportional to the thickness of the membrane, according to Fick's first law (Fig. 2).

The models discussed so far were developed and experimentally verified for membranes with a thickness on the micrometer scale. When the membrane thickness approaches the size of the pore dimension it is not certain that the models can be used to fully describe the transport, but they can be used to get a sense of which factors are important. One example of simulating the transport across thinner membranes was performed by Snyder et al. Their results indicate that thin membranes reach equilibrium faster than thick membranes and will at any time before equilibrium is reached, for both membranes, have a higher resolution of separation

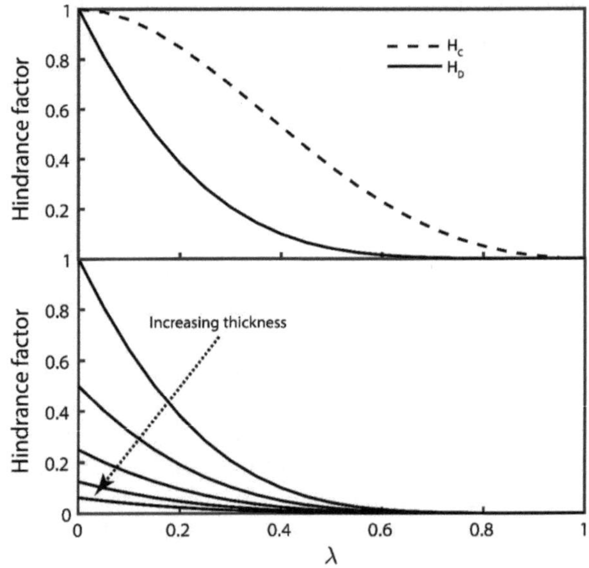

Fig. 2 (Top) Hindrance factors for diffusion or convection through a cylindrical pore calculated using the analytical expression derived by Bungay and Brenner [10]. The diffusion limited regime is more sensitive to the ratio of the molecule radius to pore radius, λ, compared to in the convection dominated regime. (Bottom) The effect of increasing the membrane thickness, with a factor two for each curve, on the diffusion dominated hindrance factor

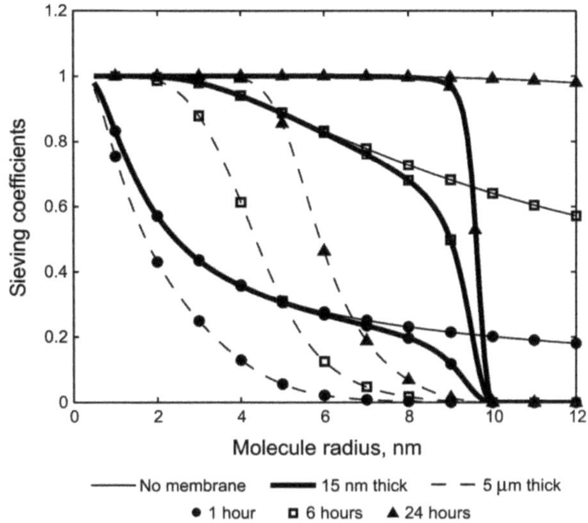

Fig. 3 Effect of the sieving coefficient, which is the ratio of the diffusion compared to free diffusion (no membrane present), for 10 nm diameter nanopores with different membrane thickness. The thin membranes have a more defined cut-off and reach equilibrium (where the concentration is the same on both sides of the membrane) faster than the thick counterparts. Reprinted with permission from [14]. Copyright 2011, Elsevier

(Fig. 3) [14]. This is explained by the fact that thicker membranes slows down the diffusion to a greater extent because of the increased distance the molecule has to travel in the constrained space, even for molecules which are considerably smaller than the opening, thereby lowering the resolution [14].

3 Different Types of Nanopore Membranes

There are several different types of membranes which can be fabricated with pore dimensions on the nanoscale. The development of straight nanopores with good uniformity started in the 1970s with track-etched membranes. Since then, a wide variety of membranes containing pores with diameters ranging from ~1–200 of nm have been developed. Other membranes which are commonly used include those fabricated by: anodization of aluminum, so called aluminum anodic oxide (AAO), template-synthesized membranes which use other membranes as a scaffold for, i.e., electroless plating, and the last group of membranes which will be addressed in this chapter are those fabricated from different types of silicon; mostly via micro-and nanofabrication techniques

3.1 *Track-Etched Membranes*

One of the first examples of fabrication of membranes containing pores with controlled size on the nanoscale arrived with the development of track-etched membranes [15–17]. These membranes allowed scientists to, for the first time, test the

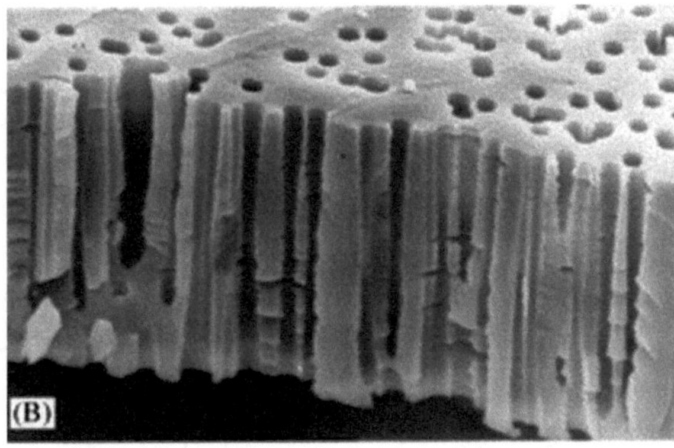

Fig. 4 Track-etched membrane with pores having a slight conical profile. Reprinted with permission from [18]. Copyright 2001, Elsevier

theories developed for hindered flow at the nanoscale; resulting in the earlier models described in the previous section.

Conventionally, track-etched membranes are made by first exposing an insulating material, such as a sheet of mica or poly carbonate (PC), to massive fission fragments from a radioactive source [16, 17] or to ion beams [18] (Fig. 4). As a result damage tracks will be generated in the insulating material, which can later be etched using an appropriate etchant for the material [16]. The etching will cause the damaged regions to develop into pores, since the etching rate is higher in these regions. Different etching strategies may be used to create pores with different profiles, e.g., conical [19]. Since the irradiation of the membrane is a random process there is a chance that several damage tracks will occur close to each other, causing the pores to overlap, especially for higher porosity samples [17].

One benefit of track-etched membranes is the versatility of the fabrication process, enabling a wide variety of pore densities and diameters to be manufactured. The pore diameter can be varied from 10 nm to tens of micrometers and the density from single pores to 10^{10} cm^{-2} [18]. The fabrication process is relatively inexpensive and allows for scalability [20]. Membranes with different pore sizes are widely commercially available.

The drawback of track-etched membranes is that the thickness of the membranes is on the order of tens of micrometers. As discussed this will, compared to thinner membranes, limit the resolution and can also increase the loss of molecules due to the increased surface area. The fact that the membranes are made from plastic such as PC or PET also makes integration with microfluidic systems more challenging, as the membranes cannot directly be bonded to common microfluidic materials such as polydimethylsiloxane (PDMS).

3.2 Aluminum Anodic Oxide Membranes

The electrochemical anodization of aluminum has for a long time been used as a way to change both the appearance and corrosion resistance of aluminum [21]. It was later discovered and confirmed that the anodized aluminum consisted of close packed, close to hexagonally shaped, cells of oxide containing a single approximately cylindrical pore in each cell [22, 23]. However, the fabricated structures were far from the theoretical perfectly hexagonal structure.

With the discovery of certain long-range order regimes in the anodization process, by Masuda and Fukuda [24], substrates with considerable higher degree of long-range order, as described theoretically, could be produced (Fig. 5). In order to obtain this highly ordered structure the anodization needs to be precisely controlled, as the ordering only occurs under certain conditions. It was also shown that increasing the anodization time increases the defect-free areas [25, 26]. The production of AAO membranes using the method developed by Masuda and Fukuda in short contains the following steps: (1) anodization in sulphuric acid, oxalic acid, or phosphoric acid at 25 V, 40 V, or 195 V respectively, (2) removal of the aluminum substrate and (3) bottom (barrier) layer removal and pore widening via etching [24, 26]. The method results in thick (~100 μm) membranes as a result of the long anodization time, but further development showed that the fabrication of thinner membranes with high order can be achieved using a two-step process [27], or via pre-texturing of the aluminium substrate [28, 29]. More recent advances in the fabrication has also contributed to new structures with tailored properties [25, 30, 31].

Some of the drawbacks with AAO are the insufficient chemical and thermal stability as well as low mechanical strength [24]. Another parameter to take into account is, like for track-etched membranes, the thickness of the membrane which is on the micrometer scale. Thinner membranes can be fabricated, but they are often quite fragile since the material is rigid compared to membranes made of polymers [32].

3.3 Template-Synthesized Membranes

Another way to create membranes for separation purposes was pioneered by the group of Martin [3, 34]. The method is based on using other membranes, such as PC membranes created via track-etching or AAO membranes, as a template material for creating so-called nanotubule membranes in gold using electrochemical plating or, more commonly, electroless plating methods [4]. Since the nanotubules are synthesized within the pores of other membranes the porosity is dependent on the template used. The electroless deposition process, in short, starts with the application of a catalyst to all surfaces to be coated. Following this the membrane is immersed in the electroless plating solution containing Au(I) and a reducing agent. The reduction of Au(I) to metallic Au will occur in the presence of the catalyst, forming a gold film which covers the membrane surface [4, 35]. The diameter of the nanotubules can be controlled by varying the plating time, allowing nanotubule diameters to be tuned down to <1 nm [3]. Membranes with other types of nanotubules materials may also be fabricated. It is, for example, possible to create silica nanotubes using sol-gel chemistry [2].

Fig. 5 AAO membranes formed using different parameters. The ordering occurs under specific regimes and the long-range order increases with time. Adapted from [33]. Copyright Journal of the Electrochemical Society 1997

3.4 Silicon and Silicon Nitride Membranes

The last material commonly used for creating membranes is the wide range of membranes fabricated in silicon or silicon nitride using micro- and nanofabrication. Compared to the previously discussed membranes it is possible to reduce the membrane thickness to less than 100 nm while retaining good mechanical stability [36]. These types of membranes can be fabricated using a wide range (Fig. 6) of principles and we will discuss some of them.

The fabrication of membranes in silicon most often starts with the deposition of the membrane material, often silicon nitride (SiN), on top of a silicon wafer. Second, a "TEM window" is created on the wafer using dry etching in combination with UV-lithography followed by anisotropic wet etching, creating a thin free-standing membrane [37]. In order to define and create the pores in the membrane a wide variety of options may be used. One is to use focused ion beam milling [36, 38, 39] to draw the pores in the free-standing membrane, or track-etching applied to silicon nitride [40]. Another method is to use some sort of lithographic technique such as: electron beam lithography [41–43], nano-imprint lithography [44], colloidal lithography [37, 45], laser interference lithography [43, 46, 47], or deep ultraviolet lithography [7]. All of the methods can produce pore diameters down to at least

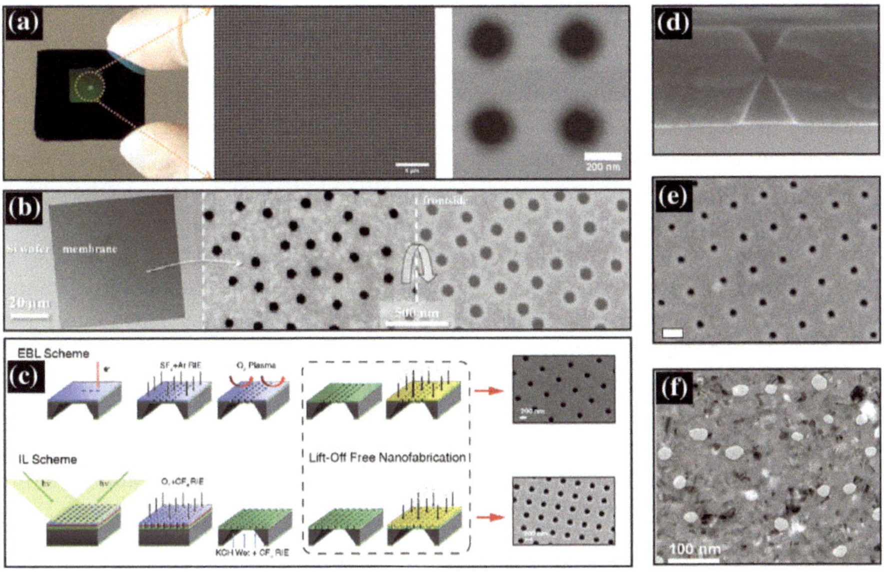

Fig. 6 Silicon membranes fabricated with different techniques. **a** Template stripping, adapted with permission from [44]. Copyright 2014, American Chemical Society. **b** Colloidal lithography, adapted with permission from [45]. Copyright 2014, Wiley. **c** Electron beam or interference lithography, adapted from [43]. Copyright 2011, National Academy of Sciences. **d** Track-etched (etched from both sides), adapted from [40]. Copyright 2009, National Academy of Sciences. **e** Focused ion beam (scale bar 50 nm), reprinted with permission from [36]. Copyright 2004, American Chemical Society. **f** pnc-Si (pores shown as white dots), adapted from [14]. Copyright 2011, Elsevier

200 nm, with some being able to produce sub 100 nm diameters. The need to fabricate multiple pores, within a reasonable time frame, limits the use of serial ion beam milling and electron beam lithography to smaller areas (μm^2), while several of the other methods, e.g., colloidal lithography, can be used to produce pores over large areas (cm^2) without increasing the fabrication time.

A somewhat different approach, compared to the methods described so far, is to create the membrane out of porous nanocrystalline silicon (pnc-Si) [48]. Membranes formed using the method described by Striemer et al. [48]. can be as thin as 15 nm, while still retaining good mechanical stability. The average pore size, which is formed via rapid thermal annealing of amorphous silicon, can be tuned between 7 and 22 nm using different annealing temperatures.

Even if silicon membranes have many benefits the continued fabrication of thin and robust membranes, with a reasonable cost, it is essential to increase the use of such membranes also outside of academia.

4 Functionalized Membranes

In order to improve or change the properties of membranes to achieve additional features, in addition to their size exclusion properties, it is essential that some surface functionalization is made post-production. This functionalization could be used in order to reduce the amount of non-specific binding of proteins to the surface [49], since membrane fouling can greatly reduce the performance of membranes and is a prominent problem in bioseparation processes [1, 50–52]. It could also be used as a way to add specificity in the transport across the membrane, facilitating increased transport of the molecule of interest [2, 53]. Other intriguing possibilities is to use functionalization in order to create "smart" membranes [54]. The term "smart" refers to materials which can, due to some external stimuli such as, e.g., pH or temperature, change their properties. This can be exploited in order to construct membranes which react to the stimuli, by opening and closing, and thereby regulating the passage of molecules. The reduction of non-specific binding and the "smart" membranes can be achieved using different polymerization strategies.

4.1 *Polymer Functionalization*

To improve and expand the use of membranes, polymers are perfect and commonly employed. The wide range of different polymers makes it possible to tailor the membranes for a variety of different uses, from pH responsive openings to protein fouling prevention.

There are different methods employed in polymer functionalization. A common approach is to use so-called "grafting to", where the polymer in question contains a material specific group, such as a thiol or silane, which will bind to the surface [55]. If a sufficient density of polymers can be attached the coils will start to overlap and stretch away from the surface, forming a polymer brush. The high-density regime can be challenging to reach using grafting to due to the self-limiting nature of the reaction, as attached chains hinder new ones from reaching the surface. These limitations can be reduced by, e.g., grafting in a "poor" solvent [49], thus reducing the polymer coil size in solution, or through the use of a "grafting from" approach [56, 57]. In grafting from the polymer is synthesized from the surface, starting from some sort of self-assembled layer of initiator molecules. The reaction may be carried out using different polymerization schemes, e.g., atom-transfer radical-polymerization [57], and can generally produce higher grafting densities compared to the grafting to approach. Other ways to functionalize surfaces with polymers include spin- or dip-coating [58] and layer-by-layer deposition of polyelectrolytes [59]. There are also several different types of polymer gels which can be used [60].

4.1.1 Smart Membranes

During the last decade there has been substantial research in constructing different types of functionalized pores which can be used to alter ionic transport [61–69] and water permeation due to some external stimuli [70–73]. These devices are achieved using different types of responsive materials which react to external stimuli such as: pH [61, 64, 69–71], temperature [66, 72], redox state [67, 73], voltage [68], and light [74, 75].

There have also been efforts in trying to achieve similar controlled membranes but for larger molecules [76]. Trying to mimic the gating of the nuclear pore complex [53] Caspi et al. showed that a PNIPAM functionalized membrane could be used as a receptor-mediated selective diffusion barrier, transporting PNIPAM modified DNA with a higher flux than unlabelled DNA, despite being a larger complex.

Thermo responsive polymers have also been used as a means to switch the adhesion of proteins and cells to surfaces [77–79]. This could potentially be used as a controlled means to capture molecules inside membranes or as a self-cleaning mechanism [80].

5 Biomolecule Separation Using Nanopore Membranes

5.1 Separation Using Template-Synthesized Membranes

Traditional ultrafiltration membranes can only separate molecules with an order of magnitude difference in size efficiently (according to the manufacturer). This limits their applications to primarily concentration of proteins and removal of molecules significantly smaller than proteins, such as ions and ligands. The poor resolution of traditional membranes, compared to nanopore membranes, can be attributed to their broad pore size distribution and long tortuous pore geometry. During the last 15 years, several types of nanopore membranes have been presented which can separate molecules with higher resolution compared to traditional alternatives.

Using DNA-functionalized gold nanotubule membranes Kohli et al. demonstrated that it is possible to create a membrane which recognizes a complementary DNA strand and facilitates its diffusion through the membrane compared to non-complementary strands [81]. The system showed a higher flux for the complementary strand even when compared to single-base mismatched strands (Fig. 7).

Lee et al. showed using a similar membrane system, composed of silica-coated AAO membranes, that enantiomeric separation of the drug 4-[3-(4-fluorophenyl)-2-hydroxy-1-[1, 2, 4]triazol-1-yl-propyl]-benzonitrile was possible [2]. The separation was achieved by the use of an antibody specific for one of the enantiomers and by tuning the binding affinity of the antibody using dimethyl sulfoxide.

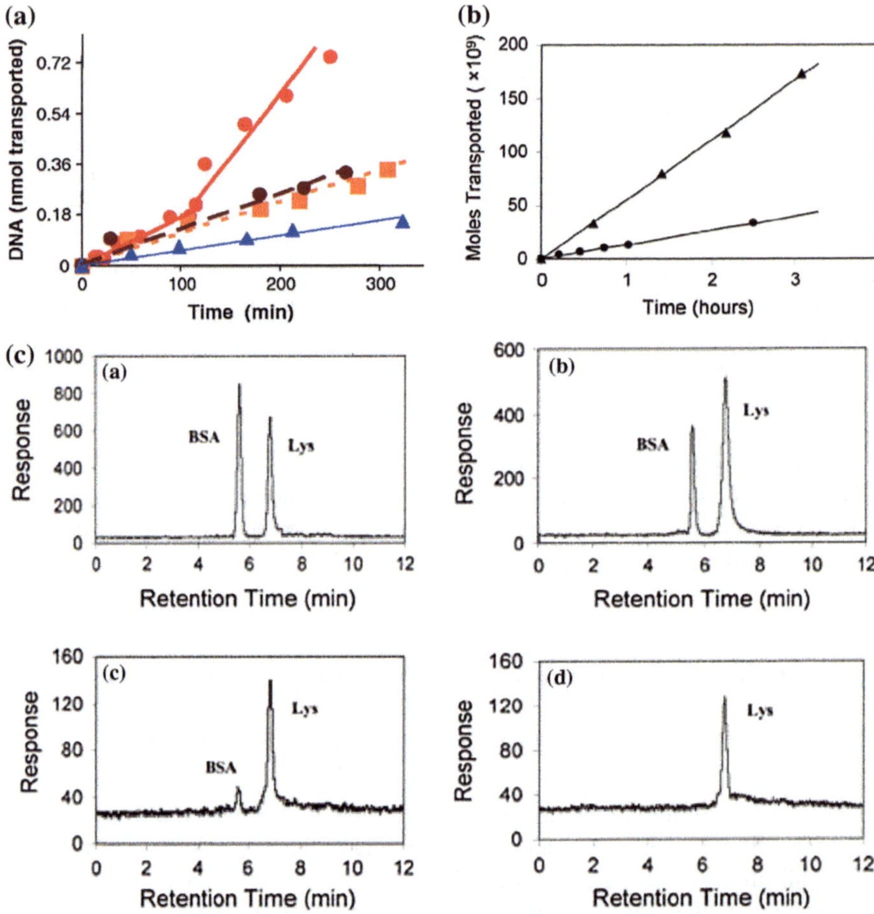

Fig. 7 **a** Transport of DNA across DNA-functionalized gold nanotubule membranes. The flux of the complementary strand (red circles) is increased compared to: single-base mismatch at the end of the strand (brown circles), seven-mismatch (blue triangles) and single-mismatch in the middle of the strand (orange squares). Reprinted with permission from [81]. Copyright 2004, American Association for the Advancement of Science. **b** Moles transported of lysozyme (triangles) and bovine serum albumin (circles) across PEG coated gold nanotubule membranes with a diameter of 40 nm. Reprinted with permission from [1]. Copyright 2001, American Chemical Society. **c** HPLC data for the separation of a mixture of BSA and lysozyme for PEG coated gold nanotubule pores with a diameter of: 45 nm (top right), 30 nm (bottom left) or 20 nm (bottom right) compared to the feed solution (top left). Reprinted with permission from [1]. Copyright 2001, American Chemical Society

Another example of separation using template-synthesized membranes, but with pore diameters ranging from 20 to 40 nm, in combination with PEG functionalization has also been demonstrated as a way to separate proteins depending on their size by Yu et al. [1].The authors achieved a selectivity ratio of more than 15 times when comparing the separation of bovine serum albumin (BSA, 66 kDa) to β-lactoglobulin A (36 kDa) and lysozyme (14 kDa) (Fig. 7). In addition, the PEG-coating increased the lifetime of the membranes, from hours to days, due to reduced non-specific binding and clogging of pores. In combination with trans-membrane electrophoresis it is possible to achieve even higher ratios with the correct combination of solution pH and protein pI [82]. It should be kept in mind that the pH and pI will play a role even without the use of electrophoresis. Highly selective transport of proteins with different pI is possible across charged membranes depending on the pH used, since if the protein and membrane have the same charge this will increase the electrostatic exclusion component. One example is the pH dependant separation of BSA from bovine hemoglobin, which has very similar molecular weight, but different pI [83–85].

5.2 *Separation Using Thin Silicon Membranes*

Due to their thickness and sponge-like structure traditional membranes for ultra-filtration and dialysis provide poor size cut-off properties, as well as a greater loss of analyte and transport rate [36, 48, 86]. With the development of thinner nanopore membranes, predominantly in silicon materials, higher resolution separation can be performed in a shorter timeframe. This is of importance not only for the use of such membranes in protein fractioning [87] but also for the development of devices combining sensing and separation of molecules.

Research on separation of proteins using silicon based membranes has mostly been conducted on pnc-Si membranes [48]. The membrane thickness in this case (15 nm) is far less than that of track-etched membranes (~6 μm) providing, as discussed earlier, a higher transport rate and improved cut-off resolution. Compared to a traditional dialysis membrane a nine times higher transport rate of small molecules (Alexa 546) was achieved (Fig. 8). The rate limiting step in the case of the thin membranes was, compared to thicker membranes, not the diffusion across the membrane but rather the diffusion to the membrane in the bulk solution [14, 48].

Other examples of protein separation using silicon based devices include the use of track-etched silicon nitride membranes (Fig. 8) [40]. Albeit thicker than the pnc-Si membranes the method used provides the possibility to tune the porosity from a single pore to highly porous and the controlled fabrication of different pore shapes is also possible.

It should be kept in mind that increased selectivity, by reducing the nanopore dimension, comes at the cost of reduced flux and throughput. This raises the question if the increased selectivity is worth the loss in flux? As a way to purify molecules this might be the case, however if used as a means to improve the performance of sensors by increasing selectivity the reduced flux might be less of an issue [88]. Decreasing the membrane thickness is also an option to overcome such hinders as this will lead to an increased flux.

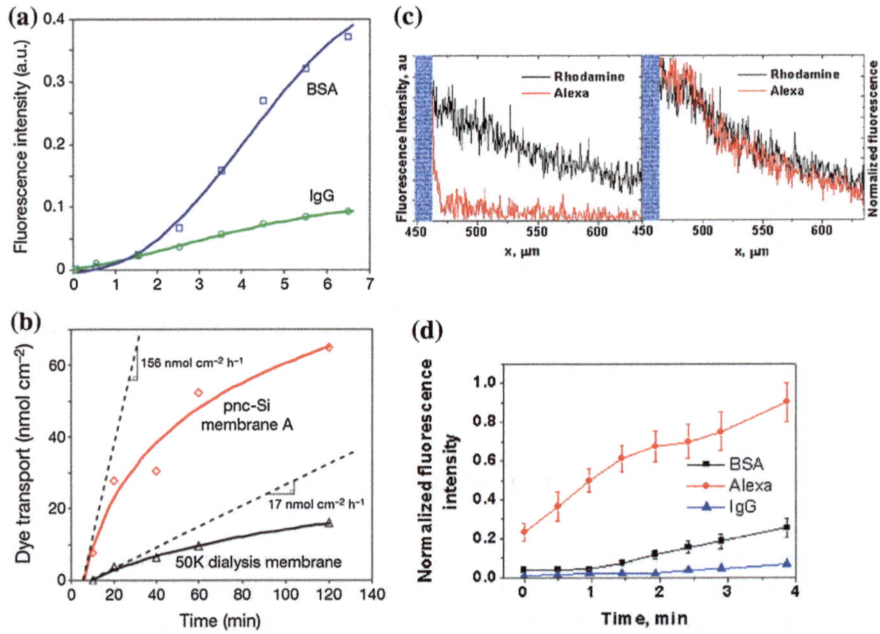

Fig. 8 **a** Fluorescence measurements of the transport of BSA and IgG across 15 nm thick pnc-Si membranes. **b** Comparison for the transport rate of small molecules (Alexa 546) across pnc-Si and dialysis membranes, showing a 9 times higher rate for the thin pnc-Si membrane. **c** Transport of charged molecules across track-etched SiN membranes without (left) or with (right) modification with a aminosilane. The modification makes the membrane less negatively charged and thus increases the passage of negatively charged molecules (Alexa 568, red line) across it. Adapted from [40]. Copyright 2009, National Academy of Sciences. **d** Transport of proteins across the track-etched SiN membranes. Adapted from [40]. Copyright 2009, National Academy of Sciences

6 Membranes Used as Biosensors

Membranes can not only be used to separate molecules but can also be combined with, or used as, sensors to detect molecular binding or translocation. Depending on the pore size the sensing principle will differ. Pores with a diameter greater than 60 nm can generally be used as optical sensors, if coated with a metal layer to support the excitation of surface plasmons, while smaller pores are more suited for electrical techniques measuring changes in the ionic conductivity as molecules enter the pore [89].

6.1 Electrical Techniques

If the dimension of the pores is small enough (a few nm), as in the case of template-synthesized membranes, the membranes may be used to detect a drop in

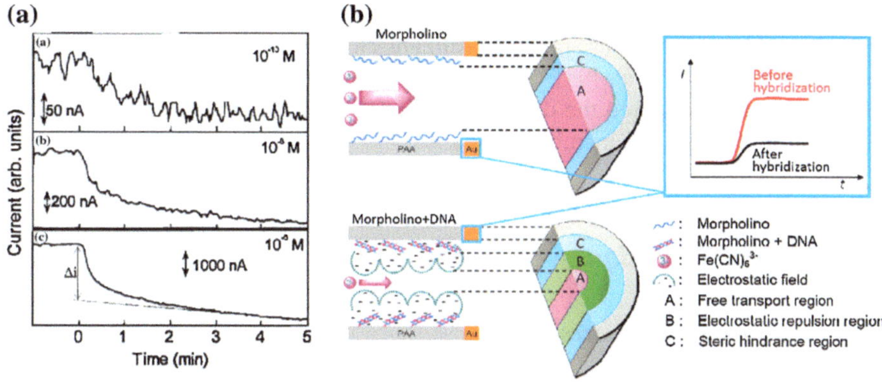

Fig. 9 **a** Transmembrane conductivity measurements as a method to detect analytes using ~2.8 nm gold nanotubules. The partitioning of analyte ($Ru(bpy)_3^{2+}$) into the pore opening lowers the conductivity through the membrane. The magnitude of the current drop across the membrane can therefore be related to the concentration of analyte using a calibration curve. Reprinted with permission from [90]. Copyright 1999, American Chemical Society. **b** A gold coated AAO membrane functionalized with a charge neutral synthetic DNA analogue (Morpholino) used as a sensor for DNA. When DNA hybridizes with the synthetic DNA analogue the membrane charge becomes more negative, because of the DNA, which increases the hindrance of the diffusion of $Fe(CN)_6^{3-}$ ions across the membrane, resulting in a reduced current. Reprinted with permission from [94]. Copyright 2010, American Chemical Society

transmembrane current when molecules, with a size similar to the pore diameter, partition into the pores [89]. The membrane in question is positioned between two chambers containing salt solution and the transmembrane current, when a constant transmembrane potential is applied between electrodes in the two compartments, is measured [90]. This should not be confused with single nanopore sensors, where the blockage causes a distinct spike in current, as the membrane consists of several pores and therefore is not a stochastic sensor but instead shows a constant current drop proportional to the concentration of the analyte (Fig. 9). To determine the concentration a series of calibration curves are first needed in order to relate the current drop to the concentration of analyte. Even if the method described can be used to detect low amounts of analyte using model compounds, its performance is worse when used on mixtures of different types of molecules. As such, the analysis of a specific analyte in a mixture of molecules would be challenging without any additional modification [91]. This, however, is something which remains true for most types of label-free sensor devices [92].

A different type of membrane sensor using electrical impedance as the signal was used to detect DNA by utilizing the increased electrostatic repulsion from surface charges once the DNA hybridizes [93]. In order to achieve this, the insides of the gold coated AAO membrane was functionalized with Morpholinos, which is a synthetic analogue of DNA where the charged phosphate groups have been replaced with neutral phosphorodiamidate, that recognizes the DNA strand of interest. The hybridization of DNA to the membrane increases the negative charge of the walls and changes the conductance through the membrane which is detected in the impedance. A similar approach utilizing the Morpholino-DNA hybridization was employed by Li et al. [94]. The difference was that the authors measured how the hybridization hindered the diffusion of $Fe(CN)_6^{3-}$ ions and thereby reduced the current flow through the membrane (Fig. 9).

6.2 *Optical Techniques*

6.2.1 Plasmonic Nanopores

One type of optical sensor principle which can be utilized in membranes is plasmonic nanopore array sensors [95]. Such membranes are fabricated by any of the different types of lithographic methods described previously for silicon, most often consisting of gold coated silicon nitride (with some exceptions [96]), and will support the excitation of surface plasmons via grating-type coupling when the pores are arranged in a periodical manner [95]. Plasmonic sensors are refractometric sensors, meaning that changes in refractive index, close to the surface as the electromagnetic field is confined, will generate a response in the form of a shift in the surface plasmon resonance position (Fig. 10). This shift will be proportional to the surface coverage of the molecules binding to the surface and enables label-free and real-time sensing.

One of the greatest benefits with using plasmonic nanopores, compared to traditional plasmonic sensors, is the possibility to have the analyte flow-through the sensing region as opposed to flow-over [7, 41, 97]. The flow-through delivery of analyte to the sensor can for some conditions greatly improve the response time of the sensor (Fig. 10). The principle of flow-through sensing with plasmonic nanopores has been explored by the groups of Sinton [38], Höök [37] and Altug [7]. The principle is also true for other types of porous membrane sensors [98].

In general membranes utilizing plasmonic sensing will be limited to nanopore diameters larger than 60 nm, since the coupling to plasmon modes is less efficient for small holes [99]. This can be compared to the sensors utilizing a change in transmembrane conductivity where smaller pores are preferable, and the use of pores with a diameter less than 10 nm is common. In order for plasmonic nanopore membranes to function both as sensors and filters the inside of the pores would need to be functionalized further with, e.g., a polymer interior [1] to enable protein separation, or for molecular separation some sort of nanocomposite could be used [100]. Another approach would be to combine a filter with a plasmonic sensing element, such a device could make use of plasmonic nanopores as the sensing element but traditional flow-over designs could also be used. Such a combined filter and sensing device was demonstrated by Breault-Turcot et al. [101]. by integrating a microporous membrane together with a traditional SPR sensor, enabling analysis in crude blood samples.

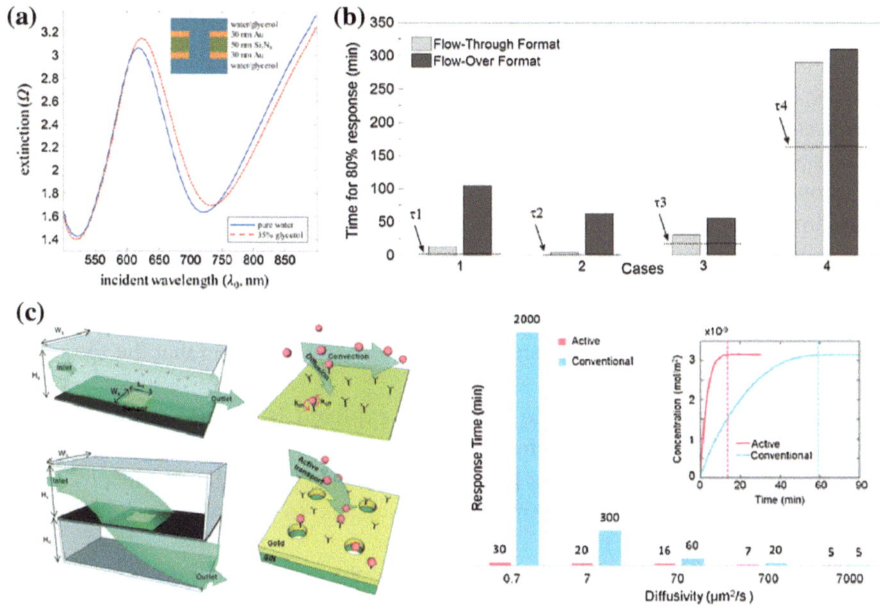

Fig. 10 **a** Spectra for a plasmonic nanopore membrane, consisting of two 30 nm gold films separated by a 50 nm SiN layer fabricated using colloidal lithography, in water (blue) and 35% glycerol (red). Reprinted with permission from [45]. Copyright 2014, Wiley. **b** The effect of flow-through versus flow-over for different reaction parameters. The parameter τ is the characteristic timescale of the binding reaction (defined as the inverse of $k_{on}C + k_{off}$). For long characteristic timescale the effect from the flow-through format is obscured (case 3 and 4). Reproduced with permission from [97]. Copyright 2010, American Chemical Society. **c** Calculated response times for flow-through (active, red) versus flow-over (conventional, blue) depending on the diffusivity of the analyte. Adapted with permission from [7]. Copyright 2013, Royal Society of Chemistry

6.2.2 Other Optical Measurements

The optical detection of transported ionic molecules was utilized by Jágerszki et al. as a means to detect DNA [102]. The template-synthesized gold nanotubule membranes used were functionalized with a peptide nucleic acid which hybridizes with the complementary DNA and increase the charge of the membrane, in a similar fashion as the Morpholino functionalization discussed previously. The increased charge of the membrane hinders the transport of the charged dye bromocresol green, which concentration can be determined via absorption measurements.

Other approaches have been to use commercially available AAO membranes as a support for hybridization assays. By immobilizing upconversion nanoparticles in the membrane lower amounts of Ebola virus oligo coated gold nanoparticles, which causes emission quenching upon hybridization, could be detected [103].

7 Conclusion and Outlook

The use of nanopore membranes for separation, sensing and fluidic control is a field which still has many aspects to explore. In this chapter we have discussed some aspects and examples of the use of such membranes in the different areas. There are still, however, many interesting applications which we have not touched upon. One is the implementation of nanopore membranes with microfluidic devices, which is an area that is of important especially for the use of membranes as sensing devices. The challenge is, in other words, to integrate horizontal microchannels with vertical nanochannels, thereby approaching 3D lab-on-a-chip devices. There exist several successful examples of such devices [104–107] and we envision that the future will bring even more.

Another interesting concept to be further explored is the use of metal-coated membranes as dielectrophoretic trapping devices [108], or as a means in order to achieve concentration of analytes [39]. There are also interesting applications with the implementation of electroosmosis as a means to control the flow-through membrane devices [109, 110].

Finally, we believe that further exploration of the use of polymers to modify and control the passage of molecules through membrane devices is essential to utilize larger pores to separate molecules. However, much still remains to be understood of the behavior of polymers in confined environments in order to truly mimic the nanopores found throughout biology [111].

References

1. Yu S, Lee SB, Kang M, Martin CR (2001) Size-based protein separations in poly(ethylene glycol)-derivatized gold nanotubule membranes. Nano Lett 1(9):495–498
2. Lee SB, Mitchell DT, Trofin L, Nevanen TK, Söderlund H, Martin CR (2002) Antibody-based bio-nanotube membranes for enantiomeric drug separations. Science 296 (5576):2198–2200
3. Jirage KB, Hulteen JC, Martin CR (1997) Nanotubule-based molecular-filtration membranes. Science 278(5338):655–658
4. Wirtz M, Parker M, Kobayashi Y, Martin CR (2002) Molecular sieving and sensing with gold nanotube membranes. Chem Record 2(4):259–267
5. de Jong J, Lammertink RGH, Wessling M (2006) Membranes and microfluidics: a review. Lab Chip 6(9):1125–1139
6. Han J, Fu J, Schoch RB (2008) Molecular sieving using nanofilters: past, present and future. Lab Chip 8(1):23–33
7. Huang M, Galarreta BC, Cetin AE, Altug H (2013) Actively transporting virus like analytes with optofluidics for rapid and ultrasensitive biodetection. Lab Chip 13(24):4841–4847
8. Deen WM (1987) Hindered transport of large molecules in liquid-filled pores. AIChE J 33(9):1409–1425
9. Paine PL, Scherr P (1975) Drag coefficients for the movement of rigid spheres through liquid-filled cylindrical pores. Biophys J 15(10):1087–1091
10. Bungay PM, Brenner H (1973) The motion of a closely-fitting sphere in a fluid-filled tube. Int J Multiph Flow 1(1):25–56

11. Renkin EM (1954) Filtration, diffusion, and molecular sieving through porous cellulose membranes. J Gen Physiol 38(2):225–243
12. Brenner H, Gaydos LJ (1977) The constrained brownian movement of spherical particles in cylindrical pores of comparable radius. J Colloid Interface Sci 58(2):312–356
13. Dechadilok P, Deen WM (2006) Hindrance factors for diffusion and convection in pores. Ind Eng Chem Res 45(21):6953–6959
14. Snyder JL, Clark A Jr, Fang DZ, Gaborski TR, Striemer CC, Fauchet PM, McGrath JL (2011) An experimental and theoretical analysis of molecular separations by diffusion through ultrathin nanoporous membranes. J Membr Sci 369(1–2):119–129
15. Bean CP, Doyle MV, Entine G (1970) Etching of submicron pores in irradiated mica. J Appl Phys 41(4):1454–1459
16. Fleischer RL, Alter HW, Furman SC, Price PB, Walker RM (1972) Particle track etching. Divers Technol Range Virus Identif Uranium Explor 178(4058):255–263
17. Quinn JA, Anderson JL, Ho WS, Petzny WJ (1972) Model pores of molecular dimension: the preparation and characterization of track-etched membranes. Biophys J 12(8):990–1007
18. Apel P (2001) Track etching technique in membrane technology. Radiat Meas 34(1–6): 559–566
19. Apel PY, Korchev YE, Siwy Z, Spohr R, Yoshida M (2001) Diode-like single-ion track membrane prepared by electro-stopping. Nucl Instrum Methods Phys Res Sect B 184(3): 337–346
20. Stroeve P, Ileri N Biotechnical and other applications of nanoporous membranes. In: Trends in biotechnology 29(6):259–266
21. Diggle JW, Downie TC, Goulding CW (1969) Anodic oxide films on aluminum. Chem Rev 69(3):365–405
22. Keller F, Hunter MS, Robinson DL (1953) Structural features of oxide coatings on aluminum. J Electrochem Soc 100(9):411–419
23. Wood GC, O'Sullivan JP, Vaszko B (1968) The direct observation of barrier layers in porous anodic oxide films. J Electrochem Soc 115(6):618–620
24. Masuda H, Fukuda K (1995) Ordered metal nanohole arrays made by a two-step replication of honeycomb structures of anodic alumina. Science 268(5216):1466–1468
25. Lee W, Ji R, Gosele U, Nielsch K (2006) Fast fabrication of long-range ordered porous alumina membranes by hard anodization. Nat Mater 5(9):741–747
26. Masuda H (2005) Highly ordered nanohole arrays in anodic porous alumina. In: Ordered porous nanostructures and applications. Springer, US, Boston, MA, pp 37–55
27. Hideki M, Masahiro S (1996) Fabrication of gold nanodot array using anodic porous alumina as an evaporation mask. Jpn J Appl Phys 35(1B):L126
28. Masuda H, Yamada H, Satoh M, Asoh H, Nakao M, Tamamura T (1997) Highly ordered nanochannel-array architecture in anodic alumina. Appl Phys Lett 71(19):2770–2772
29. Asoh H, Nishio K, Nakao M, Tamamura T, Masuda H (2001) Conditions for fabrication of ideally ordered anodic porous alumina using pretextured Al. J Electrochem Soc 148(4): B152–B156
30. Lee W, Schwirn K, Steinhart M, Pippel E, Scholz R, Gosele U (2008) Structural engineering of nanoporous anodic aluminium oxide by pulse anodization of aluminium. Nat Nano 3(4): 234–239
31. Chen W, Wu J-S, Xia X-H (2008) Porous anodic alumina with continuously manipulated pore/cell size. ACS Nano 2(5):959–965
32. Robatjazi H, Bahauddin SM, Macfarlan LH, Fu S, Thomann I (2016) Ultrathin AAO membrane as a generic template for sub-100 nm nanostructure fabrication. Chem Mater 28(13):4546–4553
33. Masuda H, Hasegwa F, Ono S (1997) Self-ordering of cell arrangement of anodic porous alumina formed in sulfuric acid solution. J Electrochem Soc 144(5):L127–L130
34. Nishizawa M, Menon VP, Martin CR (1995) Metal nanotubule membranes with electrochemically switchable ion-transport selectivity. Science 268(5211):700–702

35. Menon VP, Martin CR (1995) Fabrication and evaluation of nanoelectrode ensembles. Anal Chem 67(13):1920–1928
36. Tong HD, Jansen HV, Gadgil VJ, Bostan CG, Berenschot E, van Rijn CJM, Elwenspoek M (2004) Silicon nitride nanosieve membrane. Nano Lett 4(2):283–287
37. Jonsson MP, Dahlin AB, Feuz L, Petronis S, Höök F (2010) Locally functionalized short-range ordered nanoplasmonic pores for bioanalytical sensing. Anal Chem 82(5):2087–2094
38. Eftekhari F, Escobedo C, Ferreira J, Duan X, Girotto EM, Brolo AG, Gordon R, Sinton D (2009) Nanoholes as nanochannels: flow-through plasmonic sensing. Anal Chem 81(11): 4308–4311
39. Escobedo C, Brolo AG, Gordon R, Sinton D (2012) Optofluidic concentration: plasmonic nanostructure as concentrator and sensor. Nano Lett 12(3):1592–1596
40. Vlassiouk I, Apel PY, Dmitriev SN, Healy K, Siwy ZS (2009) Versatile ultrathin nanoporous silicon nitride membranes. Proc Natl Acad Sci 106(50):21039–21044
41. Yanik AA, Huang M, Artar A, Chang TY, Altug H (2010) Integrated nanoplasmonic-nanofluidic biosensors with targeted delivery of analytes. Appl Phys Lett 96(2)
42. Yanik AA, Huang M, Kamohara O, Artar A, Geisbert TW, Connor JH, Altug H (2010) An optofluidic nanoplasmonic biosensor for direct detection of live viruses from biological media. Nano Lett 10(12):4962–4969
43. Yanik AA, Cetin AE, Huang M, Artar A, Mousavi SH, Khanikaev A, Connor JH, Shvets G, Altug H (2011) Seeing protein monolayers with naked eye through plasmonic Fano resonances. P Natl Acad Sci USA 108(29):11784–11789
44. Kumar S, Cherukulappurath S, Johnson TW, Oh S-H (2014) Millimeter-sized suspended plasmonic nanohole arrays for surface-tension-driven flow-through SERS. Chem Mater 26 (22):6523–6530
45. Dahlin AB, Mapar M, Xiong K, Mazzotta F, Höök F, Sannomiya T (2014) Plasmonic nanopores in metal-insulator-metal films. Adv Opt Mat n/a–n/a
46. Stein K, van Henk W, van Cees R, Wietze N, Gijs K, Miko E (2001) Fabrication of microsieves with sub-micron pore size by laser interference lithography. J Micromech Microeng 11(1):33
47. van Rijn CJM (2006) Laser interference as a lithographic nanopatterning tool. MOEMS 5(1), 011012–011012-6
48. Striemer CC, Gaborski TR, McGrath JL, Fauchet PM (2007) Charge- and size-based separation of macromolecules using ultrathin silicon membranes. Nature 445(7129):749–753
49. Emilsson G, Schoch RL, Feuz L, Höök F, Lim RYH, Dahlin AB (2015) Strongly stretched protein resistant poly(ethylene glycol) brushes prepared by grafting-to. ACS Appl Mat Interfaces
50. van Reis R, Brake JM, Charkoudian J, Burns DB, Zydney AL (1999) High-performance tangential flow filtration using charged membranes. J Membr Sci 159(1–2):133–142
51. Asatekin A, Kang S, Elimelech M, Mayes AM (2007) Anti-fouling ultrafiltration membranes containing polyacrylonitrile-graft-poly(ethylene oxide) comb copolymer additives. J Membr Sci 298(1–2):136–146
52. Shannon MA, Bohn PW, Elimelech M, Georgiadis JG, Marinas BJ, Mayes AM (2008) Science and technology for water purification in the coming decades. Nature 452(7185): 301–310
53. Caspi Y, Zbaida D, Cohen H, Elbaum M (2008) Synthetic mimic of selective transport through the nuclear pore complex. Nano Lett 8(11):3728–3734
54. Stuart MAC, Huck WTS, Genzer J, Muller M, Ober C, Stamm M, Sukhorukov GB, Szleifer I, Tsukruk VV, Urban M, Winnik F, Zauscher S, Luzinov I, Minko S (2010) Emerging applications of stimuli-responsive polymer materials. Nat Mater 9(2):101–113
55. Zdyrko B, Luzinov I (2011) Polymer brushes by the "grafting to" method. Macromol Rapid Commun 32(12):859–869
56. Edmondson S, Osborne VL, Huck WTS (2004) Polymer brushes via surface-initiated polymerizations. Chem Soc Rev 33(1):14–22

57. Barbey R, Lavanant L, Paripovic D, Schüwer N, Sugnaux C, Tugulu S, Klok H-A (2009) Polymer brushes via surface-initiated controlled radical polymerization: synthesis, characterization, properties, and applications. Chem Rev 109(11):5437–5527
58. Tokarev I, Minko S (2009) Multiresponsive, hierarchically structured membranes: new, challenging, biomimetic materials for biosensors, controlled release, biochemical gates, and nanoreactors. Adv Mater 21(2):241–247
59. Bruening ML, Dotzauer DM, Jain P, Ouyang L, Baker GL (2008) Creation of functional membranes using polyelectrolyte multilayers and polymer brushes. Langmuir 24(15):7663–7673
60. Tokarev I, Minko S (2010) Stimuli-responsive porous hydrogels at interfaces for molecular filtration, separation, controlled release, and gating in capsules and membranes. Adv Mater 22(31):3446–3462
61. Zhang H, Hou X, Zeng L, Yang F, Li L, Yan D, Tian Y, Jiang L (2013) Bioinspired artificial single ion pump. J Am Chem Soc 135(43):16102–16110
62. Zhang Z, Kong X-Y, Xiao K, Liu Q, Xie G, Li P, Ma J, Tian Y, Wen L, Jiang L (2015) Engineered asymmetric heterogeneous membrane: a concentration-gradient-driven energy harvesting device. J Am Chem Soc 137(46):14765–14772
63. Liu Q, Xiao K, Wen L, Lu H, Liu Y, Kong X-Y, Xie G, Zhang Z, Bo Z, Jiang L (2015) Engineered ionic gates for ion conduction based on sodium and potassium activated nanochannels. J Am Chem Soc 137(37):11976–11983
64. Yameen B, Ali M, Neumann R, Ensinger W, Knoll W, Azzaroni O (2009) Synthetic proton-gated ion channels via single solid-state nanochannels modified with responsive polymer brushes. Nano Lett 9(7):2788–2793
65. Yameen B, Ali M, Neumann R, Ensinger W, Knoll W, Azzaroni O (2010) Proton-regulated rectified ionic transport through solid-state conical nanopores modified with phosphate-bearing polymer brushes. Chem Commun 46(11):1908–1910
66. Yameen B, Ali M, Neumann R, Ensinger W, Knoll W, Azzaroni O (2009) Ionic transport through single solid-state nanopores controlled with thermally nanoactuated macromolecular gates. Small 5(11):1287–1291
67. Elbert J, Krohm F, Rüttiger C, Kienle S, Didzoleit H, Balzer BN, Hugel T, Stühn B, Gallei M, Brunsen A (2014) Polymer-modified mesoporous silica thin films for redox-mediated selective membrane gating. Adv Func Mater 24(11):1591–1601
68. Buchsbaum S, Nguyen G, Howorka S, Siwy ZS (2014) DNA-modified polymer pores allow ph- and voltage-gated control of channel flux. J Am Chem Soc
69. de Groot GW, Santonicola MG, Sugihara K, Zambelli T, Reimhult E, Vörös J, Vancso GJ (2013) Switching transport through nanopores with pH-responsive polymer brushes for controlled ion permeability. ACS Appl Mater Interfaces 5(4):1400–1407
70. Ito Y, Ochiai Y, Park YS, Imanishi Y (1997) pH-sensitive gating by conformational change of a polypeptide brush grafted onto a porous polymer membrane. J Am Chem Soc 119(7): 1619–1623
71. Ito Y, Park YS, Imanishi Y (1997) Visualization of critical pH-controlled gating of a porous membrane grafted with polyelectrolyte brushes. J Am Chem Soc 119(11):2739–2740
72. Park YS, Ito Y, Imanishi Y (1998) Permeation control through porous membranes immobilized with thermosensitive polymer. Langmuir 14(4):910–914
73. Ito Y, Nishi S, Park YS, Imanishi Y (1997) Oxidoreduction-sensitive control of water permeation through a polymer brushes-grafted porous membrane. Macromolecules 30(19): 5856–5859
74. Park YS, Ito Y, Imanishi Y (1998) Photocontrolled gating by polymer brushes grafted on porous glass filter. Macromolecules 31(8):2606–2610
75. Liu Dunphy DR, Atanassov P, Bunge SD, Chen Z, López GP, Boyle TJ, Brinker CJ (2004) Photoregulation of mass transport through a photoresponsive azobenzene-modified nanoporous membrane. Nano Lett 4(4), 551–554

76. Lokuge I, Wang X, Bohn PW (2006) Temperature-controlled flow switching in nanocapillary array membranes mediated by poly(n-isopropylacrylamide) polymer brushes grafted by atom transfer radical polymerization†. Langmuir 23(1):305–311
77. Huber DL, Manginell RP, Samara MA, Kim B-I, Bunker BC (2003) programmed adsorption and release of proteins in a microfluidic device. Science 301(5631):352–354
78. Yu Q, Shivapooja P, Johnson LM, Tizazu G, Leggett GJ, Lopez GP (2013) Nanopatterned polymer brushes as switchable bioactive interfaces. Nanoscale 5(9):3632–3637
79. Liu H, Liu X, Meng J, Zhang P, Yang G, Su B, Sun K, Chen L, Han D, Wang S, Jiang L (2013) Hydrophobic interaction-mediated capture and release of cancer cells on thermoresponsive nanostructured surfaces. Adv Mater 25(6):922–927
80. Liu Z, Wang W, Xie R, Ju X-J, Chu L-Y (2016) Stimuli-responsive smart gating membranes. Chem Soc Rev 45(3):460–475
81. Kohli P, Harrell CC, Cao Z, Gasparac R, Tan W, Martin CR (2004) DNA-functionalized nanotube membranes with single-base mismatch selectivity. Science 305(5686):984–986
82. Yu S, Lee SB, Martin CR (2003) Electrophoretic protein transport in gold nanotube membranes. Anal Chem 75(6):1239–1244
83. Osmanbeyoglu HU, Hur TB, Kim HK (2009) Thin alumina nanoporous membranes for similar size biomolecule separation. J Membr Sci 343(1–2):1–6
84. Ku J-R, Stroeve P (2004) Protein diffusion in charged nanotubes: "On–Off" behavior of molecular transport. Langmuir 20(5):2030–2032
85. Chun K-Y, Stroeve P (2002) Protein transport in nanoporous membranes modified with self-assembled monolayers of functionalized thiols. Langmuir 18(12):4653–4658
86. Kuiper S, van Rijn CJM, Nijdam W, Elwenspoek MC (1998) Development and applications of very high flux microfiltration membranes. J Membr Sci 150(1):1–8
87. Gaborski TR, Snyder JL, Striemer CC, Fang DZ, Hoffman M, Fauchet PM, McGrath JL (2010) High-performance separation of nanoparticles with ultrathin porous nanocrystalline silicon membranes. ACS Nano 4(11):6973–6981
88. Martin CR, Siwy Z (2004) Molecular filters: pores within pores. Nat Mater 3(5):284–285
89. Bayley H, Martin CR (2000) Resistive-pulse sensing from microbes to molecules. Chem Rev 100(7):2575–2594
90. Kobayashi Y, Martin CR (1999) Highly sensitive methods for electroanalytical chemistry based on nanotubule membranes. Anal Chem 71(17):3665–3672
91. Gyurcsányi RE (2008) Chemically-modified nanopores for sensing. TrAC Trends Anal Chem 27(7):627–639
92. Reimhult E, Höök F (2015) Design of surface modifications for nanoscale sensor applications. Sensors 15(1):1635–1675
93. Wang X, Smirnov S (2009) Label-free DNA sensor based on surface charge modulated ionic conductance. ACS Nano 3(4):1004–1010
94. Li S-J, Li J, Wang K, Wang C, Xu J-J, Chen H-Y, Xia X-H, Huo Q (2010) A nanochannel array-based electrochemical device for quantitative label-free DNA analysis. ACS Nano 4 (11):6417–6424
95. Dahlin AB (2015) Sensing applications based on plasmonic nanopores: the hole story. Analyst
96. Junesch J, Sannomiya T (2014) Ultrathin suspended nanopores with surface plasmon resonance fabricated by combined colloidal lithography and film transfer. ACS Appl Mater Inter
97. Escobedo C, Brolo AG, Gordon R, Sinton D (2010) Flow-through vs flow-over: analysis of transport and binding in nanohole array plasmonic biosensors. Anal Chem 82(24):10015–10020
98. Zhao Y, Gaur G, Retterer ST, Laibinis PE, Weiss SM (2016) Flow-through porous silicon membranes for real-time label-free biosensing. Anal Chem 88(22):10940–10948
99. Xiong K, Emilsson G, Dahlin AB (2016) Biosensing using plasmonic nanohole arrays with small, homogenous and tunable aperture diameters. Analyst

100. Yamaguchi A, Uejo F, Yoda T, Uchida T, Tanamura Y, Yamashita T, Teramae N (2004) Self-assembly of a silica-surfactant nanocomposite in a porous alumina membrane. Nat Mater 3(5):337–341
101. Breault-Turcot J, Masson J-F (2015) Microdialysis SPR: diffusion-gated sensing in blood. Chem Sci
102. Jágerszki G, Gyurcsányi RE, Höfler L, Pretsch E (2007) Hybridization-modulated ion fluxes through peptide-nucleic-acid- functionalized gold nanotubes. a new approach to quantitative label-free dna analysis. Nano Lett 7(6):1609–1612
103. Tsang M-K, Ye W, Wang G, Li J, Yang M, Hao J (2016) Ultrasensitive detection of ebola virus oligonucleotide based on upconversion nanoprobe/nanoporous membrane system. ACS Nano 10(1):598–605
104. Li F, Guijt RM, Breadmore MC (2016) Nanoporous membranes for microfluidic concentration prior to electrophoretic separation of proteins in urine. Anal Chem 88(16):8257–8263
105. Hereijgers J, Desmet G, Breugelmans T, De Malsche W (2015) Strategies to integrate porous layers in microfluidic devices. Microelectron Eng 132:1–13
106. Escobedo C (2013) On-chip nanohole array based sensing: a review. Lab Chip 13(13):2445–2463
107. Chen X, Shen J (2016) Review of membranes in microfluidics. J Chem Technol Biotechnol n/a–n/a
108. Barik A, Otto LM, Yoo D, Jose J, Johnson TW, Oh S-H (2014) Dielectrophoresis-enhanced plasmonic sensing with gold nanohole arrays. Nano Lett 14(4):2006–2012
109. Snyder JL, Getpreecharsawas J, Fang DZ, Gaborski TR, Striemer CC, Fauchet PM, Borkholder DA, McGrath JL (2013) High-performance, low-voltage electroosmotic pumps with molecularly thin silicon nanomembranes. Proc Natl Acad Sci 110(46):18425–18430
110. Wu X, Ramiah Rajasekaran P, Martin CR (2016) An alternating current electroosmotic pump based on conical nanopore membranes. ACS Nano 10(4):4637–4643
111. Tagliazucchi M, Szleifer I (2015) Transport mechanisms in nanopores and nanochannels: can we mimic nature? Mater Today 18(3):131–142

Nanoporous Gold Nanoparticles and Arrays for Label-Free Nanoplasmonic Biosensing

Camille G. Artur and Wei-Chuan Shih

1 Introduction

Surface plasmons (SP) are depicted in the classical picture as a fundamental electromagnetic mode of an interface between a metal (or a semi-conductor) and a dielectric medium and involving surface collective electronic oscillations [1]. Surface plasmons were predicted and described more than 60 years ago, however, plasmonics did not form as a sub-field of photonics dealing with the manipulation of light at the nanoscale until more recently. With rapid technological advances in miniaturization, electronics and optical detection, plasmonics has taken off in the past decade and continued to flourish in both fundamental and applied fields [2, 3].

Metallic nanoparticles and nanostructures with dimensions from a few up to several hundred nanometers, support localized surface plasmon resonance (LSPR) that induce enhanced electromagnetic fields localized at the nanoparticle surface when illuminated by light. The LSPR frequency is highly dependent on the geometry but also on the permittivity of the metal itself as well as the surrounding media. Synthesis and fabrication of a large variety of nanostructures, be they in suspension in the liquid phase (metallic colloids) or on planar substrates has

C.G. Artur · W.-C. Shih (✉)
Department of Electrical & Computer Engineering, University of Houston,
4800 Calhoun Rd, Houston, TX 77204, USA
e-mail: wshih@uh.edu

W.-C. Shih
Department of Biomedical Engineering, University of Houston, Houston,
TX 77204, USA

W.-C. Shih
Program of Materials Science & Engineering, University of Houston, Houston,
TX 77204, USA

W.-C. Shih
Department of Chemistry, University of Houston, Houston, TX 77204, USA

S.-H. Oh et al. (eds.), *Miniature Fluidic Devices for Rapid Biological Detection*,
Integrated Analytical Systems, https://doi.org/10.1007/978-3-319-64747-0_2

enabled a variety of applications. On the other hand, rational design of plasmonic nanoparticles has allowed for the tuning of LSPR frequencies from ultraviolet (UV) up to the mid-infrared (MIR) portions of the electromagnetic spectrum [4].

To date, plasmonics has advanced significantly and become a highly multidisciplinary field with the potential for a wide range of technological applications, many of them having led to commercialization. Among many, two prominent directions are enhanced sensing and light manipulation which directly take advantage of the plasmonic field localization. These two applications are particularly relevant to medicine (diagnostics and therapy) and molecular biology [5, 6, 7]. A particularly useful and often adopted sensing mechanism is based on LSPR's sensitivity to local refractive index changes near the metal surface (so-called LSPR spectroscopy), similar to Surface Plasmon Resonance (SPR) sensors which however relies on propagating surface plasmons. Beyond index sensing, the electromagnetic field localization and enhancement near the nanostructures have been demonstrated to have profound impact on a variety of light–matter interactions such as surface-enhanced Raman scattering (SERS), surface-enhanced Infrared Absorption (SEIRA), surface-enhanced fluorescence (SEF), and recently, surface-enhanced near-infrared absorption (SENIRA) [8] Such surface-enhanced spectroscopy techniques have the ability to detect trace amount of analytes and in certain conditions, down to the single-molecule sensitivity [9]. A tremendous amount of effort has been devoted to designing reliable and low-cost plasmonic-based biosensors and translating them eventually into biological and biomedical applications [10].

Plasmonic nanostructures of gold have emerged as an attractive class of nanomaterials because of their corrosion resistance, relative chemical inertness, and ability to bind to a wide range of thiolated ligands and biomolecular species through the metal–S bond. Historically, gold nanospheres were proposed first as plasmon mediated photothermal therapy agent [11] but in vivo applications were limited because their plasmon resonance is near 540 nm where biological tissue is turbid. Thus, various other colloidal nanoparticles have been developed with a primary goal of shifting the resonance into the near-infrared (NIR) or the so-called "diagnostic widow". Many types and shapes have been developed and they all share the same feature: they are all made out of a solid core [12, 13, 14].

One feature LSPR/SPR-based refractive index (RI) sensing and surface-enhanced spectroscopy techniques have in common is the dramatic dependence on the distance between the target molecule and the metallic surface. The LSPR effect is a near-field, i.e., extremely localized, phenomenon and decays rapidly from the surface. The regions where the electromagnetic field is confined and its intensity greatly amplified are often called "hot spots". It is therefore desirable to optimize the sensor performance by increasing the "optically hot" surface-to-volume ratio while keeping hot spots accessible to the target molecules.

Nanoporous gold (NPG) thin films have recently captured intense attention for their large surface area (Fig. 1) and they have been recognized as plasmonic materials with LSPR frequencies exhibiting pore size dependent tunability [15, 16]. The increased total surface area permits NPG to possess a much higher density of

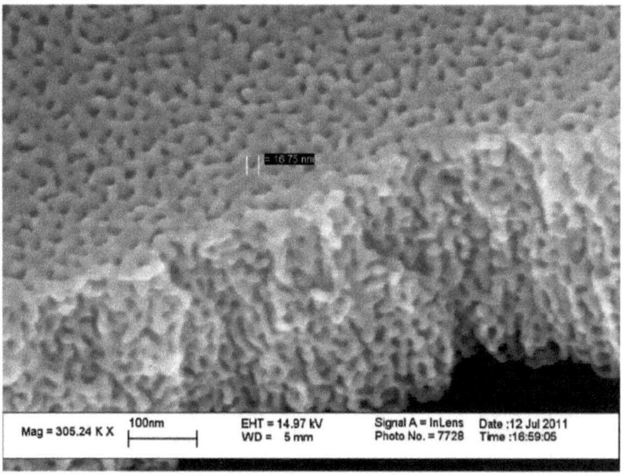

Fig. 1 Nanoporous gold with 300 nm thickness and 15–20 nm pore size from the dellayoing of a gold/silver alloy

potential hot spots for surface adsorbates and NPG thin films are "semi-transparent"; thus even adsorbates over the "internal" surfaces have a chance to be optically measured. Unlike conventional sensors featuring more or less flat sensing surfaces, NPG features an ultrahigh surface-to-volume ratio for collecting a large number of molecules inside the sensing volume.

But NPG films have also some limitations in their use as plasmonic sensor. Their LSPR band centering around 600 nm in an extinction spectrum (far field) has limited tunability of about 50 nm, which is achieved by varying the pore size from 10 to 50 nm [17]. A fundamental reason for the low tunability of NPG thin films is the relatively small pores that are merely a few percent of the optical wavelength. However, since the enhanced electromagnetic fields of LSPR excited in the ligaments are considered a highly desirable feature, a simple idea is to improve coupling between light and plasmons by forming NPG nanoparticles.

In the rest of this chapter, we will discuss the design, fabrication, and characterization of uniform, monolithic NPG nanoparticles produced by the combination of lithographic patterning and dealloying. We will demonstrate that NPG nanoparticles with large surface area, tunable plasmonic resonances, and high-density hot spots. NPG nanoparticles feature a fine porous network with pore size ~ 20 nm in some embodiments throughout its entire volume; the external shape of the nanoparticles is similar to that of nanodisks whose diameter and thickness can be easily tuned by varying the fabrication parameters. We will show how NPG disks can be integrated into microfluidic devices and will demonstrate their potential as a multifunctional platform for chemical and biological sensing.

Several label-free nanobiosensing applications will be discussed and can be classified in two groups: NPG disks based sensors with or without surface modifications. We will show that the bare-surfaces NPG disks enable quantitative detection of analytes in complex, unprocessed biological fluids such as urine. We will then demonstrate the ultrahigh sensitivity of functionalized NPG disks for sequence-specific DNA hybridization monitoring at the single-molecule level using molecular sentinel and direct sensing of small molecules such as malachite green using G-quadruplex as a capturing scaffold.

2 Nanoporous Gold Disks Fabrication and Characterization

2.1 Nanosphere Lithography and Dealloying of Au/Ag Films

To fabricate NPG disks, we take advantage of both top-down lithographic patterning and bottom-up atomic dealloying, which together demonstrate great synergy in precisely tuning the plasmonic properties of nanoporous materials. This section describes the fabrication process of the NPG disks (NPGD) substrates whose sensor applications are the core of this chapter [18].

As shown in Fig. 2, a film of gold and silver alloy approximately 120 nm thick was first deposited by DC sputtering onto a substrate (e.g., silicon wafer or glass slide) using an $Ag_{82.5}Au_{17.5}$ (atomic percentages) alloy target. A monolayer of 460–1100 nm diameter polystyrene beads (PS) was then formed on top of the alloy film an coating of over 90% of the alloy surface with close-packed PS beads can be achieved reproducibly [19] (Fig. 2a). Next, a timed oxygen plasma treatment was employed to shrink the PS beads, thus separating them from neighboring beads. The sample was then sputter-etched in argon plasma through the PS mask to transfer the bead pattern into the alloy film (Fig. 2b). Once the pattern transfer was completed, the PS beads were removed by sonication in chloroform or in isopropanol for 30 s (Fig. 2c). The alloy disks were dealloyed in concentrated nitric

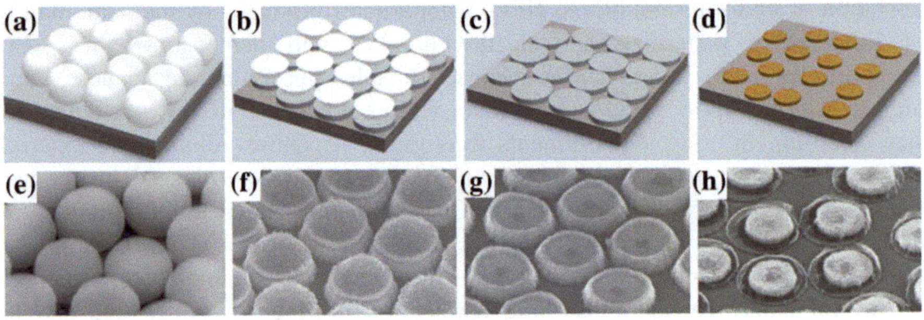

Fig. 2 **a–d** Illustration of the fabrication process used to prepare NPG disks on a silicon (or glass) substrate: (**a**) formation of a monolayer of polystyrene (PS) beads on an alloy-coated silicon (or glass) substrate; (**b**) O_2 plasma shrinkage of the PS beads and Ar sputter etching to form isolated alloy disks; (**c**) selective dissolution of PS beads by chloroform; (**d**) formation of NPG disks by dealloying. Figures (**e–h**) are SEM images taken at each step of the process with a 45° viewing angle

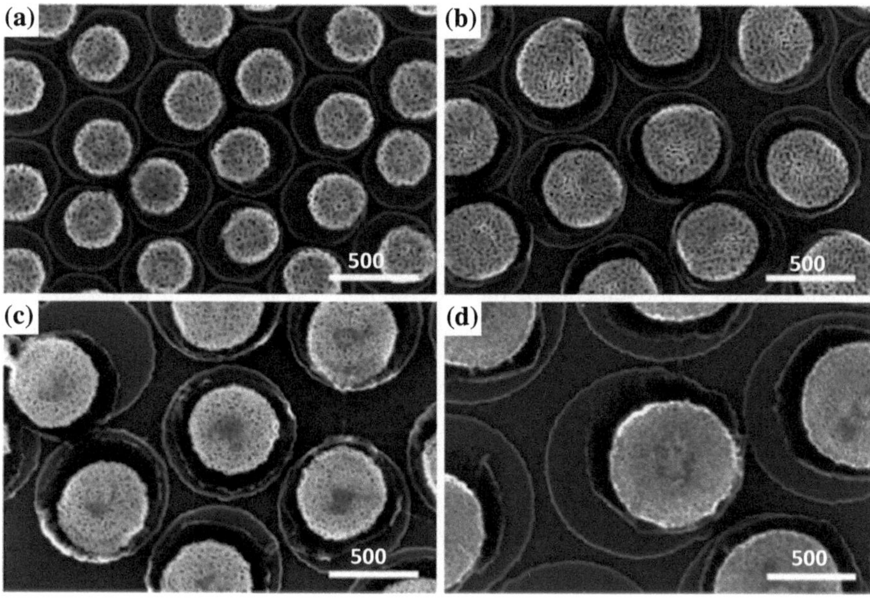

Fig. 3 SEM images of NPG disks made using 460 ± 9, 600 ± 12, 800 ± 9 and 1100 ± 14 nm PS beads on Si substrates. The corresponding final NPG disks diameters are **a** 300 ± 7, **b** 400 ± 10, **c** 500 ± 6 and **d** 700 ± 13 nm, respectively

acid (70% HNO_3 in water), followed by rinsing in deionized (DI) water (Fig. 2d) to produce an array of NPG disks. There was substantial size shrinkage during the PS bead etching step as well as the dealloying process. Scanning electron microscopy (SEM) images (Fig. 2e–h) show the corresponding nanostructures through the fabrication steps. To produce suspended colloidal NPG disks, high-density NPG disk arrays on a 3 inch Si wafer are further sonicated in DI water.

Figure 3 shows SEM images of monolayer samples of NPG disks on Si substrates made from PS beads with different diameters. The mean size and the standard deviation of different NPG disks are determined by measuring ≈100 disks using SEM images for each set of samples. The small size dispersion confirms the high fidelity of the pattern-transfer process.

Compared to the original sizes of the PS beads, there is an approximately 33–37% decrease in the NPG disk diameter, of which ≈5% occurs during the oxygen plasma treatment and up to 32% occurs during the dealloying process. Similarly the thickness decreases by ≈30%. This is consistent with the volume shrinkage effect observed during the formation of nanoporous gold by electrochemical dealloying [20]. Simulations of geometric relaxation in bicontinuous nanoporous metals revealed that surface relaxation played a significant role in the dramatic shrinkage during selective dissolution [21]. We note that NPG disks have move off-center during the fabrication process and this can be attributed to the weakening of the adhesion between the Si substrate and the deposited Au/Ag alloy during HNO_3

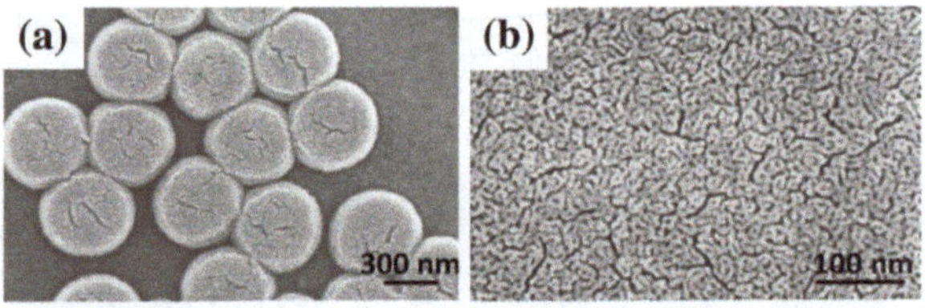

Fig. 4 **a** Top view of NPGDs ringed by gold; **b** unpatterned NPG film

dealloying through oxidation of Si to SiO_2. The strong stress generated by volume shrinkage leads to movements of the NPG disks. It is also worth noting that the weakened adhesion of the disks on the Si substrate is nevertheless sufficient to sustain rinsing with water. In contrast, NPG disks which are fabricated and strongly immobilized on a 300 nm Au film substrate as in [18] show cracks on their surface as can be appreciated on Fig. 4.

2.2 *Size and Porosity Characterization*

NPG disks prepared following the process described above can be easily released from the silicon substrate into DI water by sonication, due to their weak adhesion, to form colloidal NPG disks suspensions. Surfactant-free NPG disks are easily transferred into DI water without aggregation. The inset in Fig. 5c shows colloidal NPG disks dried on a Si wafer. Therefore, by flowing individual colloidal NPG disks in microfluidic channels, the single disk (400 nm diameter) buoyant mass is determined to be $6.04\times10^{-14}\pm7.6\times10^{-15}$ g as shown in Fig. 5d. For comparison, 400 nm diameter Au disks without porous structures immediately form aggregates in the millimeter size range in an aqueous solution upon release from the substrate.

As an effort to understand the colloidal stability of the NPG disks suspensions in water, their zeta potentials are measured and presented in Table 1. In general, when

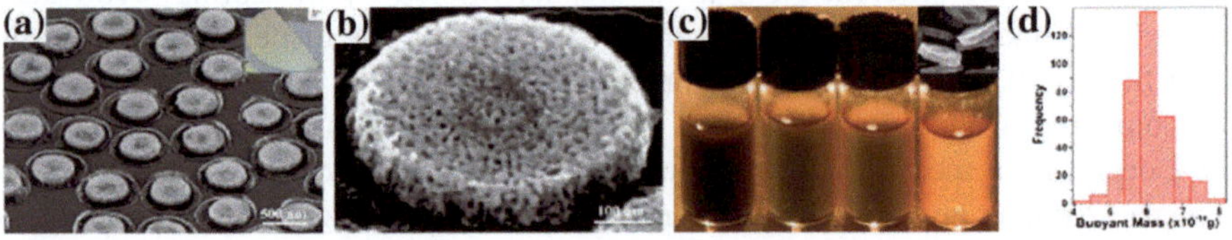

Fig. 5 SEM images of NPG disks taken at a 45° viewing angle, stored in DI water, and single disk buoyant mass measurements: **a** high-density NPG disk arrays on Si wafer before release. The inset is a 3 silicon wafer covered by a high-density monolayer of PS beads. **b** A single NPG disk with a diameter of 500 nm. **c** NPG disks having different sizes 300 ± 7, 400 ± 10, 500 ± 6, and 700 ± 13 nm (from left to right) after being released from the substrates by sonication in DI water to form colloidal NPG disk suspensions. The inset is the SEM image of NPG disks released from the substrate by sonication, dropped and dried on a Si wafer. **d** Histogram of 400 nm NPG disk buoyant mass distribution measured by flowing colloidal NPG disks in the microfluidic channel

Table 1 Average size, porosity and Zeta potentials for NPG disks 75 ± 1 nm thick. [a]NPG disks were made by using 460, 600, 800 and 1100 nm PS beads as masks and identical alloy thickness. [b]The roughness factor was obtained by using the expression $3h\beta/r$, where h, β, and r are the NPG disk thickness, two-dimensional porosity, and mean pore radius, respectively. [c]Zeta potentials were measured in DI water. [d]The full width at half maximum (FWHM) of the in-plane peaks of NPG disks obtained in air was measured by GRAMS/AI

NPG disk samples[a]	Average diameter (nm)	Average pore size (nm)	Roughness factor[b]	ζ(Mv)[c]	FWHM of the in-plane peak[d](nm)
1	300 ± 7	13.8 ± 2.2	6.56 ± 0.38	−28.5 ± 2.1	421.9
2	400 ± 10	13.7 ± 2.9	7.38 ± 0.41	−26.4 ± 3.2	460.9
3	500 ± 6	12.5 ± 2.0	7.71 ± 0.11	−19.0 ± 1.3	717.6
4	700 ± 13	12.8 ± 2.4	7.65 ± 0.27	−22.7 ± 1.2	1329.8

the absolute value of the zeta potential is larger than 25 mV, a nanoparticle suspension has a high degree of stability due to strong electrostatic repulsion between particles [22]. The zeta potentials of the 300 and 400 nm NPG disks suggests that both sizes of colloidal NPG disks have negatively charged surfaces and are quite stable in solution. Although the 500 and 700 nm diameter NPG disks possess negative surface charges but with slightly smaller zeta potentials, these larger NPG disks also exhibit practically useful long-term stability (i.e., no or minimal aggregation when stored in DI water at 4 °C for 4 months). The observed negative surface charge can be explained by the presence of deprotonated hydroxyl groups at the surface of NPG disks in aqueous solutions, which could plausibly form during the dealloying process in nitric acid. Another contributor to the NPG disks colloidal solutions stability could be the reduced inter-particles van der Waals forces due to their surface roughness which limits contact between particles.

The roughness factor of the NPG disks is estimated from the SEM images of the surface following a procedure described in [23].

2.3 *In-Plane Dealloying*

The fabrication of nanoporous gold disks described above involved pre-patterning alloy thin films into individual alloy disks *before* dealloying and thus can be classified as pre-dealloying modifications. Although highly effective, the dealloying rate using concentrated nitric can be as high as hundreds of nanometers per second, thus making the precise control and observation of the dealloying progresses challenging. In the method described in Sect. 2.1, unrestricted dealloying typically occurred from all fronts where Ag atoms were met with nitric acid and then dealloying propagated three dimensionally throughout the alloy, producing the internal nanoporous network.

In [24], a novel in situ NPG patterning method is introduced which generates hierarchical NPG structures during the dealloying process. In situ patterning is achieved by limiting the initiation point of dealloying to designated locations and restrict the reaction to only progress in-plane. After forming a film stack of an optional adhesion layer, an alloy film, and a Cr layer, the PMMA (poly(methyl methacrylate)) photoresist is deposited by spin-coating onto the surface, exposed by SEM equipped with a nanopattern generation system and developed, leaving a PMMA mask containing the via patterns on the top Cr layer. Next, the pattern is transferred from PMMA to the Cr layer by wet etching. A timed-immersion in concentrated nitric acid for dealloying follows. Due to the protection provided by the Cr layer, nitric acid could only etch the alloy through the openings, and then progressed in-plane or laterally. By controlling the dealloying time, NPG disks with various diameters are fabricated during dealloying. Experiments show that the dealloying distance is linearly proportional to dealloying time.

The smallest NPG disk that can be fabricated with in-plane dealloying is limited by the smallest via diameter and the precise dealloying time control. Since 20 nm via diameter can be routinely achieved by commercial e-beam writers, together with the wet etching undercut, 30 nm diameter via is achievable. Combining with the slowest dealloying rate 43 nm/s measured in the experiment with the conditions estimated above, the minimum diameter is 460 nm. Since the size of the NPG structure and the size of the pores are both controlled by the dealloying time, with this method, the size of the pores induced by dealloying cannot be tuned independently.

2.4 In Situ Laser-Assisted Dealloying

In [25], another in situ NPG micropatterning method is presented; the patterning is realized by localized laser heating, during which dealloying occurs at the laser focal spots due to elevated temperature. It has been shown that the dealloying rate increases with rising temperature and decreases with lowering acid concentration [26]. Therefore, dealloying can occur at elevated temperature even in diluted acid which otherwise has negligible dealloying effects at room temperature. Laser-induced localized heating can be utilized through the concomitant use of a spatial light modulator and design and fabricate NPG micropatterns on substrates immersed in an aqueous environment with diluted nitric acid which otherwise has negligible dealloying effect.

2.5 Integration of the NPG Disks into a Microfluidic Device

Microfluidic platforms hold great promise in achieving detection and identification of biomolecules in low cost as well as point-of-care applications due to their distinct advantages such as small sample and reagent consumption, fast reaction and analysis times and low cost. Among the variety of methods which have been implemented within microfluidic systems for molecular sensing, Surface-Enhanced Raman Scattering (SERS) based detection techniques have a prominent role due to their molecular specificity, label-free and multiplexed sensing capabilities. SERS harnesses the strong local electromagnetic field intensity enhancement at the surface of metal nanoparticles at plasmonic resonances to increase by several orders of magnitude the Raman scattering from a molecule near the metallic surface [9].

There are two main approaches in designing a SERS-based microfluidics detection system. One is based on colloidal metal nanoparticles acting as Raman enhancers dispersed in solution, which interact with analytes of interest and adsorb them onto the surface via mixing before SERS measurements are conducted [27, 28]. In general, SERS intensities depend on the degree of colloid aggregation that is significantly influenced by sample properties such as ionic strength and controlled by addition of aggregating agents such as potassium chloride. Clusters of different sizes and trapping different amounts of the analytes are produced and complicate the interpretation of the statistics of SERS intensities.

Another approach is that of integrating nanostructured metal surfaces as SERS-active substrates directly into the microfluidic platform. The existing nanostructure-integrated microfluidic SERS sensors primarily use silver. Although Ag-based SERS substrates provide inherently higher SERS enhancement compared to Au-based ones, the long-term stability is worse due to the gradual degradation of Ag surfaces [29]. Although the nanostructured surface approach typically provides better robustness against sample conditions, the amount of molecules that can effectively interact with the nanostructures is quite limited by diffusion—as low as $<1\%$ can be estimated using a simple boundary layer analysis in a largely laminar flow situation [30].

To address the diffusion limit mentioned above, a strategy is to increase the total usable area of the nanostructured surface, which provides more adsorption sites for biomolecules. As presented earlier, the NPG disks have large specific areas and high roughness factor (see Table 1) and we will see later that they possess a high density of plasmonic hot spots, high photothermal conversion efficiency and high SERS enhancement factor [18, 31, 32]. In the following, the design and fabrication of nanoporous gold disks monolithically integrated inside a microfluidic chip is described [33].

Figure 6 schematically illustrates the fabrication process for the NPG disk sensor, which is accomplished by oxygen-plasma bonding of two constructs: a SERS-active glass coverslip substrate with patterned NPG disks arrays and a poly-dimethylsiloxane (PDMS) layer containing a microfluidic network. The overall fabrication process includes a single low-cost transparency mask and four major steps: (a) definition of the SERS detection area, (b) construction of the NPGD arrays, (c) patterning of the microfluidic channel, and (d) plasma bonding.

The first step (a) produces a confined SERS detection area within the microchannel network, in which a photoresist mold containing microchannel structures are built using standard photolithography techniques through a transparent mask having the structures of the desired microchannel network (see Fig. 6(I)).

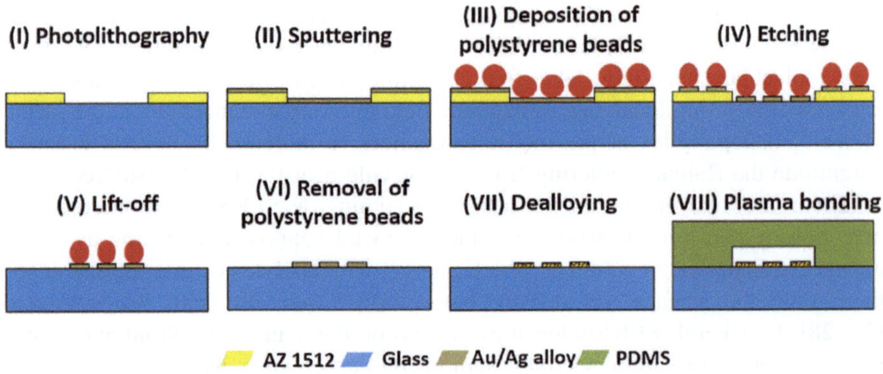

Fig. 6 Schematic view of the fabrication process for the surface-enhanced Raman scattering SERS-active nanoporous gold disk arrays within a microfluidic channel. (I) AZ 1512 in yellow is the UV sensitive photoresist. (II) In [33], the chosen alloy atomic ratio is $Ag_{82.5}Au_{17.5}$. (III) The self-assembled monolayer consist of closely packed 600 nm PS beads

Then, (b), SERS-active NPGD arrays are fabricated within the predefined region for SERS detection by generic polystyrene beads nanosphere lithography as described earlier in the chapter. Once the closely packed PS beads monolayer has been deposited (see Fig. 6III), the glass coverslip is treated with two steps of plasma etching in Fig. 6IV: oxygen etching to shrink the PS beads and argon sputter etching to etch away the portions of alloy film unmasked by the PS beads. A lift-off process is then conducted to remove the photoresist and all the nanostructures on its surface (Fig. 6V), leaving only alloy disks and PS residues within the predefined SERS detection region designed during step (a). After removing the PS residues (Fig. 6VI) by sonication in chloroform for 1 min, and dealloying of silver (Fig. 6VII) in concentrated nitric acid for 1 min, NPGD arrays with the designed geometry are formed on the coverslip. Due to the internal nanoporous network, the fabricated NPGDs have a large specific surface area for molecular adsorption sites.

The next step (c) is the patterning of the matching microfluidic channel. The PDMS microfluidic channel is fabricated using a soft lithography technique [34]. SU-8 negative photoresist is patterned through a standard photolithography technique to create a positive relief of the microchannel pattern on the surface of a silicon wafer. The pattern of the microfluidic network is designed to match the SERS detection region on the coverslip. Thus, the same transparent mask used previously in step (a) is employed here. Then liquid PDMS prepared by mixing a base and curing agent at a weight ratio of 10:1 is poured over the master, degassed, and cured in a vacuum oven. The negative PDMS cast with the microchannel pattern is then peeled off the master, and two via holes are punched for fluidic access.

The final step (d) consists in closing the microfluidic device to form enclosed channels covered by NPG disks arrays. Both the top surface of the prepared

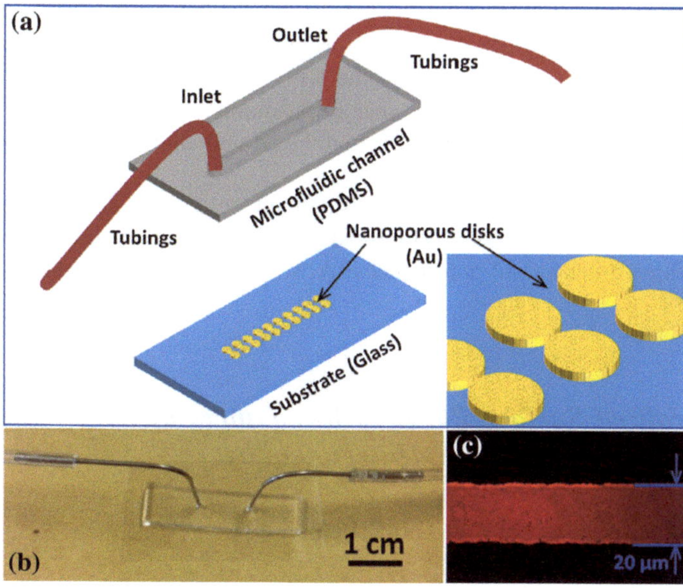

Fig. 7 microfluidic sensor architecture. **a** Preassembled parts consisting of the NPGD arrays coverslip and the PDMS microchannel network. The inset at the corner is an enlarged diagram showing the NPGDs; **b** photograph of the final sensor used in [33]; **c** a fluorescence microscopic image of the microchannel filled with Rhodamine 6G solution to confirm properly sealed microchannels

coverslip with NPGD arrays and the microchannel side of the PDMS layer are treated by oxygen plasma. After that, both surfaces are covered with methanol acting as a lubricant, and are precisely aligned under a microscope for face-to-face contact then heated at 90 °C for 5 min. Figure 7 shows a schematic of the assembled device together with a visual image of the final sensor utilized in the experiments in [33].

3 Optical Properties of NPG Disks

Semi-infinite dealloyed nanoporous gold (NPG) films with their unique bicontinuous structure of nanoscale ligaments and pore channels possess interesting optical properties, originating from the surface plasmon resonances (SPR), such as mixed states of localized/propagating surface plasmons [35, 36]. The intensity and peak position—i.e., wavelength at which absorption and scattering of light through excitation of the SPR are maximum—strongly depend on the nanopores size and the characteristic lengths and sizes of the ligaments [37]. In "as-dealloyed" NPG thin films, the LSPR band centering around 600 nm has a limited tunability of about 50 nm, achieved by varying the pore size from 10 to 50 nm [16]. In mechanically stamped NPG thin films, the grating modulation provides a propagating SPR mode coupled with NPG's original LSPR band [38]. However, the plasmonic landscape is dominated by the NPG LSPR. Furthermore, in thermally wrinkled NPG thin films, random, sparse plasmonic hot-spots form at gaps and junctions due to structural deformation, and thus do not alter the average plasmonic behavior [17].

NPG disks however, with their well-defined "exterior" disk shape of 300–700 nm with thickness around 100 nm, add a two-dimensional confinement to the "interior" three-dimensional porous network, which results in intriguing plasmonic properties.

3.1 Localized Surface Plasmon Resonances of NPG Disks and Their Size-Dependent Tunability

3.1.1 Extinction Spectra and LSPR Peaks

As shown in the extinction spectra in Fig. 8a, three peaks have been assigned to NPG LSPR ("▲"), out-of-plane resonance ("■"), and in-plane resonance ("●"). The NPG LSPR mode originates from the nanoporous structures, whereas the in-plane and out-of-plane modes are associated with the external disk shape. Size-dependent plasmonic shifts in these peaks are observed when the disk diameter is increased from 300 to 700 nm. Among these peaks, the in-plane resonance clearly dominates and only exists in NPG disks but not in semi-infinite NPG thin films.

The peak at 515 nm assigned to NPG LSPR ("▲") shows limited tunability with respect to pore size and ambient refractive index; a red shift of this peak to 540 nm in air was observed in [16] when the pore size was varied from 10 to 30 nm. In contrast, NPG disks have highly tunable plasmonic properties for all peaks as shown in Fig. 8a, due to plasmonic coupling between the nanoporous structures NPG LSPR and the patterned disk shape in-plane disk LSPR.

In addition, as seen in Fig. 8b, the out-of-plane resonance mode ("■") also exhibits a red shift with increasing disks diameter, whereas this mode resonance wavelength is fixed in the case of NPG films.

The plasmonic properties of NPG disks can be further understood by comparing with those of Au disks having the same diameter and thickness on glass substrates as presented in Fig. 8c. Compared to Au disks, the in-plane plasmonic band of NPG disks exhibit a remarkable red shift from 858 to 1100 nm which could be interpreted by plasmonic coupling [39] between the three-dimensional bicontinuous porous nanostructures and the outer geometrical size and shape. By normalizing the extinction spectra to their respective buoyant mass measured on a single-particle basis (Fig. 8d), it is evidenced that the peak height of the in-plane mode of NPG disks is about twice that of Au disks of the same external geometry. The NPG disk

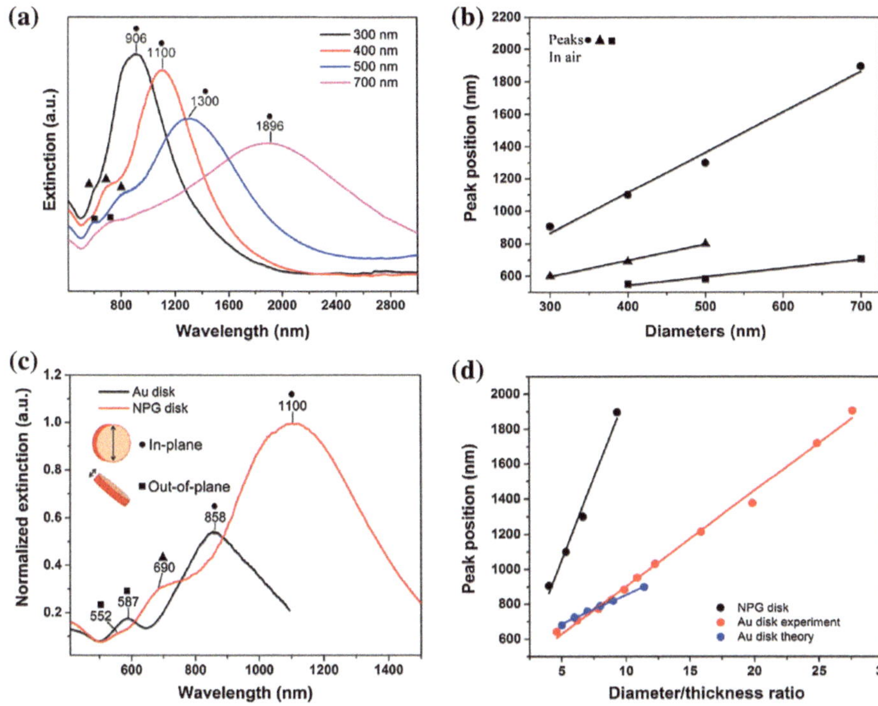

Fig. 8 Size-dependent plasmonic properties of NPG disk and comparison with Au disk: **a** extinction spectra of NPG disks with different diameters: 300, 400, 500, and 700 nm. The samples consisted of high-density NPG disk monolayers on glass substrates in air. **b** Plasmonic resonance peak positions versus NPG disk diameters in air. **c** Extinction spectra of 400 nm diameter and 75 nm thickness Au disks and NPG disks on glass substrates measured in air. Both spectra were normalized to buoyant mass. The inset shows the in-plane and out-of-plane resonance modes. **d** The in-plane dipole resonance peak positions are plotted as a function of the diameter/thickness ratio. The black circles represent NPG disks. The red and blue circles correspond to Au disks results from experiments in and theoretical calculations respectively. All extinction spectra were collected at normal incidence

also shows a much broader in-plane peak compared to the Au disk. Overall, the total extinction per buoyant mass for NPG disks is 3.3 times that of Au disks. The peak broadening can be attributed to random nanoporous structures and nanoscale Au ligaments.

Figure 8d shows the in-plane resonance peak position dependence on the diameter-to-thickness ratio (DTR). While Au disks in-plane resonance exhibits a 40 nm redshift per unit DTR increase, this redshift is 4.5 times larger for NPG disks suggestive of a much larger tunability by geometrical modifications.

The plasmonic tunability of monolithic NPG disks can also be explored by external and internal morphological changes. Specifically, increased dealloying time results in internal morphological evolution such as size growth of both the pore and ligament, thus further coarsening the porous network. Thermal annealing, in contrast, introduces both external morphological changes such as disk thickness and diameter reduction, as well as internal modifications such as pore coalescence [40].

3.1.2 Electric Field Distribution and Localized Field Enhancement

The UV-visible extinction spectra presented previously give a measure of light absorption and scattering by the NPG disks per unit wavelength and indicate where the LSPR resonance frequencies are; these spectra are however recorded in the "far field" and the local distribution of the electromagnetic field over the metallic structures and its intensity are lost. The principle of enhanced optical spectroscopies such as SERS and SEF is to harvest the locally intense electromagnetic fields that are confined to "hot spots" which are the result of the excitation of the localized surface plasmons. Analytes that are in the immediate vicinity of these hot spots will experience a local EM field whose intensity and polarization can be dramatically different from the incident field and therefore will exhibit modified interaction with light (enhanced Raman scattering, enhanced, or quenched fluorescence).

Electromagnetic modeling and simulations can be used to elucidate the local response of the nanostructures to an incident electromagnetic field.

In Fig. 9, finite difference time domain (FDTD) simulations were performed for a NPG disk and compared with a bulk Au disk having identical external shape parameters: 500 nm in diameter and 75 nm in thicknesses. The NPG disk model shown in Fig. 9a was constructed directly from the SEM image in [41]. Figure 9c displays the calculated electric field distribution for the 1300 nm incident wavelength which matches the in-plane resonance previously discussed. The maximum field enhancement is $\approx$100 in the pores around the edge whereas the Au disk produced a maximum E-field enhancement of $\approx$15, confined to either side of the disk. At 785 nm excitation wavelength (matching the NPG LSPR peak), the hot sports are uniformly distributed within the entire disk and the maximum enhancement factor is about 32 versus 5 for the Au disk. Thus, the NPG disk maintains a sixfold higher E-field enhancement compared to the Au disk.

It is worth emphasizing that the hot spots distributions in the NPG disk for 785 nm and 1300 nm incident wavelengths display remarkable differences. At 1300 nm, the hot-spot distribution appears to be concentrated near the pores around edges, supporting our previous interpretation of coupling between the in-plane resonance and the pores around edges. In contrast, the uniform hot spot distribution for 785 nm supports the interpretation that it is solely from NPG LSPR excitation. Overall, the FDTD results provide further support that the plasmonic coupling originating from the random nanoporous structure and the disk shape plays a key role in the unique plasmonic properties of NPG disks.

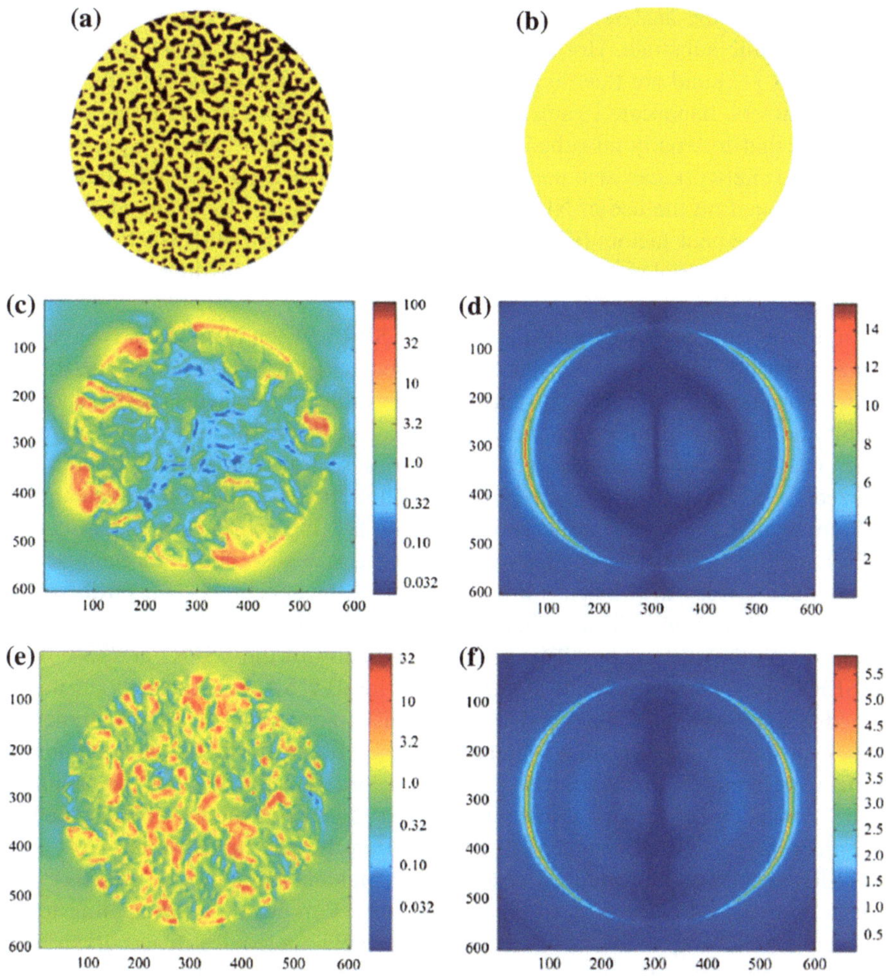

Fig. 9 E-field distribution of the NPG disk and Au disk with 500 nm diameter and 75 nm thickness: **a** and **b** are simulated models for NPG disk and Au disk, respectively. The E-field distribution was simulated using FDTD with plane wave incidence perpendicular to the disks, horizontally polarized. (**c** and **d**) E-field distribution of the NPG disk and Au disk for the 1300 nm incidence wavelength, respectively. (**e** and **f**) E-field distribution of the NPG disk and Au disk for the 785 nm incidence wavelength, respectively

3.1.3 Local Field Enhancement for SERS

Surface-Enhanced Raman Scattering (SERS) is a widely employed powerful spectroscopic technique which combines specificity—due to the uniqueness and sharp features of Raman spectra—and sensitivity, down to the single molecule, through the enhancement of the Raman signal by the excitations of plasmons at the surface of noble metal nanoparticles [9]. The enhancement of the signal is the strongest when the analyte is in the immediate vicinity of the plasmonic hot spots on the metallic substrate. Hot spots are often associated with nanoscale gaps and protrusions [42] and are thus likely to be found in nanoporous gold with 5–20 nm pore diameters, nanoscale ligament sizes and/or thickness. An additional potential benefit is that internally adsorbed molecules and those that may be traversing the nanoporous network can also participate in Raman scattering. Several studies have been published on the use of NPG films as SERS substrates and wide variations in SERS enhancement factors (EF) have been reported [43, 44, 45] which reflects the differences in material composition and morphology due to the different fabrication techniques.

In this section, the SERS effect and related SERS EF on NPG disks is explored. Benzenethiol (BT)—also called thiophenol—molecules are chosen to be the Raman marker for they have the ability to form self-assembled monolayers (SAM) and thus enable the number of molecules on individual disks to be quantified. Furthermore, their absolute intrinsic Raman scattering cross section has been measured and utilized in SERS EF evaluation [46, 47]. The chosen excitation wavelength is 785 nm which is off electronic resonance for BT molecules and is particularly well-suited for biomedical applications due to the low absorption and fluorescence of cells and tissues in the near-infrared region of the spectrum [18]. Furthermore, inspection of Fig. 9 indicates that locally intense electromagnetic field can be produced on the surface on NPG disks by excitation of the LSPRs at 785 nm.

SERS enhancement factors are evaluated for BT SAMs deposited on unpatterned nanoporous gold films (NPG), NPG disks arrays and Klarite (commercial SERS substrate from Renishaw with an advertised SERS EF of 10^6) and the normal Raman spectrum of a neat solution of BT is recorded. Table 2 gives the estimated SERS enhancement factors at 785 nm for two fingerprint modes of Benzenethiol when compared to the enhancement measured on the Klarite substrate.

SERS spectra are normalized according to incident laser power, effective illumination area, and estimation of the number of BT molecules participating to the detectable SERS signal. The number of adsorbed BT molecules in and on an NPG disk is the product of the area of the disk, the roughness R of NPG (the ratio of the chemically active surface area to the geometrical surface area), and the BT surface density. The contribution of adsorbed molecules to the total SERS signal also decreases with depth because of the decreasing laser power density and increased absorption of the Raman scattered light and an effective roughness factor for SERS must be then calculated which takes into account the skin depth of gold and silver [48]. Such considerations show that the SERS effective surface area of a NPG disk is 6.3 times larger than its geometric area.

Table 2 SERS EF estimation using Klarite's enhancement as a reference

	Klarite	NPGD	NPG film
EF of 1076 cm^{-1}	1×10^6	1.05×10^8	2.03×10^5
EF of 1575 cm^{-1}	1×10^6	1.43×10^8	2.47×10^5

An alternative way to experimentally evaluate the SERS enhancement factor on NPG disks is to compare the SERS intensities of BT adsorbed in and on NPG disks with the Raman intensity of a neat BT sample measured in the same conditions [46]. Similar SERS EF for NPG disks are obtained by this method, i.e., an enhancement factor of about 5×10^8.

3.2 Refractive Index Sensitivity of NPG Disks LSPR

Plasmon resonances are sensitive to the refractive index of the medium surrounding the metallic nanostructures and exhibit peak shifts when there is a change in refractive index which are usually quantified by the sensitivity factor $d\lambda/dn$ with the units of nm per refractive index unit (nm per RIU). The sensitivity of LSPR peak positions to the immediate environment is the basis of LSPR spectroscopy.

NPG disks can be used as plasmonic sensors due to the excellent sensitivity factor [41]. To further extend the range of index sensing into those for common solvents, the peak shifts of 400 nm NPG disks were investigated over the index range of 1.36–1.495 using pure ethanol, ethanol–toluene mixtures, and pure toluene. Figure 10a illustrates the extinction spectra of the 400 nm NPG disks in these various solvents.

The in-plane and out-of-plane resonance peaks redshift with increasing refractive indices whereas the peak shift in the NPG LSPR peak is unclear due to the overlap with the broad in-plane resonance. Overall, the sensitivity of the NPG disk in-plane peak is larger than those of spherical Au nanoparticles, Ag@Au nanoshells, SiO2@Au nanoshells, Au disks, Au nanorods, nanocages and silver nanoprisms

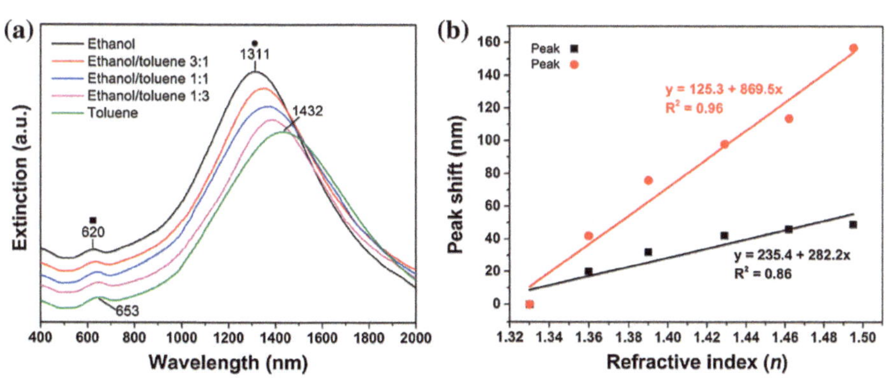

Fig. 10 **a** Extinction spectra of 400 nm NPG disks in various solvent mixtures with known refractive indices (n) varying from 1.36 to 1.495: ethanol (n = 1.36), 3: 1 ethanol–toluene (n = 1.39), 1: 1 ethanol—toluene (n = 1.429), 1: 3 ethanol–toluene (n = 1.462), and toluene (n = 1.495)0.16. **b** The peak shifts of in-plane peak (in red) and out-of-plane (in black) are plotted versus n

[49, 50, 51, 52] and comparable to nanorice [53] and nanorings [54] which range up to 800 nm per RIU.

4 Nanoporous Gold Disk Arrays in Microfluidic Sensing Applications

4.1 Microfluidic SERS Sensor with Monolithic NPG Disks Arrays for Rapid and Label-Free Dopamine and Urea Detection

The NPG disks are monolithically incorporated into the microfluidic device following the fabrication process described in Sect. 2.5. The schematic of the pre-assembled sensor is shown in Fig. 7. The height, width, and length of the microchannel are 20 μm, 20 μm and 1 cm respectively and the average diameter, thickness, and pore size of the disks decorating the microchannel are ~400, 75, and 14 nm, respectively [33]. SERS measurements are carried out on a Raman line-scan system at 785 nm [55]. The laser at the sample plane is shaped into a $1 \times 133\ \mu m^2$ line, yielding a spatial resolution of 1 μm and spectral resolution of about 8 cm^{-1}. The laser line is focused within the microchannel and scanned for hyperspectral area mapping.

4.1.1 Characterization of the Spatial Uniformity and Sensitivity of the Microfluidic SERS Sensor with no Flow

The dye rhodamine 6G (R6G) is used first as a model compound to test the robustness and uniformity of the sensor. A solution of R6G 1 mM in DI water is injected into the sensor via a syringe and let to rest for 10 min and SERS spectra are recorded from different areas onto the NPG disks surface of the microchannel. The relative intensity variations of major R6G peaks are calculated to be within 8% of the average intensity as shown in Fig. 11b, indicating the high uniformity and reproducibility of the NPGD sensor which is due to the highly uniform NPG disks arrays within the microchannel.

To further assess the sensing capability of the microfluidic sensor, concentration-dependent SERS measurements with R6G are performed, from 1 μM to 1 mM, with five recorded SERS spectra at different locations on the sample for each concentration of R6G, 10 min after injection. As shown on Fig. 12, the SERS peak intensities clearly increase with increased R6G concentrations. The intensity variations of the fingerprint mode 1366 cm^{-1} of R6G versus concentrations are used for quantitative evaluations. As shown in the inset, a highly correlated intensity-concentration relationship is obtained in the range 1 μM–1 mM. The limit of detection (LOD) is defined as the concentration at which the Raman intensity

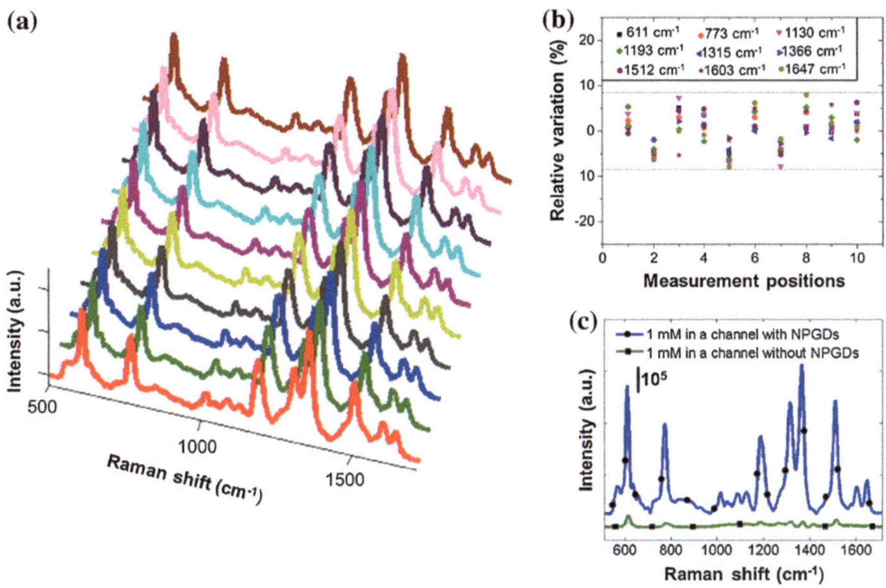

Fig. 11 R6G measurements **a** SERS spectra of 1 mM R6G detected in the SERS-active microfluidic channel at 10 different locations; **b** relative intensity variations of major peaks for the 10 locations; **c** spectra comparison of 1 mM R6G in a microchannel with NPGDs and a microchannel without NPGDs

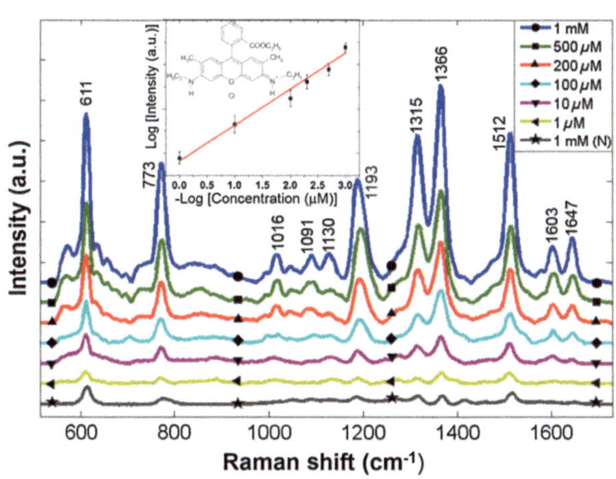

Fig. 12 Concentration-dependent SERS spectra of R6G measured in the sensor. The bottom trace was acquired from 1 mM R6G inside a microchannel without NPGD arrays. The inset indicates the variations of R6G peak intensity at 1366 cm^{-1} as a function of R6G concentration. The error bars represent the standard deviation from five measurements

value is equal to the averaged blank intensity at 1366 cm^{-1} plus three times the standard deviation.

4.1.2 SERS Detection of Rhodamine 6G with Continuous Flow

In addition to investigating the performance of the sensor without flow, its behavior under continuous flow is also studied. R6G solution at 100 μM is loaded into a syringe and injected into the sensor via a syringe pump at a flow rate of 3 μL/min. SERS measurements are recorded in situ under the continuous flow conditions with 2 s integration time per acquisition. Figure 13 shows the 1366 cm^{-1} mode SERS intensity variations within the first minute.

The starting point (t = 0 s) indicates the initial situation, where the NPGD arrays in the SERS detection region are not yet flooded by the sample solution. The signal intensity rapidly increases within the first 10 s, then steadily and slowly increases in the time range from 10 to 45 s, reaching a saturation limit in the last 10 s. After injecting the solution into the channel for 1 min, only small signal intensity variations (<10%) are observed, indicating the stability and robustness of SERS detection in continuous flow measurement. The dynamic behavior can be interpreted as follows: when more and more R6G molecules adsorb onto the surface of the NPG disks array, the SERS intensity keeps increasing within the first minute. The intensity reaches a quasi-plateau when maximum numbers of R6G molecules are adsorbed at a fixed flow rate. Further, the intensity achieved around 70% of that at the quasi-plateau state within the first 10 s. The result suggests that the sensor can provide rapid detection (i.e., <2 s) in solutions.

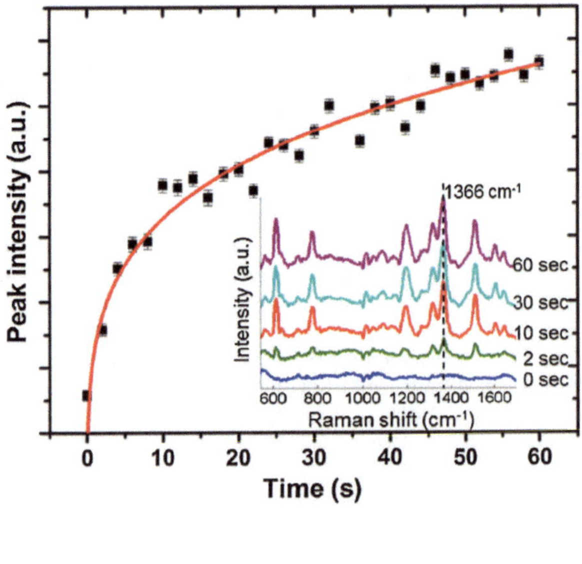

Fig. 13 Intensity variations versus time under the continuous flow condition. The 1366 cm^{-1} peak intensity of 100 μM R6G is plotted. The inset presents the measured SERS spectra at five selected time points of 0, 2, 10, 30, and 60 s

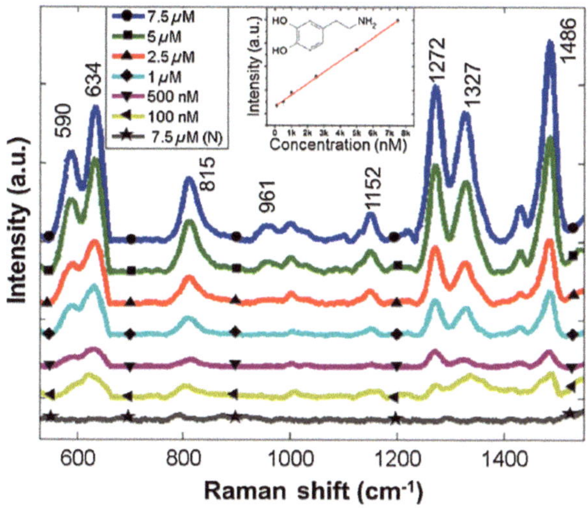

Fig. 14 Concentration-dependent SERS spectra of dopamine (DA) in the sensor. The bottom trace is acquired from 7.5 μM DA inside a microchannel without NPGD arrays. The inset shows the intensity variations at 1272 cm^{-1} with respect to DA concentration along with the molecular structure of DA. The error bars represent the standard deviation from five measurements

4.1.3 SERS Detection of Dopamine and Urea

Next, the capability of the sensor to detect dopamine, an essential neurotransmitter, is tested. Dopamine solutions of different concentrations ranging from 100 nM to 7.5 μM are introduced into the sensor by a syringe. After filling the microchannel for 10 min, SERS spectra are measured at five locations with 30 s acquisition time each. The averaged spectra are shown in Fig. 14 where the major Raman peaks of dopamine are clearly identified [56]. A good linear relationship is observed by plotting the SERS intensity at 1272 cm^{-1} versus concentrations as shown in the inset, and the limit of detection is calculated to be 32.4 nM.

4.1.4 SERS Detection of Urea at Physiological Concentrations

Artificial urine is prepared with 10 g of sodium chloride, 6 g of potassium chloride 6.4 g of sodium phosphate (monobasic, monohydrate) dissolved in 1 L of DI water and with concentrations in urea ranging from 1 to 20 mM. The corresponding SERS spectra are shown in Fig. 15, featuring the main Raman fingerprint of urea, the symmetrical C-N stretching mode around 1000 cm^{-1}. A good linear relationship is observed by plotting the SERS intensity at 1001 cm^{-1} versus concentrations as shown in the inset, and the LOD was calculated to be 0.67 mM with a power density of 0.21 mW/μm^2 and integration time of 10 s.

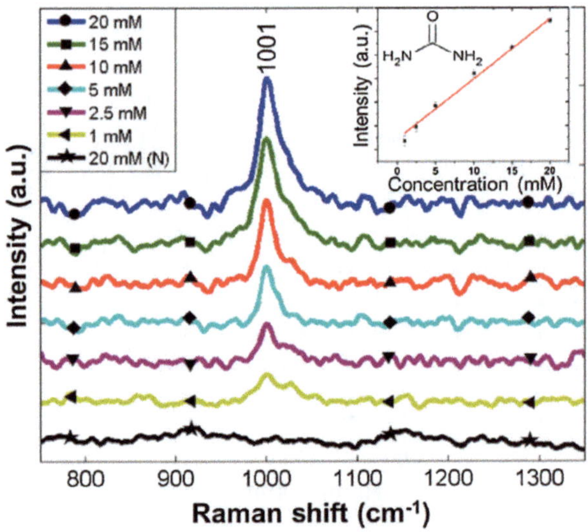

Fig. 15 Concentration-dependent SERS spectra of urea in the sensor. The bottom trace is acquired from 20 mM urea inside a microchannel without NPGD arrays. The inset shows the intensity variations at 1001 cm^{-1} with respect to urea concentration

The results suggest that the NPG disks based sensor provides a promising and versatile capability for clinical and diagnostic applications, such as kidney function monitoring and urine analysis.

4.2 *In Situ SERS Monitoring of Individual DNA Hybridization in Microfluidics*

DNA hybridization, where two single-stranded DNA (ssDNA) molecules form a duplex through non-covalent, sequence-specific interactions, is a fundamental process in biology and understanding its dynamics could help reveal molecular mechanisms involved in numerous biomolecular processes. To this end, sequence-specific detection of hybridization at the single-molecule level has become critical to many biomedical applications such as clinical diagnostics, biosensors, and drug development [57]. Current techniques to monitor diffusion and hybridization of DNA molecules include fluorescence correlation spectroscopy (FCS), single-molecule fluorescence resonance energy transfer (SMFRET), molecular beacons (MB), LSPR spectroscopy, electrochemistry and circular dichroism spectroscopy.

Label-free and amplification-free schemes are of particular interest because they could potentially provide in situ monitoring of individual hybridization events, and enable the discrimination of subtle variations due to single-base modification

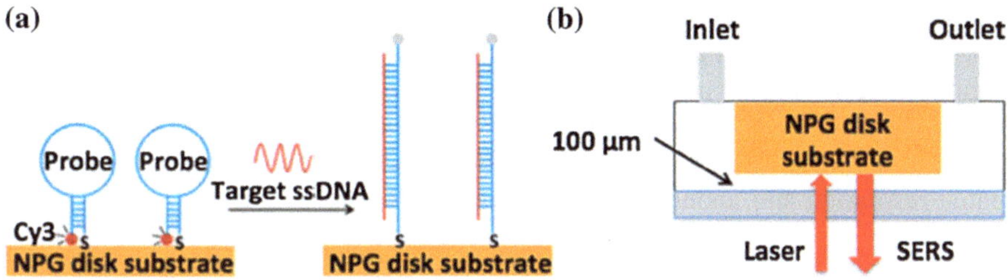

Fig. 16 Schematic of MS sensing mechanism in microfluidics

Fig. 17 SERS spectra of the MS probes on NPG disk substrates by incubation (500 pM–5 nM) and drop cast (100 pM) immobilization protocols

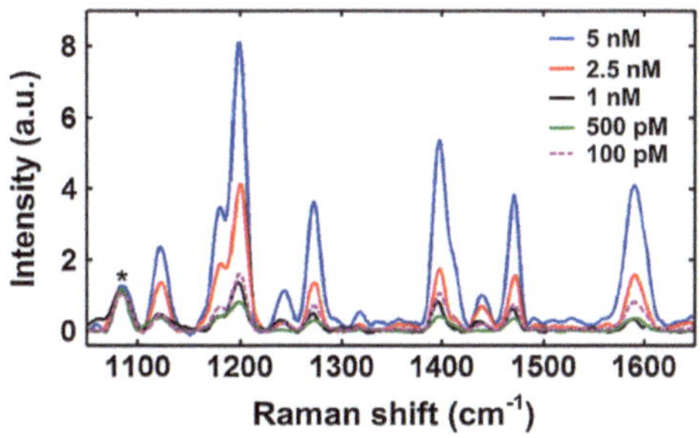

without stringency control or repetitive thermal cycling. SERS is an attractive approach for label-free multiplexed DNA/RNA detection because of its molecular specificity and potential single-molecule sensitivity and these distinct advantages have led to the development of a number of SERS sensing platforms for single DNA hybridization [58, 59].

In this section, the implementation of a microfluidic NPG disk-based sensor for in situ SERS monitoring of the same immobilized ssDNA molecules and their individual hybridization events is presented. To that end, molecular sentinels (MS, [60, 61]) are immobilized on NPG disks arrays inside a microfluidic channel, which prevents sample drying all the while keeping the sampled volume small. MS involves the design of the complementary sequence of a target ssDNA into a stem-loop "hairpin".

Figure 16 shows the schematic principle of MS use in a SERS experiment. The complementary hairpin probes have a thiol group at the 5′ end for robust immobilization on gold nanostructures and a fluorophore cyanine 3 (Cy 3) at the 3′end. Cy3 yields a strongly enhanced SERS signal when the probe is in the hairpin configuration; this signal decreases when the probe is hybridized with the target and moves away from the surface. The ssDNA target molecule is the ERBB2 gene, a critical biomarker of breast cancer and the hairpin probe consists of its complementary sequence (Fig. 17).

MS hairpin probes are immobilized onto NPG disk substrates at the bottom of a PDMS well by incubation for 40 min then the substrates are rinsed thoroughly in DI water and immersed in 6-mercapto-1-hexanol (MCH) for 10 min followed by another DI water rinse. The substrates are then mounted inside a temperature-controlled microscope microfluidic cell culture stage and a syringe pump is used to deliver target solutions of known concentration for hybridization. SERS measurements are carried out with 785 nm excitation with a 133 μm × 1 μm line-shaped focus, as described previously in the chapter, yielding about 1 μm spatial resolution and about 8 cm^{-1} spectral resolution.

4.2.1 SERS Detection of Immobilized MS Probes

Figure 17 shows SERS line spectra from different concentrations of ERBB2-sentinel probes on NPG disk substrates by incubation (500 pM–5 nM) and drop cast (100 pM), respectively. The SERS spectra show the major Raman modes of Cy3 [62] which indicates that the probe molecules are in their hairpin configuration, with the 3'-Cy3 near the gold surface. In the following experiments, the Cy3 peak height at 1197 cm^{-1} is monitored as the SERS intensity indicator. The immobilized probe density of drop cast onto NPG disk substrates is estimated from the number of probe molecules pipetted onto the NPG disk surface for the 100 pM solution or calibrated against the SERS intensity obtained from concentration-dependent SERS measurements obtained from MS drop cast substrates ([63] Supplementary Information). The average MS probe density is found to be about 2 molecules/μm^2 for NPG disk substrates incubated in 1 nM probe solution.

4.2.2 In Situ Monitoring of DNA Hybridization with Varying Target ssDNA Concentrations

In the first series of experiments, 5 nM molecular sentinel solution is immobilized by incubation on the NPG disks along with target concentrations from 5 to 20 nM. The SERS monitoring begins when the substrate is mounted into the microfluidic chamber and spectra are recorded every 10–15 min.

Figure 18a shows the Cy3 intensities at 1197 cm^{-1} from the line spectra after introducing the target ssDNA molecules; the SERS intensity of the peak begins to decrease after introduction of the 20 nM target solution then reaches a plateau at about 170 min indicating the end point of hybridization, at 40% of the starting peak intensity. Measurements over another 40 min indicate that no further hybridization occurs, i.e., 40% of the immobilized probes do not react with the target ssDNA molecules. A plausible explanation for the incomplete consumption of all immobilized probes is inefficient mass transfer of target ssDNA molecules to the NPG disk surface. According to the adsorption kinetics model of biomolecules [30], only a tiny fraction of target ssDNA molecules are able to react with probes in the current diffusion-limited configuration.

Figure 18b shows the hybridization and plateau phase of experiments with different target concentrations and non-complementary ssDNA molecules together

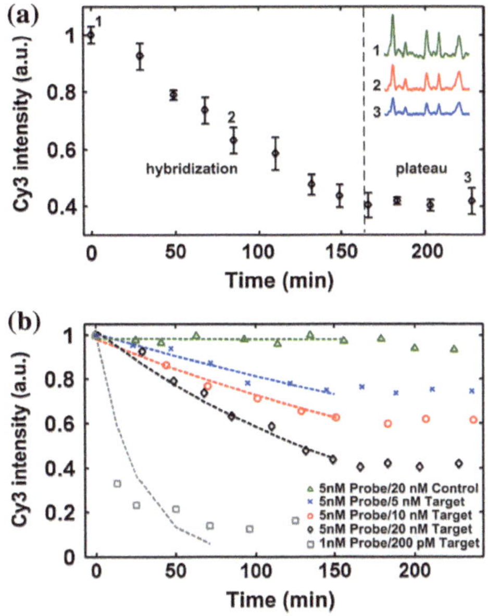

Fig. 18 In situ hybridization monitoring using SERS line spectra: **a** 5 nM MS probe hybridization with 20 nM target, **b** 5 nM MS probe hybridization with 5, 10, 20 nM target (cross, circle and diamond) and 20 nM non-complementary ssDNA (triangle); 1 nM MS probe hybridization with 200 pM target (square). Dashed curves are the exponential fits for the hybridization phase

with exponential fits of the data. The time constants increase with target concentrations, suggesting that target concentration can be determined by monitoring the decrease rate of Cy3 intensity. Alternatively, the final intensity value was also indicative of the target concentration. In the negative control experiment, 20 nM non-complementary ssDNA molecules do not cause a statistically meaningful SERS intensity change (≈5%). Furthermore, the stable SERS signal indicates that there is no photobleaching during experiments and the probe immobilization is robust. Thus, any signal decrease after adding target ssDNA molecules is attributed to hybridization.

To explore the detection limit in terms of number of target DNA molecules for the NPG disks sensor, the incubated concentration of MS probes is reduced to 1 nM for immobilization, resulting in a probe density of about two molecules/μm^2. After adding 200 pM target solution, the Cy3 SERS intensity decreases significantly in the first 10 min and reaches a plateau phase 90 min later. About 80% overall intensity decrease is observed.

Next, instead of the overall time trace extracted from the line spectra as shown in Fig. 18, individual time traces from point spectra are extracted by taking advantage of the spatial resolution of the line-scan Raman system. Ideally, there are 133 time traces using the point-spectrum, each scattered from a 1 μm^2 spot. Since the probe density is estimated to be about 2 molecules per μm^2 for substrates incubated in 1 nM MS probe solutions, and the average SERS intensities is 200 CCD counts, each 100 CCD counts is considered emanating from a single MS probe. Equivalently, each intensity decrease of 100 CCD counts during hybridization is attributed

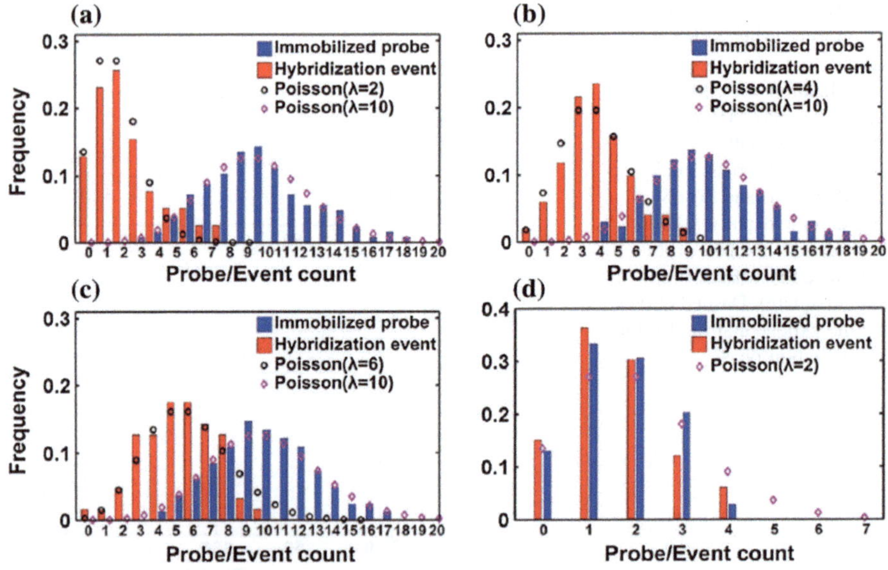

Fig. 19 Statistical analysis of individual time traces at probe/target concentrations of **a** 5 nM/5 nM, **b** 5 nM/10 nM, **c** 5 nM/20 nM and **d** 1 nM/200 pM. See text for details

to a single hybridization event. And an interval of 100 CCD counts between centers of bins is chosen in the following statistical analyses.

Figure 19 displays the histograms of immobilized probe count and hybridization event count by studying individual time traces. The point spectra showing high SERS intensities at different peak locations different from Cy3, likely from impurities in the solution, are excluded from the statistical study. The blue bars in Fig. 19 represent the frequency of the immobilized probe count on 1 μm^2 NPG disk surface before hybridization. These histograms (blue bars) can be better fit by a Poisson distribution than by a Gaussian with an average of 10 and 2 (shown as magenta diamonds) for substrates incubated in 5 nM and 1 nM probe solutions, respectively, in good agreement with the interpretation that 100 CCD counts represent a single probe. The red bars represent the frequency of hybridization event count. There are more hybridization events at higher target concentrations in 5 nM incubation experiments, which is consistent with the intensity time traces in Fig. 18b. Similarly, the histograms of hybridization event count can be better fit by a Poisson distribution (black circle in Fig. 19a–c, magenta diamond in Fig. 19d with averages of 2, 4, 6, and 2 for 5 nM, 10 nM, 20 nM and 200 pM target concentrations, respectively.

4.2.3 In Situ Monitoring of DNA Hybridization with 20 pM Target ssDNA Concentration

In these experiments, the probe molecules at 100 pM are immobilized by drop cast on the NPG disks resulting in about two probe molecules per μm^2 before hybridization. A protocol identical to the previous experiment is followed except that a 20 pM target solution is used.

Figure 20b–d shows the full-frame SERS images just before adding the target, during hybridization and at the last measurements (time points 1, 2, and 3 in Fig. 20a, respectively. As shown in Fig. 20f, the histogram (blue bars) of the immobilized probe count agrees well with a Poisson distribution with an average of 2 and a similar distribution is observed in the histogram of hybridization event count as discussed later. Four representative intensity patterns are observed and shown in Fig. 20e. The observation of quantized intensity decreases in individual time traces provide further support that individual hybridization events are observed.

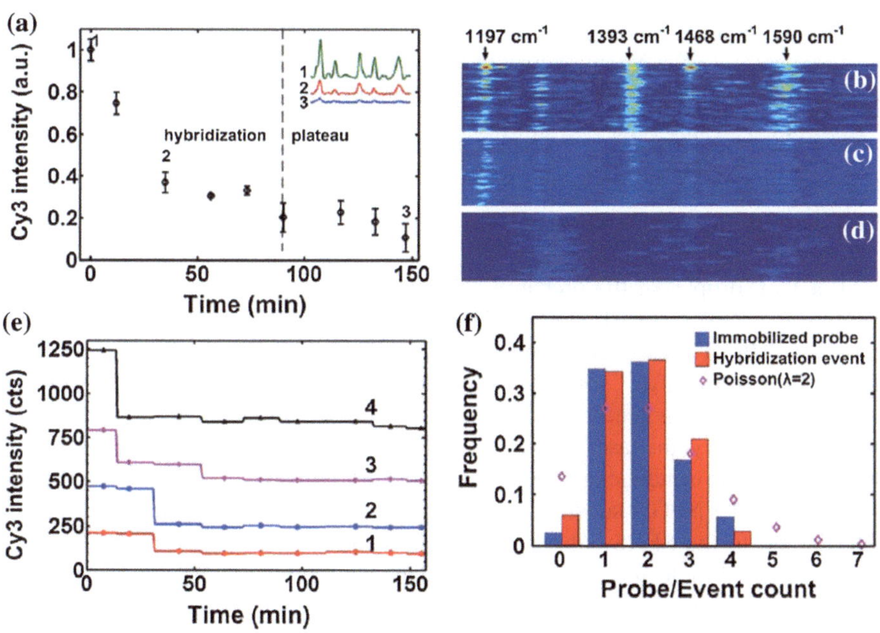

Fig. 20 **a** Overall Cy3 intensity trace with 20 pM target DNA; SERS image at **b** t = 0 min, **c** t = 40 min and **d** t = 150 min; the horizontal axis represents wavenumbers. Each row in the SERS image is a single point-spectrum. The major bands of Cy3 are labeled; **e** representative intensity patterns 1–4 corresponding to the hybridization counts 1–4 in **f** (red bars); (**f**) histogram analysis of individual time traces from 64 1 μm^2spots

5 Versatility and Performance: Nanoporous Gold Disk Arrays in Various Biomolecular Sensing Applications

5.1 Reagent and Separation-Free Measurements of Urine and Creatinine Concentration by Stamping SERS on NPG Disks Arrays

In this section, a novel method for the detection and concentration measurement of biological molecules by SERS is introduced. This technique, called stamping SERS (S-SERS) enables label-free, multiplexed molecular sensing and large-area and high-resolution molecular imaging. It also provides several benefits such as reagent and separation-free, low cost, high sensitivity and reproducibility [64, 65]. The performance of S-SERS is evaluated by the detection and quantification of creatinine spiked in three different liquids: creatinine in water, mixture of creatinine and urea in water, and creatinine in artificial urine within physiologically relevant concentration ranges. Finally, creatinine concentration measurements are realized for samples collected from a mouse model of nephritis, showing the potential for rapid, cost-effective, and reliable urine analysis for non-invasive diagnosis and monitoring of renal function.

Large-area, uniform, and reproducible NPGD arrays patterned on Au-coated silicon substrate are utilized as SERS substrates and SERS measurements are carried out with line-scan 785 nm laser excitation.

5.1.1 Stamping of the Analyte onto the SERS Substrate

First, a 1 μL droplet of the prepared solution containing the target molecules (i.e., creatinine) is first pipetted onto a PDMS thin film (~ 100 μm thick) laid flat on a glass coverslip. The droplet is then dried on the PDMS substrate, forming a film of target molecules after solvent evaporation. After that, a NPGD substrate ($\sim 0.5 \times 0.5$ cm^2) is gently stamped onto the PDMS surface bearing dried target molecules. Finally, the laser is focused at the PDMS surface to detect SERS signals arising from the sandwiched target molecules. Compared to directly drying the droplet onto the NPGD substrate, where target molecules are permanently chemically bound to the SERS substrate, the related issues like competitive adsorption among different molecules, surface and molecule affinity variability and uncertainty are reduced to some extent by the S-SERS technique.

The reproducibility of S-SERS creatinine is first tested; as shown in Fig. 21 the relative intensity variations of the major creatinine peaks are calculated to be within 12% of the average intensity, when repeatedly assayed from different locations on the sandwiched area.

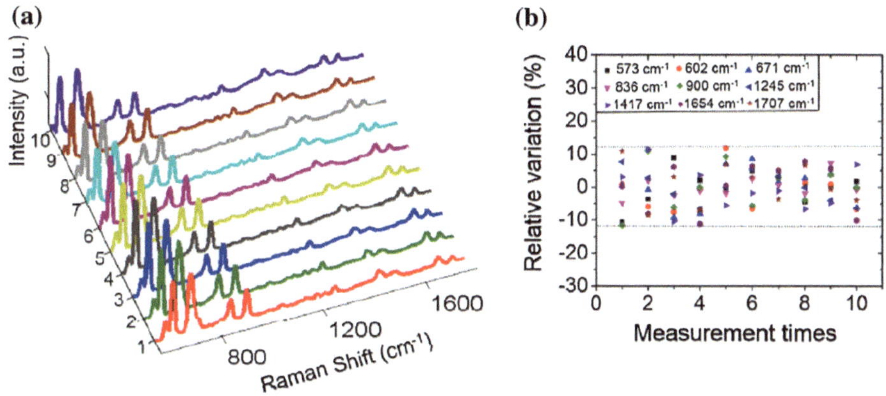

Fig. 21 Reproducibility of S-SERS: **a** SERS spectra of 100 μM creatinine detected by S-SERS at ten different locations, and **b** relative intensity variations of major peaks for the ten locations

5.1.2 SERS Detection of Creatinine in Water and Multiplexed SERS Concentration Measurements

Creatinine concentration-dependent SERS spectra are recorded for concentrations ranging from 100 nM to 100 μM in DI water. As expected, the SERS intensity of the major creatinine Raman peaks increases with increasing concentrations and the relationship is linear between the intensity variations of the 836 cm^{-1} peak and creatinine concentration within the whole [100 nM–100 μM] range. The limit of detection of creatinine in water is then estimated to be 13.2 nM based on the signal-to-noise ratio at 100 nM [65].

Next, the multiplexed sensing capabilities of urea and creatinine of the S-SERS sensor are tested. Since urea is a dominant chemical constituent in urine samples, a reagent- and separation-free technique needs to provide selectivity based on intrinsic molecular fingerprints. As presented in Fig. 22, measurements are performed on samples by mixing 100 μM creatinine and 100 mM urea solutions at five different volume ratios. Both concentrations ranges correspond to physiological concentrations in urine and are hence relevant for practical urine analysis. The results show that concentration-dependent creatinine SERS can be obtained even in the co-presence of $\sim 100\times$ higher concentration of urea.

5.1.3 SERS Detection of Creatinine in Nephritic Mouse Urine Samples

Urine samples are collected from nephritic mice with anti-GBM disease [66]. The creatinine concentrations in these samples from different mice were first determined using a commercial colorimetric assay kit based on a coupled enzyme reactions with a specified detection limit of about 0.15 mg/dl. The entire assay from warming

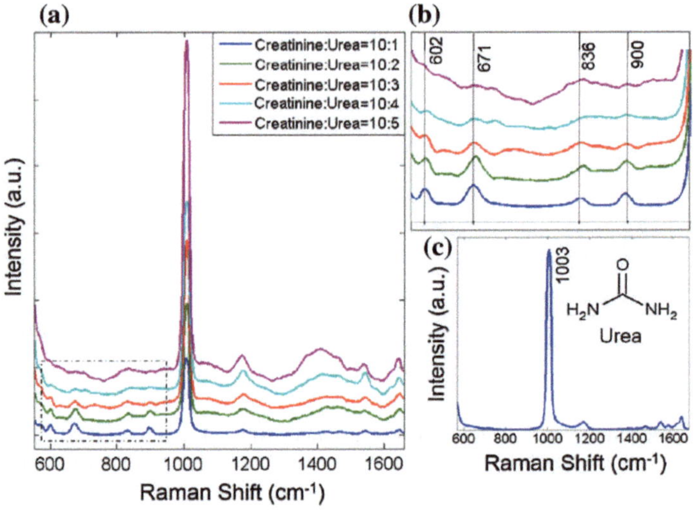

Fig. 22 SERS spectra from creatinine and urea water mixture **a** concentration-dependent SERS spectra of 100 μM creatinine and 100 mM urea mixed at different volume ratios. **b** zoomed-in view of the dashed window in (**a**), showing the intensity variation of creatinine. **c** Raman spectrum of urea in water

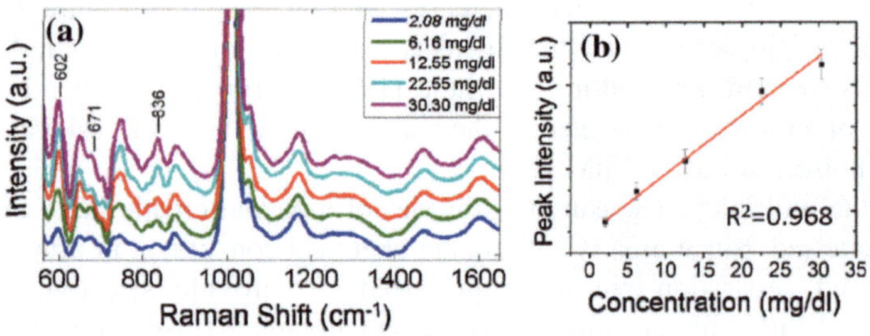

Fig. 23 Performance of S-SERS on urine from diseased mouse models: **a** Concentration-dependent SERS spectra of creatinine in urine samples collected from a nephritic mouse model of anti-GBM disease. **b** Variations of creatinine peak intensity at 836 cm^{-1} versus creatinine concentration

up reagents to room temperature to calculating creatinine concentration takes 1 h at the very least.

Figure 23a shows averaged SERS spectra of the different mouse urine samples with the concentrations determined by the assay test. A good linear relationship is observed by plotting the peak intensity at 836 cm^{-1} versus creatinine concentrations. When one compares the creatinine peak intensity at ~836 cm^{-1} from real urine to that of creatinine water solution at the same concentration, one finds that the intensity from real urine is about six-fold lower than that from the creatinine water solution which may come from local refractive index differences between water and urine, causing light attenuation and scattering. It is remarkable that S-SERS is still able to capture the creatinine-specific signature despite the co-existence of >70 different proteins in these nephritic urine samples [66].

5.2 NPG Disks Functionalized with G-Quadruplex Moieties for Sensing Small Molecules

In this section, a device for label-free SERS detection of malachite green (MG) is designed. NPG disks are functionalized with stabilized Guanine-quadruplex (G4) moieties with a highly specific topologic structure and effectively and selectively capture MG molecules for high sensitivity SERS measurements [67].

G4 are readily formed by the $\pi - \pi$ stacking of two or more G-tetrads prevalent in guanine rich (G-rich) oligonucleotides. These tetrads consist of a planar arrangement of four guanine bases held together by a cyclic array of Hoogsteen hydrogen bonds. The central core of the G4 is negatively charged due to the orientation of the carbonyl group of each G-base toward the center of the G-tetrad [68]. This configuration is believed to favor the conjugation of cations and small cationic organic molecules to the G4 structure [69, 70] and to enable ligand-specific binding into the G4 scaffold by varying its loop size and conformation [71] and the interaction between MG molecules and specific G4 scaffolds has been studied [72].

Malachite green (MG) is a well-known triphenylmethane dye with good antibacterial, antifungal, and antiparasitic properties. MG is commonly employed as an active ingredient in veterinary drugs to resist fungal and parasitic infections in aquaculture due to its low cost and availability but its use has however been restricted or prohibited due its potential carcinogenicity [73]. As a result, the European Union has mandated a sensor detection limit of 2 parts per billion (ppb) ($\sim$5.48 nM) for total MG and its reduced leuco-form. In addition, the U.S. Food and Drug Administration (FDA) prohibits the use of MG in aquaculture, hence a robust and reliable analytical technique is needed for effective detection of MG in aquaculture products, especially the fish meat sold for consumption.

5.2.1 Design of the Label-Free SERS Sensing Platform

NPG disks with 360 nm diameter are fabricated and form a semi-random array on the surface of a silicon wafer. SEM micrographs reveal the final nanoporous network with pore sizes of $\sim$15 nm in each individual disk [67]. NPG disks are then functionalized with G4 moieties stabilized by K^+ ions and 6-mercaptohexanol (MCH) is applied to block the nonspecific binding of small molecules. Exposure of the G4-decorated NPG disks to MG molecules causes the electron-rich phenyl rings of MG to effectively bind on the face of the G4 moieties to form G4-MG

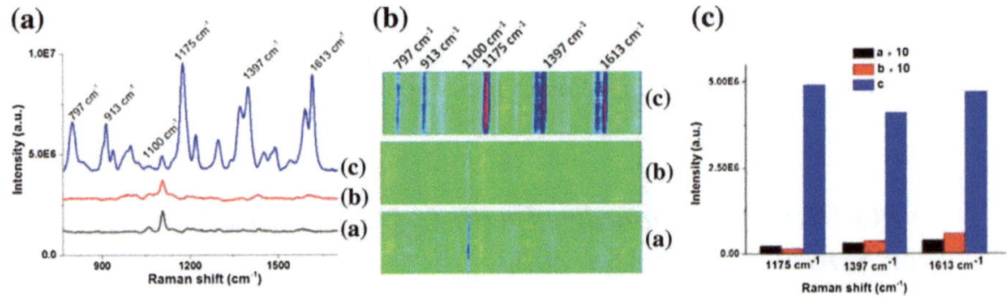

Fig. 24 **a** SERS spectra at 785 nm excitation and power 22.1 mW and **b** CCD image of different modified NPG disks; (**a**) MCH functionalized NPG disks immersed in MG solution (i.e., no G4); (**b**) G4-functionalized NPG disks immersed in buffer solution (i.e., no MG); **c** G4-functionalized NPG disks immersed in MG solution. (**c**) The corresponding SERS intensity from panel B at 1175 cm^{-1}, 1397 cm^{-1}, and 1613 cm^{-1}. The intensities from panels a and b were multiplied by 10 to be visible. The G4 concentration and MG concentration were 1.0 μM and 50 μM, respectively

conjugations through $\pi - \pi$ stacking and electrostatic interaction between MG and the G4 scaffold further improve the stability of the G4-MG complexes.

As can be seen in Fig. 24, remarkable SERS signal can be observed from the captured MG molecules on the G4-functionalized NPG disks. In contrast, no SERS signal, except for that from MCH molecules at 1100 cm^{-1}, is detected in the absence of MG molecules due to the lack of MG molecules being captured by G4 moieties. Similar results are obtained in the absence of G4 moieties, which suggests that MG molecules cannot be immobilized effectively onto the surface of MCH-incubated NPG disks without G4 as the capturing scaffold as well as the effectiveness of MCH as a blocking-agent. In contrast, the MG SERS intensity shows no significant variation in the presence of G4 before and after incubating with the MCH solution (Supplementary Information in [67]), which indicates that the MCH molecules cannot replace MG molecules that are captured by G4 moieties. In other words, the MG molecules are conjugated onto the surface of NPG disks through an interaction with G4 moieties rather than nonspecific binding on the surface of NPG disks, a key factor for high specificity.

It is found through G4 concentration studies of the evolution of the SERS sensitivity that the SERS intensity is enhanced sharply with increasing G4 concentration in the range of 100 pM–100 nM due to an increase of capturing sites formed on the NPG disks. However, the intensity increase slows down and reaches a plateau in the range from 100 nM to 1 μM. To achieve best performance, 1 μM G4 concentration is used in the following section.

5.2.2 Sensitivity for MG Detection

SERS measurements are obtained for a series of MG concentrations and presented in Fig. 25 The SERS intensity increases with the MG concentration from 50 pM to 5 μM, at which it reaches a plateau indicative of a reduction in binding events of

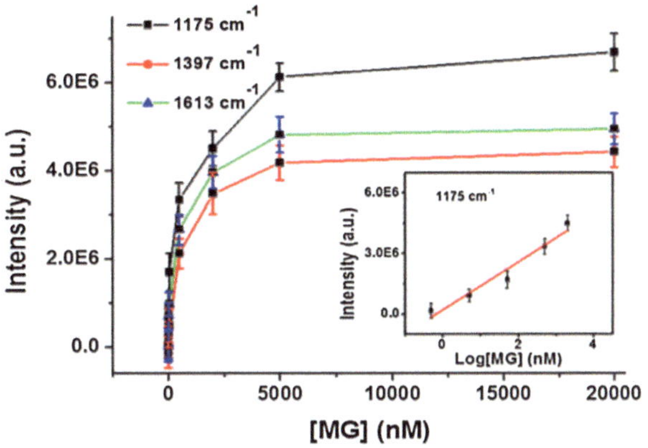

Fig. 25 SERS intensity for three fingerprint peaks of MG as a function of MG concentration. Inset shows a linear relationship between the SERS intensity at 1175 cm^{-1} and the logarithm of MG concentration at the range from 0.5 nM–2 μM. The error bars are calculated from at least three measurements on random spots on the same substrate

MG molecules onto the surface of G4-functionalized NPG disks. The prominent peak at 1175 cm^{-1} can be observed at 50 pM of MG which is 100 times more sensitive than the sensor detection limit mandated by the European Union for MG detection. The high sensitivity may be attributed to three factors: (1) the high enhancement of local electromagnetic fields from high-density hot-spots on NPG disks; (2) the high affinity of G4 moieties formed by the unique G-rich DNA to MG molecules; or (3) the nanoporous structure of NPG disks provides a high surface-to-volume ratio for loading more G4 moieties, thus capturing more MG molecules onto the surface of NPG disks (Fig. 25).

5.2.3 Sensor Performance in Complex Samples and Real-World Situations

The performance of the NPG disks sensor is evaluated on samples consisting of fish meat spiked with different MG concentrations as a model. Fish is purchased and turned into a homogenate, 2 g of which are then mixed with 10 mL PBS buffer solution and incubated at room temperature overnight. The solution is then filtered through a 0.22 μm membrane filter and various MG concentrations are used to spike the resulting supernatant.

As shown in Fig. 26, sensor performance similar to the above section is obtained for the complex mixture. Therefore, the proposed sensing system may be extended to other target assays in the biological and environmental fields while taking into account the possible need to modify the G4 sequence to accommodate for the target analyte.

Although the above-mentioned SERS sensor exhibits excellent sensitivity and good specificity, a laboratory Raman system has been employed for the SERS measurements. To demonstrate real time, onsite measurement in the field, the feasibility of the SERS detection with a portable Raman system with an optical fiber probe is tested, first on MG solutions in PBS and then on fish homogenates spiked with MG.

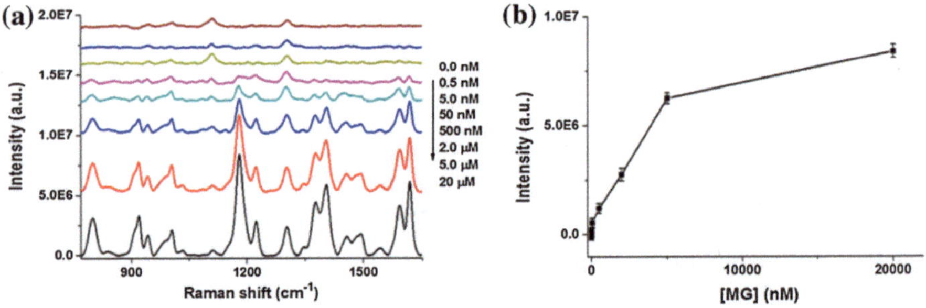

Fig. 26 **a** SERS spectra in fish samples spiked with various MG concentrations; **b** the corresponding SERS intensity change at 1175 cm^{-1}

Figure 27 summarizes the results and shows that the filter probe gives results similar to the laboratory Raman system. The lowest detectable concentration of MG with this system is 5.0 nM, which suggests that this NPG sensor can be effectively employed with a portable Raman system for onsite measurements for industrial and environmental applications. The difference between the linear relationships reported in Fig. 26 and that in Fig. 27 is due to the lower sensitivity of the portable Raman system compared to the line-scan system.

5.3 *Label-Free, Zeptomole Cancer Biomarker Detection by Surface-Enhanced Fluorescence on NPG Disks Arrays*

In this section, a label-free, spacer-free biosensor that makes use of distant-dependent detection of surface-enhanced fluorescence (SEF) is constructed on NPGD arrays and tested for ultrasensitive detection of ERBB2 cancer gene DNA targets [74].

SEF—also called metal-enhanced fluorescence (MEF)—occurs when fluorophores are within nanoscale proximity from surfaces of metallic nanostructures. Through the use of specifically designed metal nanostructures, increased fluorescence enhancement factors up to 500 times has been reported [75] and enhancements in the order of 5–50 times are typically observed [76, 77, 78]. When fluorophores are near metal surfaces, electronic energy transfer from the molecule may occur which generally leads to fluorescence quenching. Therefore, the fluorescence measured on "naked" metallic substrates is typically the outcome of the competition between SEF and metal-induced quenching and it is now understood that quenching occurs for nanoparticle-fluorophore distances smaller than 5 nm whereas enhancement has been reported mostly for distances between ~5 and 20 nm [79, 80]. Based on these results, most existing SEF demonstrations,

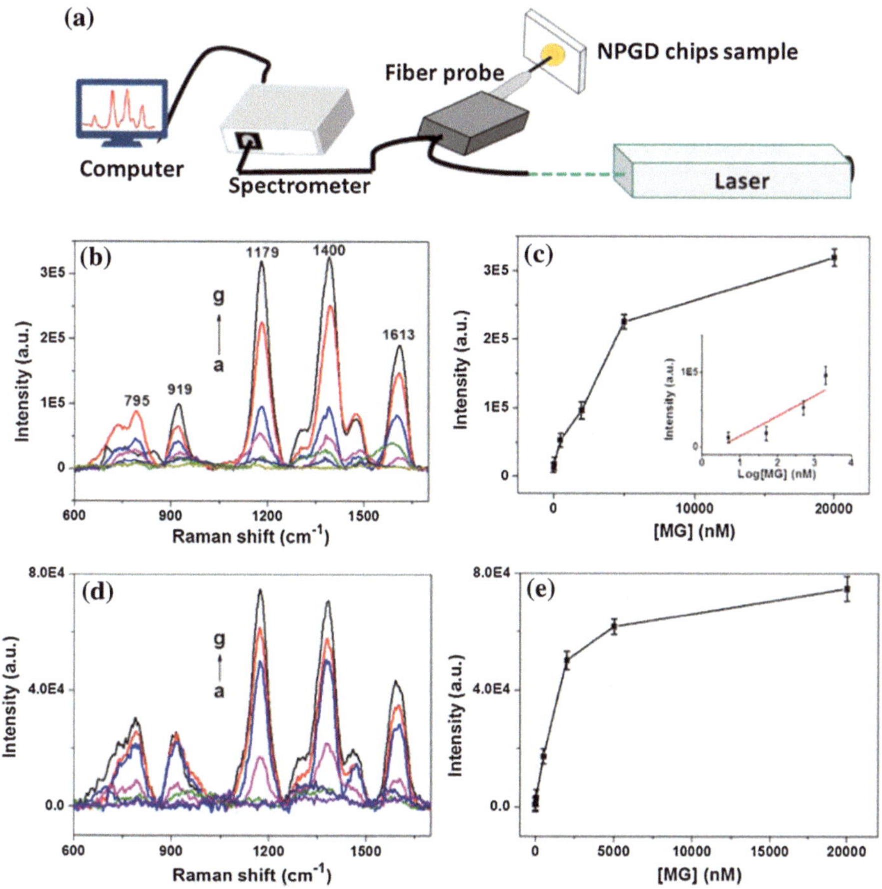

Fig. 27 **a** Portable Raman system with an optical fiber probe; **b** SERS spectra of various MG concentrations in PBS buffer solution; from a–g are 0 nM, 5 nM, 50 nM, 500 nM, 2 μM, 5 μM, and 20 μM, respectively; **c** plot of SERS intensity at 1179 cm^{-1} versus MG concentration, inset: linear range of SERS intensity at 1179 cm^{-1} versus MG concentration 5.nM–2 μM. **d** SERS spectra of fish samples spiked with various MG concentrations; from a–g are 0 nM, 5 nM, 50 nM, 500 nM, 2 μM, 5 μM, and 20 μM, respectively; **e** plot of SERS intensity at 1179 cm^{-1} versus MG concentration. Laser wavelength, 785 nm; acquisition time, 0.5 s; accumulation time, 30 s

including those on thin-film NPG materials, involve spacers to place fluorophores at the "ideal" distance from the metal [81, 82].

Here the label-free NPGD sensor is based on the scheme presented in Sect. 4.2 with a distance modulation scheme based on targeted DNA hybridization to a "hairpin" single-stranded DNA (ssDNA) probe featuring a Cyanine 3 (Cy3) fluorophore on the 3′-end and a sulfur on the 5′-end. The distance between the Cy3 molecule and NPGD can be altered. By taking advantage of the distance modulation, the sensitivity of the NPGD sensor to target DNA with 28 complementary base pairs is demonstrated. The hairpin probe sequence which is employed is complementary to the ERBB2 gene, a target breast cancer DNA biomarker.

5.3.1 Fluorescence Enhancement on NPG Disks Arrays

Fluorescence emission spectra were acquired using home-built inverted fluorescence microscopy systems with 532 and 785 nm CW laser sources. 400 nm diameter NPG disks arrays on glass are fabricated according to the method described in Sect. 2 and the dyes rhodamine 6G (R6G, 526/560 nm), Cy3 (550/570 nm), and IRDye 800 (780/800 nm) are used to investigate the SEF properties of NPG disks, 400 nm Au disks, flat gold, and glass substrates.

For R6G concentration ranging from 100 nM to 1 μM, a net total fluorescence gain up to 50 is measured on NPGD versus fluorescence on glass substrate whereas for lower R6G concentrations, between 20 nM and 5 nM, quenching of fluorescence intensity is observed [74]. It is worth mentioning here that these values should not be interpreted as the SEF enhancement factor on NPGD because the total fluorescence signals are the outcome of combined quenching and enhancement effects which are likely affected by the average surface density of the adsorbed dye molecules.

To characterize the fluorescence enhancement properties of NPGDs, fluorescence emission spectra are collected from three dyes on three different gold substrates: the NPGDs are of 400 nm diameter, 75 nm thickness, and 13 nm average pore size, the non-porous gold disks are of the same external size and the flat gold film is 75 nm in thickness. All fluorescence spectra are normalized to the peak value of the one obtained from the glass substrate. The results are presented on Fig. 28.

NPGD substrates exhibit the strongest total fluorescence for all three dyes, as compared to the other gold substrates (gold disk and gold film). The net gain obtained on NPGD substrates is ~ 7, 7, and 1.75 times compared to the glass substrate for R6G, Cy3, and IRDye 800, respectively. The flat gold substrate

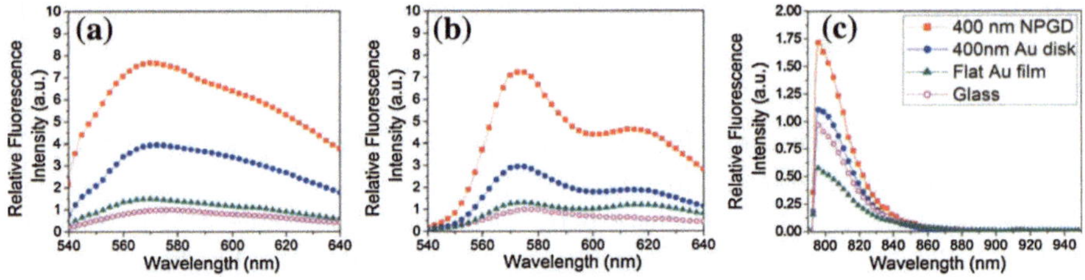

Fig. 28 Comparison of fluorescence spectra from 3 dyes on 4 substrates: **a** R6G and **b** Cy3 dyes dried on NPGD and gold disk samples acquired using a home-built fluorescence microscope with 532 nm excitation, **c** IRDye 800 dye fluorescence measured using a home-built fluorescence microscope with 785 nm excitation. An amount of 5 μL of the 100 nM aqueous dye solutions was drop-casted on each substrate

provides the lowest fluorescence with net gain slightly larger than 1 for R6G and Cy3, and ~0.6 for IRDye, suggesting quenching dominates over enhancement. The non-porous gold disks provide medium fluorescence with net gain ~3–4 times for R6G and Cy3, and 1.15 times for IRDye. The larger net gain on NPGDs is attributed to higher field enhancement within the high-density plasmonic hot spots, which are unique features only seen in NPGDs and this higher enhancement is measured despite the fact that the LSPR peak for the non-porous Au disks is closer to the excitation/emission wavelengths for either the visible or NIR dyes.

5.3.2 Zeptomole Detection of the ERBB2 Cancer DNA Biomarker

A "hairpin" ssDNA probe is employed to detect ERBB2 breast cancer biomarker. The probe features a Cy3 on the 3′-end and a sulfur on the 5′-end that enables effective binding to gold surfaces. If the probe alone is bonded, the Cy3 molecule is positioned at the close proximity of the NPGD surface, which promotes quenching. In contrast, a hybridized dsDNA would have the Cy3 molecule on the opposite end of the sulfur and far away from the NPGD surface after binding. As compared to the un-hybridized hairpin probe on NPGD, higher fluorescence intensity is expected for the dsDNA configuration, providing an effective sensing mechanism. (See Fig. 29) In the previous drop casting method, the likelihood for the fluorophore molecules to be situated in the enhancing hot spots of the NPGD is highly dependent on the local dye-NPGD molecular interactions and dye concentration. In contrast, due to its inherent self-assembly mechanism and robust sulfur-gold covalent binding, the thiolated probe structure ensures a uniform molecular coverage on the gold surface. These features of the hairpin probe minimize the randomness of its molecular attachment to the NPGD substrate and, thus, increases the precision of molecular coverage on the three-dimensional structure of NPGDs.

In Fig. 30, the total Cy3 fluorescence intensity dependence on ssDNA target sequence concentration is presented. Different amounts (0–5 μL) of the ssDNA

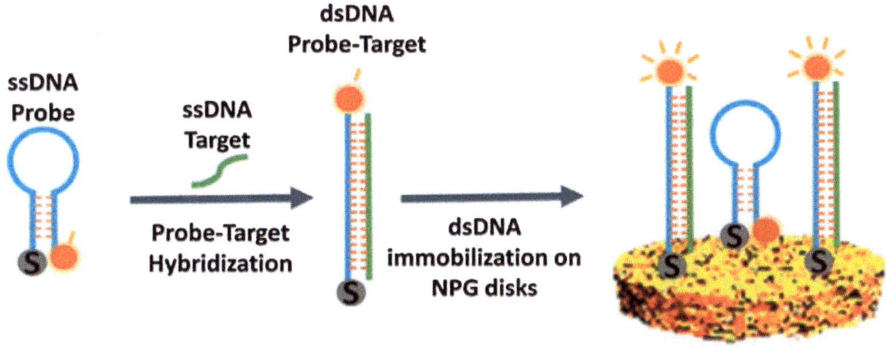

Fig. 29 Schematic of the hairpin probe and probe-target (dsDNA) configurations of the ERBB2 cancer gene, and their immobilization on NPGDs

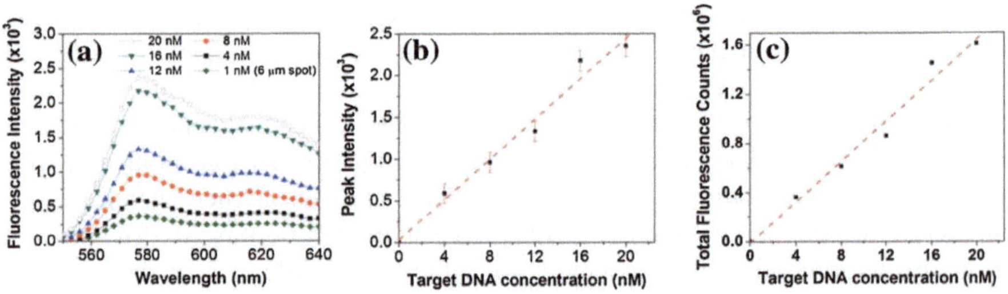

Fig. 30 Fluorescence spectra, peak intensity and total fluorescence counts: **a** with the probe signal used as reference baseline, the difference spectra at different target DNA concentration are plotted as labeled. Corresponding **b** peak intensity and **c** total intensity counts for each concentration step of the target DNA

target sequence (40 nM) are mixed and hybridized with a fixed amount of the hairpin probe (5 μL, 40 nM) at 50 °C for 60 min. With added phosphate buffer solution to a total 10 μL volume, different final concentrations of hybridized dsDNA and non-hybridized probe molecules are obtained. After cooling, 5 μL of each sample is dispensed onto the NPGD substrate surrounded by a PDMS well (2 mm diameter, 4 mm height) and allowed to incubate for 60 min. The surface is then washed with deionized (DI) water and incubated in 0.1 mM mercapto-1-hexanol (MCH) for 30 min to eliminate nonspecific binding, followed by DI water rinse. In Fig. 30, the difference spectra are shown by subtracting the baseline spectrum (0 nM target) from others. At 4 nM target concentration, the amount of dsDNA molecules is estimated to be ~2.4 zeptomole with a signal-to-noise ratio (SNR) of ~123 based on the laser spot size and the surface density calculation. In principle, based on the signal-to-noise ratio, the ultimate limit of detection (LOD) in quantity would be ~0.06 zeptomole for SNR ~3.

Next, the LOD in concentration depends on the measurement spot size. At any fixed target molecules concentration, more fluorescence can be acquired from a larger spot with more molecules. Therefore, lower concentration LOD beyond 4 nM can be achieved. Since we are only measuring the ensemble average of the enhanced population over the laser spot, the detection limit can be only be estimated according to the expected coverage at a given target concentration; the statistical limitation of the signal averaging of the enhanced molecules deters single-molecule detection.

6 Conclusion

Continual scientific discoveries and technological advances in nanoplasmonic over the past two decades have stimulated exciting development of biosensing assays and platforms. The pursuit of measuring trace analytes and biomolecules from a tiny sample volume has driven the field towards high surface-to-volume ratio metallic nanostructures. This is of particular interest to the fields of medical diagnostics, environmental toxin detection, and molecular biology. On one hand, the ability to detect trace disease biomarker at unprecedentedly low concentration can lead to early detection of various diseases such as cancer where the best treatment is early detection. On the other hand, many environmental trace toxins, although do not pose imminent health threat, can have profound negative impact after prolonged exposure and accumulation in biological subjects. Therefore, successful detection at the low concentration can be extremely important. In addition, nanoplasmonic sensing either by refractive index changes or enhanced light–matter interactions can often be implemented in a label-free fashion where the target analytes are in their most original form. Furthermore, the ability to test small volume is a practical advantage for clinical and other applications where sample collection is costly. It is therefore desirable to focus our current efforts on the design of reliable, robust, and low-cost nano-biosensors which can be integrated efficiently into routine laboratory as well as point-of-care testing and environmental monitoring. The central subject of this chapter deals with recent advances in the science and technology of nanoporous gold nanoparticles and arrays with a particular emphasis on their biosensing applications enabled by the unique combinations of large surface-to-volume ratio, three-dimensional high-density hot spots, and architectural integrity and compatibility with other planar fabrication processes.

NPG disks are easy to fabricate, low-cost and very versatile. Their three-dimensional porous network yields a unique surface-to-volume ratio and multiplies the sensing sites that analytes can reach. Their high plasmonic tunability enables one to tailor them to specific needs by changing their interior and exterior morphology. Varying their diameter is achieved straightforwardly thanks to their top-down nano-patterning and the porosity can be modified through thermal or chemical post-treatment. NPG disks can be furthermore monolithically integrated into a microfluidic device and undergo surface modifications to increase their sensing specificity.

Several label-free nanobiosensing examples discussed in the chapter can be classified into two groups: with or without advanced surface functionalization. In the former, we have shown that ultrahigh sensitivity can be achieved in sequence-specific DNA hybridization monitoring at the single-molecule level using molecular sentinel, and direct sensing of small molecules such as malachite green using G-quadruplex as a capturing scaffold. Using advanced surface functionalization greatly improves sensitivity and specificity at the price of limited multiplexing. In the latter, we have demonstrated that bare-surfaced NPG disks can successfully quantify key analytes in unprocessed biological fluids such as urine, with the potential of highly multiplexed assays. We believe that nanoplasmonic biosensing based on NPG nanoparticles and arrays has exciting exciting commercialization potential for biomedical diagnostics and environmental monitoring applications.

References

1. Dror S, William C (2010) Modern introduction to surface plasmons. Cambridge University Press, Cambridge, UK
2. Zia R, Schuller SA, Chandran A, Brongersma ML (2006) Plasmonics: the next chip-scale technology. Mater Today 9(7–8):20–27
3. Polman A, Atwater HA (2005) Plasmonics: optics at the nanoscale. Mater Today 8:56
4. Lal S, Link S, Halas NJ (2007) Nano-optics from sensing to waveguiding. Nat Photonics 11(11):641–648
5. Tokel O, Inci F, Demirci U (2014) Advances in plasmonic technologies for point of care applications. Chem Rev 114(11):5728–5752
6. Vo-Dinh T, Fales AM, Griffin GD, Khoury CG, Liu Y, Ngo H, Norton SJ, Register JK, Wang H-N, Yuan H (2013) Plasmonic nanoprobes: from chemical sensing to medical diagnostics and therapy. Nanoscale 5:10127–10140
7. Sotiriou GA (2013) Biomedical applications of multifunctional plasmonic nanoparticles. WIREs Nanomedicine Nanobiotechnoly 5:19–30
8. Shih W-C, Santos GM, Zhao F, Zenasni O, Arnob MMP (2016) Simultaneous chemical and refractive index sensing in the 1–2.5 μm near-infrared wavelength range on nanoporous gold disks. Nano Lett 16:4641–4647
9. Le Ru E, Etchegoin P (2008) Principles of Surface-Enhanced Raman Spectroscopy and related plasmonic effects. Elsevier
10. Brolo AG (2012) Plasmonics for future biosensors. Nat Photonics 6(11):709–713
11. Pitsillides CM, Joe EK, Wei X, Anderson RR, Lin CP (2003) Selective cell targeting with light-absorbing microparticles and nanoparticles. Biophys J 84:4023–4032
12. Huang X, Jain PK, El-Sayed IH, El-Sayed MA (2006) Determination of the minimum temperature required for selective photothermal destruction of cancer cells with the use of immunotargeted gold nanoparticles. Photochem Photobiol 82(2):412–417
13. Hirsch LR, Stafford JR, Bankson JA, Sershen SR, Rivera B, Price R, Hazle JD, Halas NJ, West JL (2003) Nanoshell-mediated near-infrared thermal therapy of tumors under magnetic resonance guidance. Proc Natl Acad Sci 100(23):13549–13554
14. Loo C, Lowery A, Halas N, West J, Drezek R (2005) Immunotargeted nanoshells for integrated cancer imaging and therapy. Nano Lett 5(4):709–711
15. Biener J, Nyce GW, Hodge AM, Biener MM, Hamza AV, Maier SA (2008) Nanoporous plasmonic metamaterials. Adv Mater 20(6):1211–1217
16. Lang X, Qian L, Guan P, Zi J, Chen M (2011) Localized surface plasmon resonances of nanoporous gold. Appl Phys Lett 98(9):093701
17. Liu H, Zhang L, Lang X, Yamaguchi Y, Iwasaki H, Inouye Y, Xue Q, Chen M (2011) Single molecule detection from a large-scale SERS-active $Au_{79}Ag_{21}$ substrate. Sci Rep 1:112
18. Qi J, Motwani P, Gheewala M, Brennan C, Wolfe JC, Shih W-C (2013) Surface-enhanced Raman spectroscopy with monolithic nanoporous gold disk substrates. Nanoscale 5(10): 4105–4109
19. Ruan W-D, Lu Z-C, Ji N, Wang C-X, Bing Z, Zhang J-H (2007) Facile fabrication of large area polystyrene colloidal crystal monolayer via surfactant-free Langmuir-Blodgett technique. Chem Res Chin Univ 23(6):712–714
20. Parida S, Kramer D, Volkert CA, Rosner H, Erlebacher J, Weissmuller J (2006) Volume change during the formation of nanoporous gold by dealloying. Phys Rev Lett 97(3):035504
21. Crowson DA, Farkas D, Corcoran SG (2007) Geometric relaxation of nanoporous metals: the role of surface relaxation. Scr mater 56(11):919–922
22. Read JS (1988) Introduction to the principle of ceramic processing. Wiley
23. Seker E, Berdichevsky Y, Begley MR, Reed ML, Staley KJ, Yarmush ML (2010) The fabrication of low-impedance nanoporous gold multiple-electrode arrays for neuralelectrophysiology studies. Nanotechnology 21(12):125504

24. Zhao F, Zeng J, Santos GM, Shih W-C (2015) In situ patterning of hierarchical nanoporous gold structures by in-plane dealloying. Mater Sci Eng B 194:34–40
25. Li J, Zhao F, Shih W-C (2016) Direct-write patterning of nanoporous gold microstructures by in situ laser-assisted dealloying. Opt Express 24(20):23610–23617
26. Qian L, Chen M (2007) Ultrafine nanoporous gold by low-temperature dealloying and kinetics of nanopore formation. Appl Phys Lett 91(8):083105
27. Strehle KR, Cialla D, Rosch P, Henkel T, Kohler M, Popp J (2007) A reproducible surface-enhanced Raman spectroscopy approach. Online SERS measurements in a segmented microfluidic system. Anal Chem 79(4):1542–1547
28. Quang LX, Lim C, Seong GH, Choo J, Do KJ, Yoo S-K (2008) A portable surface-enhanced Raman scattering sensor integrated with a lab-on-a-chip for field analysis. Lab Chip 8(12): 2214–2219
29. Sun J, Xianyu Y, Jiang X (2014) Point-of-care biochemical assays using gold nanoparticle-implemented microfluidics. Chem Soc Rev 43(17):6239–6253
30. Dee KC, Puleo DA, Bizios R (2003) An introduction to tissue-biomaterial interactions. Wiley
31. Santos GM, Zhao F, Zeng J, Shih W-C (2014) Characterization of nanoporous gold disks for photothermal light harvesting and light-gated molecular release. Nanoscale 6(11):5718–5724
32. Qi J, Zeng J, Zhao F, Lin SH, Raja B, Strych U, Willson RC, Shih W-C (2014) Label-free, in situ SERS monitoring of individual DNA hybridization in microfluidics. Nanoscale 6(15):8521–8526
33. Li M, Zhao F, Zeng J, Qi J, Lu J, Shih W-C (2014) Microfluidic surface-enhanced Raman scattering sensor with monolithically integrated nanoporous gold disk arrays for rapid and label-free biomolecular detection. J Biomed Opt 19(11):111611
34. Li M, Li S, Cao W, Li W, Wen W, Alici G (2012) Continuous particle focusing in a waved microchannel using negative DC dielectrophoresis. J Micromech Microeng 22(9):095001
35. Ding Y, Chen M (2009) Nanoporous metals for catalytic and optical applications. MRS Bull 34(08):569–576
36. Yu F, Ahl S, Caminade A-M, Majoral J-P, Knoll W, Erlebacher J (2006) SPP and LSPR in NPG membranes. Anal Chem 78(20):7346–7350
37. Wittstock A, Biener J, Erlebacher J (2012) Nanoporous gold: from an ancient technology to a high-tech material. R Soc Chem
38. Ryckman JD, Jiao Y, Weiss SM (2013) Three-dimensional patterning and morphological control of porous nanomaterials by gray-scale direct imprinting. Sci Rep 3
39. Halas NJ, Lal S, Link S, Chang W-S, Natelson D, Hafner JH, Nordlander P (2012) A plethora of plasmonics from the laboratory for nanophotonics at Rice University. Adv Mater 24(36): 4842–4877
40. Zeng J, Zhao F, Qi J, Li Y, Li C-H, Yao Y, Lee RT, Shih W-C (2014) Internal and external morphology-dependent plasmonic resonance in monolithic nanoporous gold nanoparticles. RSC Adv 4(69):3688–36682
41. Zhao F, Zeng J, Arnob MMP, Sun P, Qi J, Motwani P, Gheewala M, Li C-H, Paterson A, Strych U, Raja B, Willson RC, Wolfe JC, Lee TR, Shih W-C (2014) Monolithic NPG nanoparticles with large surface area, tunable plasmonics and high-density internal hot spots. Nanoscale 6(14):8199–8207
42. Camden JP, Dieringer JA, Zhao J, Van Duyne RP (2008) Controlled plasmonic nanostructures for surface-enhanced spectroscopy and sensing. Acc Chem Res 41(12):1653–1661
43. Kucheyev SO, Hayes JR, Biener J, Huser T, Talley CE, Hamza AV (2006) Surface-enhanced Raman scattering on nanoporous Au. Appl Phys Lett 89(5):053102
44. Gloria D, Gooding JJ, Moran G, Hibbert BD (2011) Electrochemically fabricated three dimensional nano-porous gold films optimised for surface enhanced Raman scattering. J Electroanal Chem 656(1):114–119
45. Li Z, Yang Y, Xia Y, Huang W, Zheng J, Li Z (2012) Fabrication of nano-network gold films via anodization of gold electrode and their application in SERS. J Solid State Electrochem 16(4):1733–1739

46. Aggarwal RL, Farrar LW, Diebold ED, Polla DL (2009) Measurement of the absolute Raman scattering cross section of the 1584-cm^{-1} band of benzenethiol and the surface-enhanced Raman scattering cross section enhancement factor for femtosecond laser-nanostructured substrates. J Raman Spectrosc 40(9):1331–1333
47. Gui JY, Stern DA, Frank DG, Lu F, Zapien DC, Hubbard AT (1991) Adsorption and surface structural chemistry of thiophenol, benzyl mercaptan, and alkyl mercaptans. Comparative studies at silver (111) and platinum (111) electrodes by means of Auger spectroscopy, electron energy loss spectroscopy, low energy electron dif. Langmuir 7(5):955–963
48. Jiao Y, Ryckman JD, Ciesielski PN, Escobar CA, Jennings KG, Weiss SM (2011) Patterned nanoporous gold as an effective SERS template. Nanotechnology 22(29):295302
49. Sun Y, Xia Y (2002) Increased sensitivity of surface plasmon resonance of gold nanoshells compared to that of gold solid colloids in response to environmental changes. Analitycal Chem 74(20):5297–5305
50. Hanarp P, Käll M, Sutherland DS (2003) Optical properties of short range ordered arrays of nanometer gold disks prepared by colloidal lithography. J Phys Chem B 107(24):5768–5772
51. Hu M, Chen J, Marquez M, Xia Y, Hartland GV (2007) Correlated rayleigh scattering spectroscopy and scanning electron microscopy studies of Au-Ag bimetallic nanoboxes and nanocages. J Phys Chem C 111(34):12558–12565
52. Jain PK, Huang X, El-Sayed IH, El-Sayed MA (2008) Noble metals on the nanoscale: optical and photothermal properties and some applications in imaging, sensing, biology, and medicine. Acc Chem Res 41(12):1578–1586
53. Wang H, Brandl DW, Le F, Nordlander P, Halas NJ (2006) Nanorice: a hybrid plasmonic. Nano Lett 6(4):827–832
54. Larsson EM, Alegret J, Käll M, Sutherland DS (2007) Sensing characteristics of NIR localized surface plasmon resonances in gold nanorings for application as ultrasensitive biosensors. Nano Lett 7(5):1256–1263
55. Qi J, Shih W-C (2012) Parallel Raman microspectroscopy using programmable multipoint illumination. Opt Lett 37(8):1289–1291
56. Park SG, Lee NS, Lee SH (2000) Vibrational analysis of dopamine neutral Bae based on density functional force field. Bull Korean Chem Soc 21(10):1035–1038
57. Sassolas A, Leca-Bouvier BD, Blum LJ (2008) DNA biosensors and microarrays. Chem Rev 108(1):109–139
58. Lu Y, Liu GL, Kim J, Mejia YX, Lee LP (2005) Nanophotonic crescent moon structures with sharp edge for ultrasensitive biomolecular detection by local electromagnetic field enhancement effect. Nano Lett 5(1):119–124
59. Kang T, Yoo SM, Yoon I, Lee SY, Kim B (2010) Patterned multiplex pathogen DNA detection by Au particle-on-wire SERS sensor. Nano Lett 10(4):1189–1193
60. Wang H-N, Dhawan A, Du Y, Batchelor D, Leonard DN, Misra V, Vo-Dinh T (2013) Molecular sentinel-on-chip for SERS-based biosensing. Phys Chem Chem Phys 15(16):6008–6015
61. Wang H-N, Fales AM, Zaas AK, Woods CW, Burke T, Ginsburg GS, Vo-Dinh T (2013) Surface-enhanced Raman scattering molecular sentinel nanoprobes for viral infection diagnostics. Anal Chim Acta 786:153–158
62. Cao YC, Jin R, Mirkin CA (2002) Nanoparticles with Raman spectroscopic fingerprints for DNA and RNA detection. Science 297(5586):1536–1540
63. Shih WC (2014) Label-free in situ SERS monitoring of individual DNA hybridization in microfluidics. Nanoscale 6(5):8521–8526
64. Li M, Lu J, Qi J, Zhao F, Zeng J, Yu JC-C, Shih W-C (2014) Stamping surface-enhanced Raman spectroscopy for label-free, multiplexed, molecular sensing and imaging. J Biomed Opt 19(5):050501
65. Li M, Du Y, Zhao F, Zeng J, Mohan C, Shih W-C (2015) Reagent-and separation-free measurements of urine creatinine concentration using stamping surface enhanced Raman scattering (S-SERS). Biomed Opt Express 6(3):849–858

66. Xie C, Sharma R, Wang H, Zhou XJ, Mohan C (2004) Strain distribution pattern of susceptibility to immune-mediated nephritis. J Immunol 172(8):5047–5055
67. Qiu S, Zhao F, Zenasni O, Li J, Shih W-C (2016) Nanoporous gold disks functionalized with stabilized G-quadruplex moieties for sensing small molecules ACS Appl Mater Interfaces 8(44):29968–29976
68. Bhasikuttan AC, Mohanty J (2015) Targeting G-quadruplex structures with extrinsic fluorogenic dyes: promising fluorescence sensors. Chem Commun 51(36):7581–7597
69. Biffi G, Di Antonio M, Tannahill D, Balasubramanian S (2014) Visualization and selective chemical targeting of RNA G-quadruplex structures in the cytoplasm of human cells. Nat Chem 6(1):75–80
70. Olejko L, Cywinski PJ, Bald I (2015) Ion-Selective formation of a guanine quadruplex on DNA origami structures. Angew Chem Int Ed 54(2):673–677
71. Koirala D, Dhakal S, Ashbridge B, Sannohe Y, Rodriguez R, Sugiyama H, Balasubramanian S, Mao H (2011) A single-molecule platform for investigation of interactions between G-quadruplexes and small-molecule ligands. Nat Chem 3(10):782–787
72. Bhasikuttan AC, Mohanty J, Pal H (2007) Interaction of malachite green with guanine-rich single-stranded DNA: preferential binding to a G-Quadruplex. Angew Chem Int Ed 46(48): 9305–9307
73. Srivastava S, Sinha R, Roy D (2004) Toxicological effects of malachite green. Aquat Toxicol 66(3):319–329
74. Santos GM, Zhao F, Zeng J (2015) Label-free, zeptomole cancer biomarker detection by surface-enhanced fluorescence on nanoporous gold disk plasmonic nanoparticles. J Biophotonics 8(10):855–863
75. Geddes CD, Parfenov A, Roll D, Gryczynski I, Malicka J, Lakowicz JR (2003) Silver fractal-like structures for metal-enhanced fluorescence: enhanced fluorescence intensities and increased probe photostabilities. J Fluoresc 13(3):267–276
76. Gartia MR, Hsiao A, Sivaguru M, Chen Y, Liu LG (2011) Enhanced 3D fluorescence live cell imaging on nanoplasmonic substrate. Nanotechnology 22(36):365203
77. Chen Y, Munechika K, Ginger DS (2007) Dependence of fluorescence intensity on the spectral overlap between fluorophores and plasmon resonant single silver nanoparticles. Nano Lett 7(3):690–696
78. Ranjan Gartia M, Eichorst JP, Clegg RM, Logan Liu G (2012) Lifetime imaging of radiative and non-radiative fluorescence decays on nanoplasmonic surface. Appl Phys Lett 101 (2):023118
79. Anger P, Bharadwaj P, Novotny L (2006) Enhancement and quenching of single-molecule fluorescence. Phys Rev Lett 96(11):113002
80. Campion A, Gallo AR, Harris CB, Robota HJ, Whitmore PM (1980) Electronic energy transfer to metal surfaces: a test of classical image dipole theory at short distances. Chem Phys Lett 73(3):447–450
81. Lang XY, Guan PF, Fujita T, Chen M (2011) Tailored nanoporous gold for ultrahigh fluorescence enhancement. Phys Chem Chem Phys 13(9):3795–3799
82. Lang XY, Guan PF, Zhang L, Fujita T, Chen M (2010) Size dependence of molecular fluorescence enhancement of nanoporous gold. Appl Phys Lett 96(7):073701

Microfluidic Mixing for Biosensors

N. Scott Lynn Jr.

Abstract Among other factors, the performance of an affinity-based biosensor is dependent on the rate at which analyte is transported to, and captured by, its active sensing surface. The efficiency of analyte delivery can be increased via the use of microfluidics, albeit not without detraction, as microfluidic biosensors are often subjected to severe diffusion limitations when used for the detection of biologically relevant analytes. Such conditions lead to the formation of a boundary layer, void of analyte, which acts to resist the rate at which analyte is captured. It is often proposed to mix the fluid in the sensing chamber, where the exchange of depleted solution with fresh analyte can potentially increase sensor performance. The nature of analyte transport in a mixed channel is complex, however, and simply mixing the contents of a microchannel does not guarantee success. In this chapter, we review developments in the characterization (and prediction of) analyte transport in both mixed and unmixed channels. Our discussion focuses on the conditions under which mixing will (and will not) be beneficial and furthermore, the magnitude of performance increase that can be expected. Special attention is given to flow in the staggered herringbone mixer (SHM): a passive chaotic micromixer often used to enhance the performance of a biosensor. We review relevant experimental works on the topic and compare the results from several studies with the behavior expected from theory. Finally, we note several challenging aspects regarding the detection of circulating tumor cells which, due to their large size, are subject to additional transport mechanisms with respect to smaller analytes.

Keywords Biosensor · Microfluidics · Mixing · Chaos
Transport phenomena · Circulating tumor cell

N.S. Lynn Jr. (✉)
Institute of Photonics and Electronics, Academy of Sciences of the Czech Republic, Chaberská 57, 18251 Prague, Czech Republic
e-mail: lynn@ufe.cz

S.-H. Oh et al. (eds.), *Miniature Fluidic Devices for Rapid Biological Detection*, Integrated Analytical Systems, https://doi.org/10.1007/978-3-319-64747-0_3

1 Introduction

Affinity-based biosensors represent an increasingly prevalent analytic tool for detection, research, and (bio)analytic purposes. Reports on such devices span a period of nearly three decades—their use has become omnipresent across both research and commercial settings—where end-use applications include medical diagnostics and drug discovery, as well as agricultural, environmental, and food monitoring [1].

The scope of these devices is extremely broad: transduction mechanisms can proceed via optical [2–4], electrochemical [5, 6], micromechanical [7], piezoelectric [8], thermometric [9], and magnetic means [10]. In addition, the list of biorecognition elements is just as large and includes antibodies, enzymes, and engineered proteins; nucleic acids and aptamers; and tissues, cells, and microorganisms. The bulk of these devices are surface-based, whereby the biorecognition elements are immobilized to a region on, or near, the active transduction element(s).

For most of these devices, the sensor output (i.e., the sensor signal) is proportional to the rate of analyte captured by the active regions of the sensor surface. In general, this condition remains independent of both the transduction mechanism and the nature of the biorecognition element. To increase the efficiency of analyte delivery, biosensing devices are often paired with a microfluidic flow cell: a trend that grew rapidly after the advent of low-cost, user-friendly, and time-efficient microfabrication methods [11, 12]. The inherently small dimensions of a microchannel allow for reduced sample volume requirements (ranging from mL down to μL) and perhaps more importantly, enhanced sensor response times [13].

The majority of microfluidic biosensors are constructed in a fairly simple manner, whereby a rectangular microchannel is situated over a planar sensing surface having immobilized biorecognition elements.[1] A schematic of this generalized layout is shown in Fig. 1, and Table 1 lists the pertinent geometrical and operational parameters. The majority of biosensing processes are operated with pressure-driven flow, as optimal conditions for electrokinetic flows (e.g., DC or AC electroosmosis [14]) are not often compatible for surface-based biosensors. Due to the small characteristic lengths, pressure-driven microchannel flows are associated with a low Reynolds number, $\mathrm{Re} = \rho UH/\mu < 100$, where flows are uniaxial and void of turbulence.

Under certain (and often encountered) conditions, microfluidic biosensors can exhibit inconvenient behavior. Biosensors having a relatively large sensing region size with fast interaction kinetics are often diffusion-limited, where the resulting analyte depletion layer acts to resist analyte delivery. In such cases, an increase in sensor response can be obtained with a higher sample flow rate (i.e., higher rate of analyte delivery), albeit at the sacrifice of efficient sample use: a 10× increase in sensor response frequently necessitates a 1000× increase in flow rate [15]. Benefits can also be obtained by simple changes to the flow cell geometry, for example, by decreasing the channel height [16]; however, such changes are limited by large

[1]The area for signal transduction is often the same as that for biocomponent immobilization, which is assumed herein.

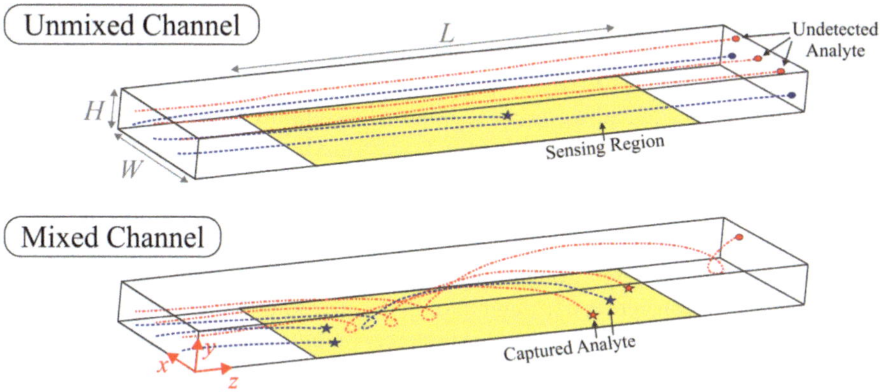

Fig. 1 The uniaxial flow in microfluidic channels often leads to a large percentage of analyte that does not interact with the sensing surface. The inclusion of mixing can increase the efficiency of analyte capture

Table 1 List of the dimensional parameters and values for the experimental sensors used in this study

Dimensional Parameter	Symbol	Units
Geometrical Parameters		
Microchannel height	H	mm
Microchannel width	W	mm
Length of sensing region	L	mm
Operational Parameters		
Volumetric flow rate	Q	$mm^3\ s^{-1}$
Fluid viscosity	μ	$g\ mm^{-1}\ s^{-1}$
Fluid density	ρ	$g\ mm^{-3}$
Diffusivity of analyte	D	$mm^2\ s^{-1}$
Kinetic Parameters		
Association rate constant	k_1	$mm^3\ mol^{-1}\ s^{-1}$
Dissociation rate constant	k_2	s^{-1}
Average surface density of bioreceptors	Γ_o	$mol\ mm^{-2}$
Dependent Parameters		
Average surface density of captured analyte	Γ	$mol\ mm^{-2}$
Analyte collection flux	J	$mol\ mm^{-2}\ s^{-1}$

increases in viscous resistance. Methods to remove the analyte boundary layer, either by fluid sheathing [17] or by fluidic removal of the depleted fluid [18], come at the cost of increased complexity.

The problems associated with diffusion-limited biosensors originate from the uniaxial flow profile inherent to rectangular microchannels at low Re. This problem can be alleviated by mixing the fluid above the sensing region. The exchange of depletion layers with fresh solution can potentially increase both the rate of analyte capture and the efficiency of sample use. The use of mixing to enhance mass (or heat) transfer in laminar flows is a long-studied problem within both chemical and mechanical engineering disciplines [19, 20].

Microfluidic mixing, however, is a relatively complex process, and extensive lengths often need to be taken to ensure proper mixing conditions. The literature on microfluidic mixing is quite extensive and, in general, can be separated into two broad categories. The first category encompasses *active* mixers, which utilize an external energy source to manipulate and mix fluid, typically in a controllable manner. Active mixers operate through a variety of mechanisms, among which include electroosmotic, electrophoretic, magnetic, and electrothermal effects. The second category encompasses *passive* mixers, which rely on the geometry of the microchannel to mix fluid and thus require no additional power, albeit often in a noncontrollable manner. Passive mixing strategies are generally based on either channel modifications to introduce non-axial flow (e.g., mixing grooves), or channel arrangements such that the flow is systematically split and recombined. There are currently a number of reviews dedicated to both groups of micromixers, to which we refer the reader [21–25].

Upon scanning the literature, one finds that only a small percentage of the literature on microfluidic mixing is dedicated to improving biosensor performance. The vast majority of these studies are focused on the staggered herringbone mixer (SHM), a passive mixer that mixes fluid in a chaotic fashion [26]. Of these studies, there are a wide range of reported values regarding level of improvement that can be attributed to the inclusion of mixing. For example, the performance improvement via the use of the SHM have ranged from 0% [27] to 26% [28] for the detection of streptavidin, whereas a similar mixer offered a 170% improvement for the detection of circulating tumor cells (CTCs) [29].

Therefore, before beginning an attempt to modify an existing biosensor with the inclusion of a microfluidic mixer (taken individually, both processes are quite complicated), it is important to ask the following two questions:

(A) Is an improvement in sensing performance expected?
(B) If so, what is the expected magnitude of improvement?

This chapter aims to provide the answers to both of these questions. We start by reviewing the concepts related to the mass transfer of analyte in an unmixed microchannel, specifically, the convective and diffusive transport of analyte and its (reactive) capture. We use these concepts to estimate the conditions under which a biosensor can be expected to be diffusion- or reaction-limited.[2] We then review, from a theoretical perspective, how the inclusion of mixing might serve to increase rates of analyte transport and furthermore, how to predict such rates. Finally, we review the literature on the use of the staggered herringbone mixer for biosensing purposes and compare the results from several experimental studies with the behavior expected from theory.

[2]As discussed later, the distinction between the two is very important: a reaction-limited biosensor will never benefit from the inclusion of mixing.

Before starting, however, it is worth noting that this topic also maintains relevance within a variety of other fields. In addition to (bio)sensing applications, control over the mass transfer of a dissolved species to the wall of a microchannel is sought after in a variety of other applications, including microfluidic fuel cells [18], the controlled deposition of nanostructured materials [30], continuous flow (heterogeneous) catalytic microreactors [31], and membrane absorbers [32].

2 Analyte Transport in Unmixed Channels

The capture of analyte by a microfluidic biosensor maintains a complex dependency on a number of factors, including the architecture of both the microchannel and sensing regions, the flow rate of sample solution, the diffusivity of analyte, and several (kinetic) parameters related to the interaction between biorecognition element and analyte. Fortunately, both convection and diffusion are well-understood phenomena on the microfluidic scale, where fluid is restricted to the laminar regime (Re < 100).

A number of theoretical, computational, and experimental reports have analyzed the problem related to unmixed sensors. Particularly useful to this chapter are those by Myszka et al. [33] and Squires et al. [15]. In this section, we summarize the problem related to unmixed sensors and furthermore, provide a simple guide to predict analyte transport to a generalized biosensor as shown in Fig. 1.

2.1 *Governing Equations*

The capture of an analyte in a microchannel is best described in mathematical terms. For steady, laminar, incompressible flow, the velocity vector field $\boldsymbol{v} = \boldsymbol{v}(x, y, z)$ is represented by both the Stokes equation

$$-\nabla p + \mu \nabla^2 \boldsymbol{v} = 0 \tag{1}$$

and the equation of continuity,

$$\nabla \cdot \boldsymbol{v} = 0. \tag{2}$$

where $p = p(x, y, z)$ is the fluid pressure, and a no-slip condition $\boldsymbol{v} = 0$ is assumed on the channel walls. For pressure-driven flows in rectangular channels (low Re, sufficiently far from the inlet), the x- and y-components of the fluid velocity vanish (i.e., the flow becomes uniaxial), and the solution to both Eqs. (1) and (2) can be solved via a Fourier series solution as

$$v_z = \frac{2U\left[1 - \left(\frac{2y}{H}\right)^2 + 4\sum_{n=0}^{\infty} \frac{(-1)^{n+1}}{m_n^3 \cosh\left(\frac{m_n W}{H}\right)} \cosh\left(\frac{2m_n x}{H}\right)\cos\left(\frac{2m_n y}{H}\right)\right]}{\frac{4}{3} - \frac{8H}{W}\sum_{n=0}^{\infty} m_n^{-5} \tanh\left(\frac{m_n W}{H}\right)}, \quad (3)$$

where $m_n = \pi(2n+1)/2$, and the domain consists of the region $-W/2 < x < W/2$ and $-H/2 < y < H/2$. Often microchannels are fabricated with dimensions such that $W \gg H$, whereby edge effects can be ignored and the fluid velocity profile becomes only a function of y, where the velocity profile then follows

$$v_z = \frac{3U}{2}\left(1 - 4\frac{y^2}{H^2}\right). \quad (4)$$

The concentration of analyte in the microchannel $c = c(x, y, z, t)$ is described by the unsteady convection–diffusion equation,

$$\frac{\partial c}{\partial t} + \boldsymbol{v} \cdot \nabla c = D\nabla^2 c. \quad (5)$$

To account for the capture of analyte a standard ligand–receptor mechanism is often assumed, $A_s + B \rightleftharpoons AB$, where A_s represents the aqueous analyte present at the sensor surface (with concentration $c_s = c_s(x, z, t)$), B represents the immobilized (free) receptor (with surface density $\beta = \beta(x, z, t)$), and AB the bound analyte/receptor complex (with surface density $\gamma = \gamma(x, z, t)$).

Under proper conditions the analyte will interact only with the sensing region (where receptors are immobilized), whereby the remaining surfaces will resist the adsorption of analyte and can be considered to be passivated. Assuming elementary reaction kinetics and ignoring surface diffusion, the boundary conditions for Eq. (5) along each respective surface can thus be written as

$$\text{sensing region: } \hat{n} \cdot D\nabla c = k_1 c_s \beta - k_2 \gamma, \quad (6)$$

$$\text{passivated surfaces: } \hat{n} \cdot D\nabla c = 0, \quad (7)$$

where $\hat{n}$ is the normal unit vector along the sensing region. Equation (6) represents the balance between the transport of aqueous analyte toward the sensor surface (or more specifically, its diffusive flux $D\nabla c$) with the rate of analyte capture on said surface. The rate of analyte capture can be written in terms of an analyte collection flux $j = j(x, z, t)$, where

$$j = \frac{\partial \gamma}{\partial t} = k_1 c_s \beta - k_2 \gamma. \quad (8)$$

This value is especially pertinent to biosensing, as the response of a biosensor is often proportional to γ (the density of captured analyte), rather than the absolute number of captured analyte molecules.

Unfortunately, Eq. (8) cannot be used directly to calculate this flux, as the distribution of c_s is unknown without the solution to Eqs. (5)–(7). An analytical solution to these equations is not possible due to the parabolic nature of the pressure-driven velocity profile. Approximate solutions can be obtained by assuming a linear velocity profile, yet they are only valid for "fast" flows having small boundary layers. Discrete solutions can be obtained via computational methods (typically via finite-element or finite-volume methods); however, even modern computational packages can be quite cumbersome and furthermore, the problem has an extensive parameter space (seen in Table 1).

2.2 *Macroscopic Approach: Averaged Rates of Transport*

Rather than examination of analyte transport in a microscopic manner (i.e., the solutions of Eqs. (5)–(7) over the entire sensing chamber), it is useful to examine the problem from a macroscopic point of view. In this approach, one assumes that the sensor response can be sufficiently estimated via knowledge of $\Gamma = \Gamma(t)$: the *average* surface density of captured analyte over the sensing region.

Given the (experimentally measurable) parameters shown in Table 1, estimations of Γ, or more appropriately $d\Gamma/dt$, can be obtained in a relatively simple manner by assuming the process of analyte capture occurs in two steps: (i) the convective and diffusive transport of analyte from the bulk solution to the sensor surface and (ii) the affinity-based capture of said analyte by immobilized bioreceptors. A reaction mechanism for such a process can be written as

$$\mathrm{A}_b \underset{K_m}{\overset{K_m}{\rightleftharpoons}} \mathrm{A}_s + \mathrm{B} \underset{k_2}{\overset{k_1}{\rightleftharpoons}} \mathrm{AB} \tag{9}$$

where A_b represents aqueous analyte far from the sensing region (with concentration C_o). The rate constants for the first step in Eq. (9) are represented by the macroscopic mass transfer coefficient K_m, defined as

$$J = K_m(C_o - C_s), \tag{10}$$

where $C_s = C_s(t)$ is the average analyte concentration along the sensor surface, and $J = J(t)$ is the average analyte capture flux over the sensor surface.[3] As shown later, K_m is related to the shape and size of the analyte boundary layer, and is dependent on many of the parameters listed in Table 1.

[3]The rate at which the first step in Eq. (9) proceeds is dependent only on the magnitude of the difference between C_s and C_o, and hence the presence of K_m in both the forward and reverse steps.

The reaction mechanism in Eq. (9) can be used to formulate a set of two coupled ordinary differential equations, obtained via a mass balance over A_s and AB, respectively:

$$\frac{dC_s}{dt} = K_m C_o - K_m C_s - k_1 C_s(\Gamma_o - \Gamma) + k_2 \Gamma, \tag{11}$$

$$\frac{d\Gamma}{dt} = k_1 C_s(\Gamma_o - \Gamma) - k_2 \Gamma, \tag{12}$$

where Γ_o is the average surface density of immobilized bioreceptors.

We first examine the behavior of these two equations in the limits where mass transport occurs both very fast and very slow with respect to the rate of reactive capture. The former can be taken as the case where $K_m \to \infty$, whereby the solution to Eq. (11) leads to $C_s \to C_o$. This represents the reaction-limited regime (sometimes referred to as being "well-mixed"), where Eq. (11) can be used to estimate the rate of analyte collection as $d\Gamma/dt = k_1 C_o(\Gamma_o - \Gamma) - k_2 \Gamma$.

The opposite situation represents the diffusion-limited regime, where mass transfer is slow and $K_m \to 0$; in the initial stages of such an assay, the amount of captured analyte is small ($\Gamma \approx 0$) and it follows that $C_s \to 0$. These conditions lead to the formation of an analyte boundary layer with variable shape and size (see Fig. 2). When such boundary layer has reached a steady state, the rate of analyte collection can be estimated as $d\Gamma/dt = K_m C_o$.

Equations (11) and (12) can be simplified by noting that the capture of analyte will often have a relatively small impact on the shape and size of the analyte boundary layer.[4] Under such conditions a quasi-steady approximation can be assumed, where $dC_s/dt \approx 0$, and Eq. (11) can be solved explicitly for C_s. Substitution into Eq. (12) then leads to

$$\frac{d\Gamma}{dt} = \frac{k_1 C_o(\Gamma_o - \Gamma) - k_2 \Gamma}{1 + k_1(\Gamma_o - \Gamma)/K_m}. \tag{13}$$

The conditions under which a biosensor can be considered to be quasi-steady have been discussed by a variety of authors; more information can be found in the works by Squires et al. [15], Eddowes [34], and Glaser [35].

As seen in the denominator of Eq. (13), the relative magnitude of K_m with respect to $k_1(\Gamma_o - \Gamma)$ aids to provide information on the limiting step in the reaction mechanism described by Eq. (9). The ratio of these two values is a dimensionless number known as the Damköhler number Da, defined here as $\text{Da} = k_1 \Gamma_o / K_m$, as we are often interested in the behavior of a sensor at the beginning of an assay. It follows that conditions leading to $\text{Da} \gg 1$ and $\text{Da} \ll 1$ represent a sensor that is diffusion- and reaction-limited, respectively.

This macroscopic approach is often referred to as a two-compartmental model and, as evidenced by its wide use in the biosensing community, can be quite useful; for conditions such that $\text{Da} \lesssim 1$, Eq. (13) is often used to extract bioanalytical

---

[4]More specifically, the characteristic time for the analyte boundary layer to reach equilibrium is often much smaller than the characteristic time for the analyte to reach equilibrium.

data (usually k_1 and k_2) from time-series biosensor signals [33, 36]. In addition, and especially pertinent to this chapter, Eq. (13) can be used to predict how changes in K_m will affect the sensor response.

We now have enough information to answer the first question: (A) is an improvement in sensing performance expected with the inclusion of mixing? In general—although, as seen later, not always true—analyte transport in mixed microchannels will be more efficient with respect to unmixed channels of similar dimension, where the process of mixing serves to increase the magnitude of K_m. However, this increase might be redundant: if an unmixed biosensor is operated under conditions such that Da $\ll 1$, then the inclusion of mixing will have no effect on the analyte collection rate.

Conversely, the performance of diffusion-limited biosensors can be increased significantly under mixed conditions: for an unmixed sensor with Da $\gg 1$, an increase in the analyte collection rate will be directly proportional to the increase in K_m.

This leads us to reformulate our second question: (B) how does a mixing process affect K_m? Before attempting to answer that question, however, we must know the magnitude of K_m for an unmixed channel. In the next two sections, we give a brief review on predicting such values. Due to the extensive parameter space of the problem, it is convenient to work in dimensionless units.

2.3 Dimensionless Behavior

Analyte transport to a biosensor surface is best characterized by the Sherwood number, the dimensionless analog to the mass transfer coefficient. From the macroscopic perspective discussed in Sect. 2.2, the average Sherwood number can be calculated as

$$\mathrm{SH} = \frac{JH}{(C_o - C_s)D} = \frac{K_m H}{D}. \tag{14}$$

It should be noted that we used the microchannel height as the characteristic length (rather than L), which allows for the direct comparison of SH between mixers having different lengths. As with the mass transfer coefficient, SH is an indicator of analyte transport efficiency to the sensing surface.

There are two dimensionless parameters that are important for the characterization (and prediction) of analyte transport to a biosensor. The first is the Péclet number, $\mathrm{Pe} = UH/D$, which represents the relative magnitude between the rate of convective versus diffusive analyte transport on the scale of the microchannel. The second is the sensor aspect ratio $\eta = L/H$. For (steady) diffusion-limited analyte transport, these two parameters have a large influence on the size (δ) and shape of the analyte boundary layer; the former influences the rate of transport, whereas the latter determines the characteristics of transport.

Following the discussion by Squires et al. [15], the effect of Pe and η on the shape and size of a diffusion-limited analyte boundary layer is conveniently illus-

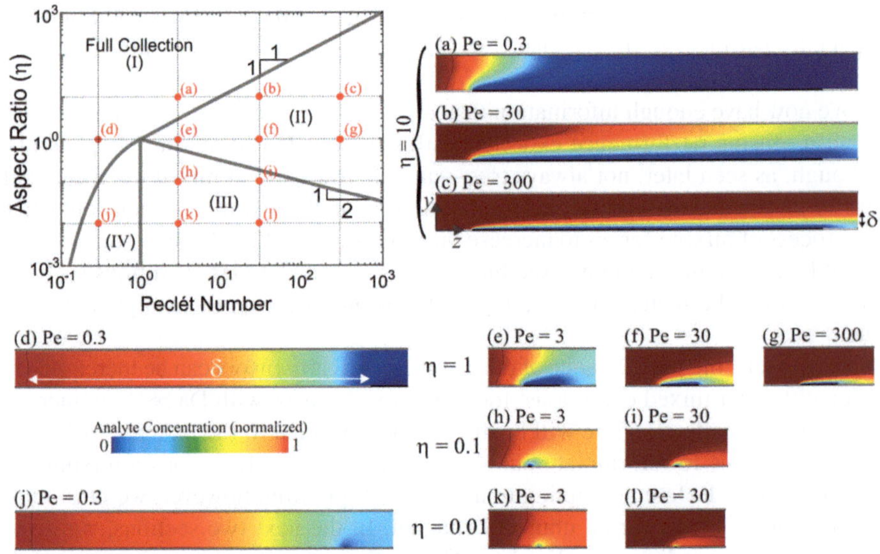

Fig. 2 Phase diagram for the steady, diffusion-limited analyte transport to a biosensor. The images shown in (**a**)–(**l**) are the steady analyte concentration profiles (across z and y) in a sufficiently wide microchannel, where the concentration of analyte along the sensing region was set to zero. These profiles were obtained using 2D finite-element simulations (COMSOL)

trated through the use of 2D finite-element simulations. Results from such simulations are shown in Fig. 2, where analyte transport in an unmixed channel can be classified into four "phases", each of which regards a different boundary layer shape. The relative positions of the boundaries between each phase can be distinguished by both Pe and η, which can be used to define two additional dimensionless numbers.

The first of these is the Graetz number, $\text{Gr} = \text{Pe}/\eta$, which provides information on the development of the boundary layer in relation to the channel height. Conditions such that $\text{Gr} \lesssim 1$ (i.e., $\text{Pe} \lesssim \eta$) correspond to a completely developed analyte boundary layer,[5] where the sensor will interact with all of the analyte flowing past. This relationship thus sets the boundary between phase (I) and (II); illustrations of this can be seen in the differences between (a), (b), and (c) in Fig. 2.

Another useful dimensionless parameter is the sensor Péclet number, $\text{Pe}_s = 6\eta^2\text{Pe}$, which represents the relative magnitude between convection versus diffusion adjusted for the size of the sensing region. For sufficiently fast flows (i.e., $\text{Pe} \gg 1$), diffusion-limited sensors with $\text{Pe}_s \gg 1$ will have boundary layers that are small with respect to both H and L, whereas sensors with $\text{Pe}_s \ll 1$ will have boundary layers that are small with respect to H, yet large with respect to L. This term can be used to classify the boundary between phase (II) and (III).

[5] Applicable for the geometries here, where a single wall acts to capture analyte. A value of $\text{Gr} \gg 1$ indicates the boundary layer has not reached the top of the channel.

Finally, the boundary between phase (III) and (IV) is determined by the value $Pe \approx 1$. The boundary between phase (I) and (IV) is slightly more complicated, and we refer the reader to previous works [15]. Nevertheless, we note that biosensors are rarely operated under conditions such that $Pe \leq 1$.

Before we discuss how the rate of analyte transport to a biosensor (i.e., SH, or K_m) can be predicted through knowledge of the relevant dimensionless parameters (i.e., Pe, η), we take a short pause to reconsider another form of question (A): is mixing worth the trouble? An qualitative answer to that question lies in the boundary layer profiles shown in Fig. 2. Intuition tells us that mixing will be most advantageous to sensors in phase (II); for example, there is a lot of "fresh" analyte just out of reach by the sensor in Fig. 2c. It is harder to deduce how mixing might affect sensors in phase (III); we leave this for later discussion (Sect. 3.1). Intuition also tells us that mixing will be of little use for sensors sufficiently far into phase (I): if $Pe \gg 1$ mixing might result in a local increase in analyte capture near the leading edge of the sensing region, yet it is clear that the averaged rate of analyte capture over the entire sensing region will remain unchanged; sensors operated at $Pe \lesssim 1$ are dominated by diffusion, and thus there will be no change in the rate of analyte collection by *any* portion of the sensor.[6]

2.4 *Predicting Analyte Transport in Unmixed Sensors*

Semi-quantitative estimates of the rate of diffusion-limited transport can be obtained through knowledge of the approximate boundary layer size: sensors in phase (I) have a boundary layer size of $\delta \approx H/\mathrm{Pe}$ (measured along the channel axis), which according to Eq. (14),[7] leads to a relatively accurate estimate of $\mathrm{SH} \approx \mathrm{Pe}/\eta$; sensors in phase (II) have a boundary layer size of $\delta \approx L \cdot \mathrm{Pe}_s^{-1/3}$, which leads to an estimate of $\mathrm{SH} \approx \eta^{-1}\mathrm{Pe}_s^{1/3}$; and sensors in phase (III) have a boundary layer size that scales as $\delta \approx L \cdot \mathrm{Pe}_s^{-1/2}$ (estimations of SH in this phase are more complicated).

Several authors have examined transport to unmixed sensors in a more quantitative fashion. Pertinent to this chapter are the works by Newman [37] and Ackerberg [38], who developed analytical solutions for SH as a function of Pe_s in the region of $\mathrm{Pe}_s \gg 1$ and $\mathrm{Pe}_s \ll 1$, respectively; these equations are listed in Table 2. Understandably, these equations are only valid under conditions relevant to phase (II) and (III). A slight correction of these equations in the respective limits of $\mathrm{Pe} \approx \eta$ and $\mathrm{Pe} \approx 1$ can be written as

$$\mathrm{SH}^{-a} = (\mathrm{Pe}/\eta)^{-a} + \mathrm{SH}_o^{-a}, \tag{17}$$

[6] In this region, the rate of analyte capture is proportional to the flow rate, no matter how fast the fluid is stirred.

[7] Where the analyte flux in the axial direction can be estimated as $C_o D/\delta$, and thus $J \approx C_o DH/\delta L$.

Table 2 List of dimensionless parameters related to transport in unmixed channels and the equations used to calculate them. Equations (15) and (16) are taken from the works of Newman [37] and Ackerberg [38]. These equations are specific to sensors such as that shown in Fig. 1, where the channel is sufficiently wide ($W \gg H$) and the sensing region spans the entire width of the channel

Dimensionless parameter	Equation
Channel Peclét number	$\text{Pe} = Q/WD = UH/D$
Sensor aspect ratio	$\eta = L/H$
Sensor Peclét number	$\text{Pe}_s = 6\eta^2\text{Pe}$
Sherwood number:	
$\text{Pe}_s > 1$	$\text{SH}_o = \frac{1}{\eta}\left(0.8075\text{Pe}_s^{\frac{1}{3}} + 0.7058\text{Pe}_s^{\frac{-1}{6}} - 0.1984\text{Pe}_s^{\frac{-1}{3}}\right)$ (15)
$\text{Pe}_s < 1$	$\text{SH}_o = \frac{1}{\eta}\left[\pi G\left(1 - 0.04633\text{Pe}_s G\right)\right]$, (16) where $G^{-1} = \ln\left(4\text{Pe}_s^{\frac{-1}{2}}\right) + 1.0559$

where $a = 5/2$, and SH_o is taken from the results in Table 2. The term Pe/η represents the estimated Sherwood number in phase (I). Figure 3 illustrates the dependence of the Sherwood number on the Péclet number in unmixed sensors; it can be seen that Eq. (17) gives a very good match to data obtained from 2D FE simulations. Equations (15) and (16) have also been shown to match data from experiment [39]. In regards to the biosensing literature, the first term in Eq. (15) represents the mass transport coefficient used by the BIAevaluation software to extract kinetic data from SPR sensorgrams [36].

The use of the dimensionless numbers listed in Table 2 should not be deterring, as these parameters are often directly related to those commonly used in experiment. For example, the data shown in Fig. 3 represent how the rate of analyte capture by a biosensor (proportional to SH) is dependent on the flow rate (proportional to Pe) for sensors of varying length (proportional to η).

The use of the results in this section is thus fairly straightforward. The parameters directly affecting the mass transfer constant (U, H, L, D) are typically known beforehand and, if not, can be easily measured or estimated. These values can be used to calculate Pe, η, and Pe_s, which in turn can be used to calculate SH with the equations shown in Table 2. Equation (17) should be used when conditions are such that $\text{Pe} \approx \eta$ or $\text{Pe} \approx 1$; otherwise, the equations in Table 2 are fine when used alone. These values of SH can then be used directly with Eq. (14) to calculate the mass transfer coefficient K_m. This value, along with knowledge of both the kinetic parameters k_1 and k_2 as well as the density of bioreceptors Γ_o,[8] can be used to calculate the Damköhler number. If desired, these values can then be used with Eq. (13) to estimate the analyte collection rate as a function of time.

[8]Measurement of these values can be accomplished using a variety of methods, the most popular of which the SPR method.

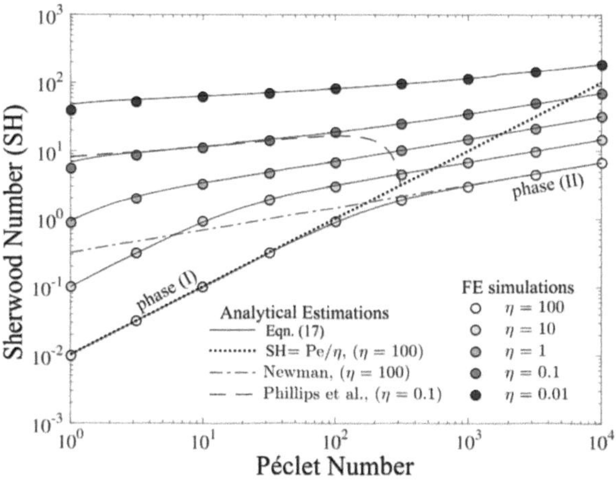

Fig. 3 The dependence of the Sherwood number on the Péclet number for biosensors having variable η. The black lines represent values of SH calculated from Eq. (17), where values of SH_o were calculated from either Eqs. (15) or (16) (depending on the value of Pe_s). The black-dotted line pertains to estimates within phase (I) ($SH = Pe/\eta$). The red-dotted-dashed line represents the solution given by Newman [37] (Eq. (15)). The blue-dashed line represents the solution given by Ackerberg [38] (Eq. (16)). The results shown as the symbols were obtained using 2D finite-element simulations (COMSOL), details of which are given in the work by Lynn and Homola [40]

3 Analyte Transport in Mixed Channels

In mixed channels, the velocity field is no longer uniaxial, as the effect of stirring induces a fluid movement transverse to the channel axis (i.e., in the x- and y-directions). Figure 4 illustrates how this additional fluid motion aids to stir the contents flowing through a "lid-driven" micromixer. Although this example is theoretical (a single moving wall on a microchannel presents several technical challenges), there are experimental devices that have similar flow profiles (e.g., an electroosmotically stirred microchannel [41]) and furthermore, this type of fluid motion has been shown to emulate the flow profile of the staggered herringbone mixer [42].

The fluid streamlines shown in Fig. 4 are indicative of analyte transport at very high Péclet numbers. From a biosensing perspective, this fluid motion would seemingly increase the rate of analyte capture. Yet, either of the two mixers shown in Fig. 4, used alone, would not provide optimal mixing conditions. Fluid streamlines in each device rotate around one or more stagnation points, thus only the analyte near the outer edges of each "vortex" will interact with a sensing region.

It is often a goal to design a micromixer (used for both mixing *and* biosensing purposes) such that fluid mixing proceeds in a chaotic fashion. The requirements to reach such a condition are fairly complex, and thus we point the reader to several useful sources [43–45]. In general, a chaotic mixing state can be achieved by periodic alteration of the flow profile along the channel axis, where alternating flow profiles do not have overlapping stagnation points. In doing so, the fluid is systematically

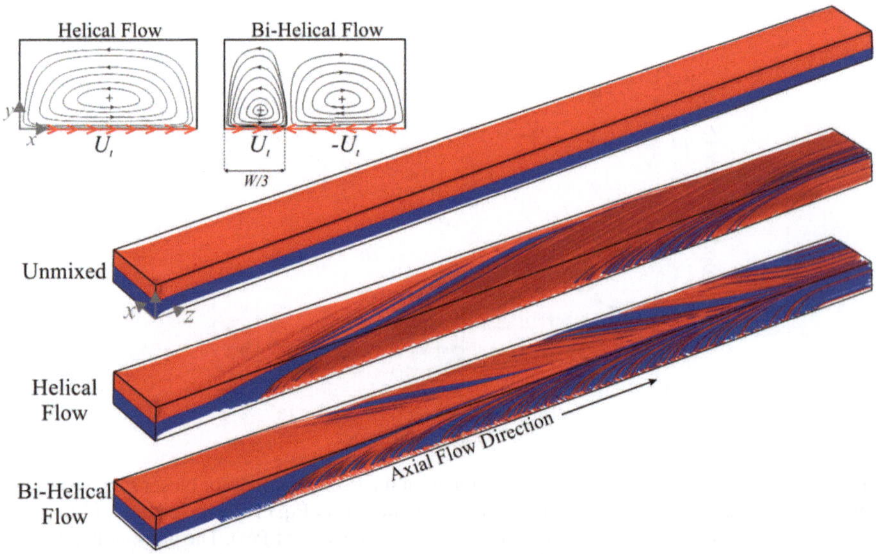

Fig. 4 Fluid mixing in a "lid-driven" microchannel, where three of the channel walls obey the no-slip condition and the channel floor moves with a velocity U_t (in the x-direction). For pressure-driven flow in the Stokes regime (Re $\ll$ 1), the fluid velocity in the entire chamber can be sufficiently estimated by the superposition of axial Poiseuille flow (v_z, described by Eq. (3)) and a transverse flow profile $\boldsymbol{v}_t$ created from the moving wall. Cross-sectional streamlines (regarding $\boldsymbol{v}_t$) for two types of fluid motion are shown in the upper left. Stagnation points for each flow profile are shown as the crosses. These simulations were performed with COMSOL, where $W = 2H$ and $U_t = U$

stretched and folded (in a manner similar to a baker's transformation) such that fluid streamlines (and solutes) experience the entire cross-sectional space. For the mixers in Fig. 4, this can be achieved by alternating the helical and bi-helical flow profiles or more optimally, alternating sequences of the bi-helical flow with its mirror image (across x) [42].

A full discussion on chaotic mixing in laminar flows would require much more than the space available in this chapter. Thus, for now, we will disregard specific flow profiles and rather classify a mixer according to (i) it being either chaotic or non-chaotic and (ii) its characteristic transverse velocity U_t. Outside of the example shown in Fig. 4, U_t can be taken as the average magnitude of the fluid velocity in the x-direction.

3.1 *Examination from a Local Perspective*

Understandably, the equations listed in Table 2 are not valid when the fluid in the sensing chamber is mixed, and thus a different approach is required.

In order to characterize (and predict) the effect of mixing on analyte transport, it is useful to examine the problem from a local perspective. In this approach, we assume—for both unmixed and mixed channels—that rates of analyte transport to the entire sensing region can be sufficiently represented by $j_z = j_z(z)$, the analyte flux averaged across the width of the channel (along the x-direction). A *local* Sherwood number can then be defined as

$$\mathrm{Sh}_z = \frac{k_m H}{D} = \frac{j_z H}{(c_b - c_s)D}, \tag{18}$$

where $k_m = k_m(z)$ is the local mass transfer coefficient, and $c_b = c_b(z)$ is the local ("mixing-cup") analyte concentration,[9] defined as

$$c_b(z) = \frac{\int_y \int_x c(x, y, z) v_z(y) dx dy}{UHW}. \tag{19}$$

The advantage of analysis from a local perspective is illustrated in Fig. 5a, which plots Sh_z as a function of the dimensionless axial distance $\bar{z} = z/\mathrm{Pe}H$ for both mixed and unmixed channels; from Sect. 2.3, $\bar{z}$ represents the inverse Graetz number. Under conditions spanning either phase (I) or (II), all sensors in an *unmixed* rectangular channel will exhibit the following trends:

- In the entrance region, where boundary layers remain small and $\bar{z} \ll 1$, the local Sherwood number will scale as $\mathrm{Sh}_z \propto \bar{z}^{-1/3}$.
- In the fully developed region, where $\bar{z} \gtrsim 0.1$ and the boundary layer has filled the channel, the local Sherwood number will asymptotically approach a value of $\mathrm{Sh}_\infty \approx 2.5$.

As noted by Kirtland et al. [46], sensors in *mixed* rectangular channels will exhibit similar trends:

- The rate of analyte transport in a mixed channel will be equivalent to that of an unmixed channel for axial distances of $z < z_\infty \approx WU/U_t$, where U_t is the characteristic transverse fluid velocity.
- For axial lengths of $z > z_\infty$, the local Sherwood number will deviate to an asymptotic value of $\mathrm{Sh}_\infty \geq 2.5$. For sufficiently high values of Pe, this asymptotic Sherwood number will scale as $\mathrm{Sh}_\infty \propto \mathrm{Pe}_t^{1/3}$, where $\mathrm{Pe}_t = \mathrm{Pe}(U_t/U)$ is the transverse Péclet number.
- For *chaotically* mixed fluids in the asymptotic region ($z > z_\infty$), Sh_z will remain constant with increases in $\bar{z}$.
- For *non-chaotically* mixed fluids in the asymptotic region ($z > z_\infty$), there will be a decrease in Sh_z with increases in $\bar{z}$.

[9] The mixing-cup concentration c_b represents the concentration one would obtain by collecting the microchannel effluent with a small cup.

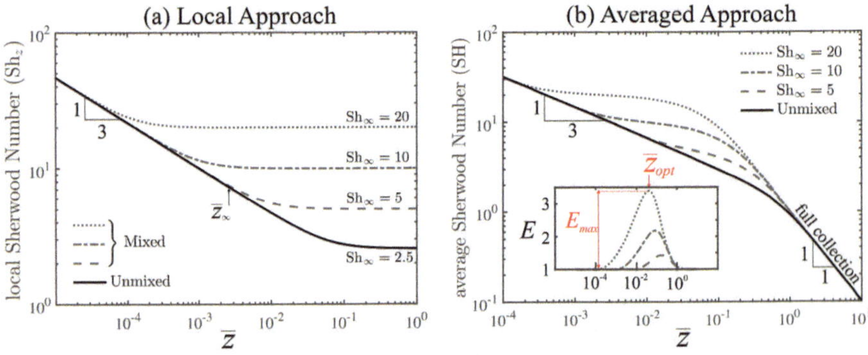

Fig. 5 Characteristic differences between unmixed and chaotically mixed sensors. **a** The local Sherwood number as a function of the dimensionless axial distance. Data for both the mixed and unmixed sensors were calculated from Eq. (20) using the respective values of Sh_∞. **b** The macroscopic Sherwood number as a function of the dimensionless axial distance ($L/\mathrm{Pe}H$). These data were calculated from those shown in (**a**) via the use of Eqs. (14), (22), and (21). The inset shows E: the ratio of SH between the mixed and unmixed channels

For both unmixed and chaotically mixed fluids, the local Sherwood number can be sufficiently estimated in an analytical fashion as

$$\mathrm{Sh}_z = (\bar{z}^{-5/3} + \mathrm{Sh}_\infty^5)^{1/5}, \tag{20}$$

where it follows that $\mathrm{Sh}_\infty = 2.5$ for an unmixed channel. It should be noted that Eq. (20) was not derived from first principles; rather, it represents a convenient method to estimate Sh_z analytically; this equation matches well to both data from FE simulations in unmixed channels (< 2% error) as well as random walk simulations in mixed channels [46, 47].

To put things in perspective, the data shown in in Fig. 5a might be representative of data taken from a variety of experimental situations, for example, (i) concerning an adjustable, chaotically mixed channel used for the detection of a single analyte at a single flow rate, increases in Sh_∞ would be a result of an increase in the magnitude of stirring (i.e., a higher U_t); or (ii) concerning a passive chaotic mixer (constant U_t/U), increases in Sh_∞ would be a result of either increasing the flow rate or similarly, the detection of analytes with progressively smaller diffusivities.

More information on the topic, including the behavior of non-chaotically mixed channels, can be found in [46, 48]. We also point the reader to a review of other topics related to transport phenomena within chaotic flows [49].

3.2 Prediction of Analyte Transport in Mixed Channels

Despite the convenience of an examination from a local approach, the results shown in Fig. 5a can be misleading to those not familiar with dimensionless analysis. For example, an increase in Sh_∞ by 10× does not lead to a 10× increase in the rate of mass transfer (either averaged across the sensor or measured at a specific axial distance corresponding to $\bar{z}$). As per the discussion regarding Eq. (13), it remains desirable to have information related to the average Sherwood number, which is directly related to experimental measurements. For such purposes, the local Sherwood number can be converted to a local flux as

$$j_z(z) = \frac{\mathrm{Sh}_z(z) C_o D}{H} \exp\left(- \int_0^z \frac{\mathrm{Sh}_z(z)}{\mathrm{Pe}H} dz \right), \tag{21}$$

where it follows that the average analyte flux can be calculated as

$$J = \frac{1}{L} \int_0^L j_z(z) dz, \tag{22}$$

which can be used directly with Eq. (14) to calculate SH (or likewise, to calculate $K_m = J/C_o$). Figure 5b displays the average Sherwood number as a function of the dimensionless axial distance for sensors of varying length; these data were calculated from those shown in Fig. 5a via Eqs. (21)–(22). The data for the unmixed sensor are equivalent to that given by Eq. (17).

The approach shown in Fig. 5 thus represents a convenient method to predict rates of mass transport to a sensing surface in chaotically mixed channels. Such predictions are relatively straightforward and follow a linear progression:

(i) The asymptotic Sherwood number Sh_∞ can either be calculated numerically [46, 47] or for $\mathrm{Pe} \gg 1$, be estimated as $\mathrm{Sh}_\infty \approx (\mathrm{Pe} U_t H / UW)^{1/3}$ [49].
(ii) The local Sherwood number $\mathrm{Sh}_z(z)$ can then be estimated via Eq. (20).
(iii) The average analyte flux J can then be estimated via Eqs. (21) and (22).
(iv) If one has knowledge of the kinetic parameters, the average mass transfer coefficient $K_m = J/C_o$ can be used with Eq. 13 to estimate the rate of analyte collection as a function of time.

As demonstrated in Sect. 4.2, this approach can be used to provide accurate predictions of the transport behavior observed in experiment: specifically, in biosensors using both slanted [28] and herringbone grooves [50].

3.3 Scaling Behavior for Optimal Conditions

In order to discuss optimal mixing conditions for a biosensor, it is useful to compare how the rate of analyte capture in a mixed channel differs from that in an unmixed

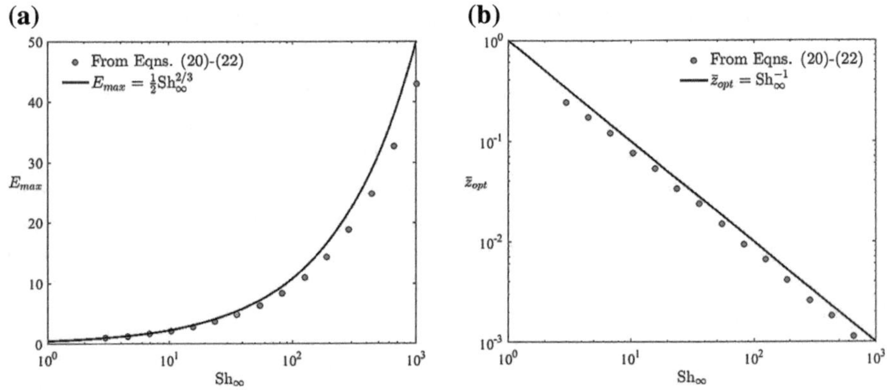

Fig. 6 Scaling behavior for optimal mixing conditions. The values E_{max} and $\bar{z}_{opt}$ correspond to the positions of the maximum mixing enhancement E as shown in Fig. 5. The symbols were calculated from Eqs. (20)–(22) for both mixed channel (variable Sh_{∞}) and unmixed channels ($\mathrm{Sh}_{\infty} = 2.5$)

channel. This comparison can be made through E, the ratio of SH (or equivalently K_m) between a mixed and unmixed sensor of similar dimension.[10] This ratio—shown in the inset of Fig. 5b—is useful in highlighting several characteristics of how a chaotic mixing process affects the performance of a biosensor.

It is clear from the results in Fig. 5b that there is an optimal dimensionless axial distance, $\bar{z}_{opt}$, such that the effect of mixing will be maximized with an enhancement E_{max}. If a mixer is too short (i.e., $z < z_{\infty}$), the analyte collection rate will be equivalent to an unmixed channel; the same is true when a mixer is too long (i.e., $\bar{z} > 1$). From the perspective of maximizing J, optimal mixing conditions correspond to the distance at which a mixed channel reaches the limit of full collection (specifically, at the point where SH starts to scale as SH $\propto \bar{z}^{-1}$). It follows that this mixer length is also optimal in terms of the analyte collection efficiency.

We now have enough information to give a partial answer to question (B): what is the expected performance improvement that mixing can provide? A general answer to this question can be obtained by examining the scaling behavior of E_{max}, shown in Fig. 6a. It can be seen that, per Eqs. (20)–(22), this enhancement scales with the asymptotic (local) Sherwood number as $E_{max} \propto A_o \mathrm{Sh}_{\infty}^{2/3}$, where $A_o \approx 0.5$. Knowing that the latter scales as $\mathrm{Sh}_{\infty} \propto (\mathrm{Pe}U_t/U)^{1/3}$, it follows that the maximum sensing enhancement scales as $E_{max} \propto A_o(\mathrm{Pe}U_t/U)^{2/9}$.

It should be noted, however, that under many circumstances it may be difficult to realize the conditions leading to E_{max}. Referring to the results in Fig. 6b, the axial position corresponding to E_{max} is seen to scale as $\bar{z}_{opt} \propto \mathrm{Sh}_{\infty}^{-1}$. In dimensional terms, the optimal length of a sensor will thus scale as $L_{opt} \propto H(U/U_t)^{1/3}\mathrm{Pe}^{2/3}$.

---

[10]For quasi-steady, diffusion-limited conditions, the limit of detection for a mixed biosensor (LOD_m) is related to that of an unmixed biosensor (LOD) of similar dimension as $\mathrm{LOD}_m = \mathrm{LOD} \cdot E^{-1}$.

Applying these scaling results to one of our previous experimental studies [50], where a passive mixer was used to enhance the SPRi-based detection of bacteria ($H = 27\,\mu m$, $U_t/U \approx 0.1$, and Pe $= 10^6$), we find that the maximum enhancement will follow $E_{max} \approx 6$; however, this maximum will require a mixer length of $L_{opt} \approx 581$ mm! Such a length would likely present numerous experimental difficulties for a SPR-based sensor. Nevertheless, nonoptimal mixers can still provide benefits: under the same conditions, a mixer length of $L = 15$ mm was shown to have an enhancement of $E = 2.4$ [50].

It is also important to keep in mind that the value E represents the enhancement in analyte transport under diffusion-limited conditions. This value thus represents the upper bound of what can be experimentally realized: following Eq. 13, the actual level of enhancement due to mixing may be restricted by reaction limitations.

4 Experimental Applications: The Staggered Herringbone Mixer

The staggered herringbone mixer is perhaps the most notable example of a passive microfluidic mixer [26]. Illustrated in Fig. 7a, the SHM consists of herringbone-shaped grooves fabricated onto at least one wall of the microchannel. Under pressure-driven flow, these grooves generate two counter-rotating helical flows similar to that shown in Fig. 4, where alteration of the herringbone asymmetry along the channel length serves to mix the fluid in a chaotic fashion. For biosensing purposes, the SHM grooves are often placed on the wall opposite of the sensing region. Under proper conditions, the transverse bi-helical fluid motion can aid to increase the rate of analyte transport to a biosensor.

Since the seminal work on the topic, there have been a large number of analytical, numerical, and experimental studies focused on optimizing the traditional SHM geometry, a selected portion of which can be found in [51–56]. Additional works have focused on the optimization of mixing in SHM-like channels [57–60]. For readers of this chapter, however, caution must be taken in interpreting (or applying) the results of these studies, as they examine how changes to the SHM geometry affect the rate of mixing within the channel. For sensing applications, we are rather more interested in the rate of mass transfer to a sensing surface. These two end-use purposes (mixing and sensing) thus remain distinct; optimization of the first does not guarantee the same result for the second.

4.1 Optimization of the SHM Geometry for Biosensing

To date, only a limited number of studies have examined how changes in the SHM geometry affect analyte transport to a sensing surface. The initial study by Kirtland et al. was pertinent in establishing many of the important concepts on the topic [46]; however, those results approximated flow in the SHM as being lid-driven [42], similar to that in Fig. 4, where axial fluid velocities remained unchanged along the length of the sensor. Consideration of the more complex SHM geometry requires a more detailed approach.

The complexity of analyte transport within a SHM is illustrated in Fig. 7a. Axial fluid velocities are higher in the thin channel regions between grooves, and lower in the regions under the grooves (where the effective channel height is larger). Transverse fluid velocities are even more intricate. The 3D flow profile leads to interweaved boundary layers and furthermore, a variable analyte flux profile that mirrors the groove design. Analyte flux is highest near the apex of each groove structure when the flow direction is aligned with the "points" of the herringbones, and lowest in the regions under the grooves.[11]

From a numerical perspective, there is an inherent difficulty in studying analyte transport in the SHM. Accurate computation of fluid velocity fields in the SHM is possible with relatively modest computational resources; however, the computation of concentration fields is susceptible to numerical artifact, and accurate simulations require the use of unrealistic mesh densities [62]. For this reason, the contour and flux profiles shown in Fig. 7a cannot be used to obtain quantitative information.

A workaround to this computational problem was shown by Forbes and Kralj, who examined how changes in the herringbone geometry influence the frequency at which fluid streamlines come into proximity with the SHM surfaces [61]. As shown in Fig. 7b, they found that this streamline interaction frequency with *all* of the SHM surfaces is maximized when the hydraulic resistance of the grooves are equal to that of the channel. Thus, the sizes for both the channel (H) and groove height (h_g) can be used to analytically estimate the optimum groove width (w_g) without the need of computational simulation. Such a conclusion is certainly advantageous for SHM biosensors having all surfaces functionalized with bioreceptors. It is unclear if this approach is applicable toward the sensors considered in Fig. 7a, where only the surface opposite the grooves is functionalized. Although there is surely a relationship between the frequency of streamline interaction and the rate of analyte transport, the exact nature of that relationship is unknown.

An alternative method to estimate rates of analyte transport in the SHM is via the use of particle tracing simulations. These simulations, also used in the study by Kirtland et al. [46], are often applied to study mass transfer in complex domains [63]. We recently used these simulations to study analyte transport in the SHM [47]; examples of those results are shown in Fig. 7c. As expected, we found that (when averaged over each 1/2-cycle) the local Sherwood number deviates to an asymptotic value Sh_∞ that is dependent on both the Peclét number and more importantly, the geometry of the SHM. In general, we found that for a set microchannel height, increases in Sh_∞ can be obtained by decreasing W and increasing h_g, whereas changes in the groove pitch

[11]When flow is reversed, the analyte flux is highest where the grooves meet the channel sidewalls. This arrangement is nonoptimal due to the lower axial velocities near the channel edges.

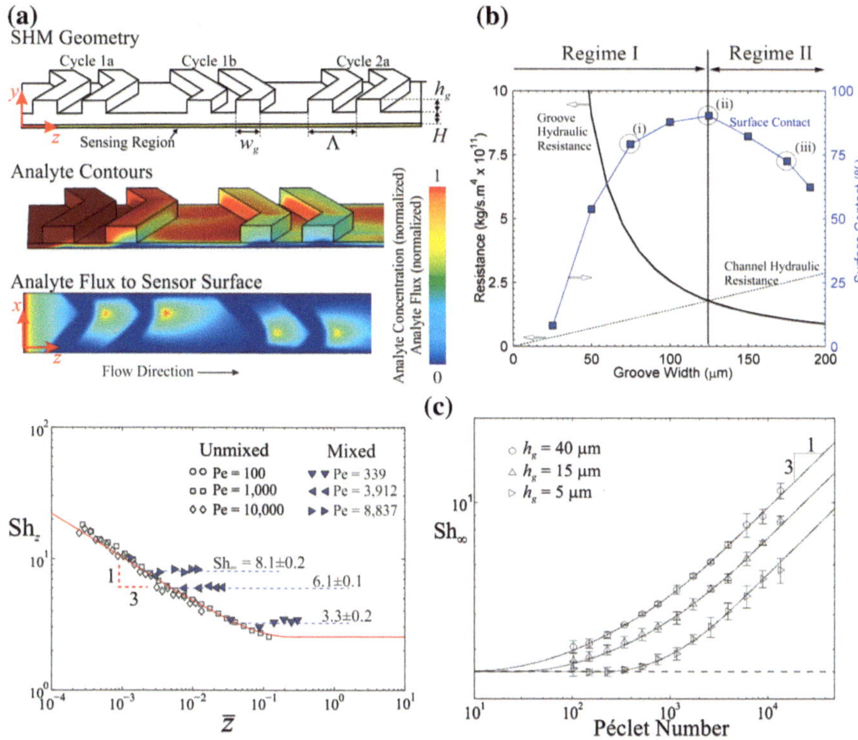

Fig. 7 **a** Schematic of the SHM geometry used for sensing purposes. Steady diffusion-limited analyte contours (normalized to C_o) and analyte flux contours (normalized by the maximum value) are shown for the first cycle (obtained via COMSOL) **b** Results from Forbes et al. [61]: the streamline surface contact with all surfaces of the SHM is maximized when the hydraulic resistance of the grooves (solid line) is equal to that of the channel (dashed line). **c** Results from Lynn et al. [47]: particle tracing simulations can be used to characterize the dimensionless behavior of a SHM; such behavior is similar to that in Fig. 5a (left). Increasing the groove depth (h_g) leads to increases in Sh_∞ (right), where increases in Pe for all channels lead to increases in Sh_∞ (Dimensional parameters in (**c**) were $H = 20\,\mu\mathrm{m}$, $W = 200\,\mu\mathrm{m}$, and $\Lambda = 150\,\mu\mathrm{m}$). The results in (**b**) were adapted from [61] with permission from the Royal Society of Chemistry. The results in (**c**) were adapted with permission from [47]. Copyright 2015 American Chemical Society

(Λ) and the number of grooves per 1/2-cycle (N_g) had little to no effect (the lack of dependence on Λ was also shown in [61]).

Using these simulations, we found that for an SHM with constant H and h_g there exists an optimal groove width such that Sh_∞ is maximized [47]. However, in contrast to the findings by Forbes and Kralj [61], this optimum groove width was not at the position of equal hydraulic resistance (as shown in Fig. 5b). To further complicate matters, we found that optimal groove widths (to maximize Sh_∞) were nonoptimal in terms of maximizing transverse fluid velocities.

Unfortunately, these previous studies cannot be used to formulate a "handbook" for the optimization of SHM-based biosensors. Nonetheless, there are a few guidelines that, while might not be optimal, can help improve a biosensor's performance. Perhaps, the most important design tip is to simply increase the groove depth: a value of $h_g/H \approx 2$ seems to maximize Sh_∞ for all SHM geometries [47]. The other tip is to avoid the use of a single herringbone pattern over sufficiently wide channels: channels with symmetric herringbone patterns (with a symmetric width W_s) have a smaller effective width and thus serve to increase transport. As shown below, symmetric SHM designs are often used in experimental biosensors.

4.2 Experimental Use of the SHM for Biosensing

In the decade after the seminal work [26], there were only a few experimental studies utilizing the SHM for biosensing purposes. Most of the early reports pertained to the detection of analytes having relatively small sizes. To our knowledge, the first experimental use of a groove-based micromixer for biosensing purposes was reported by Golden et al., who used a mixer consisting of alternating sequences of slanted and v-grooves situated over a surface having immobilized bioreceptors [28]; below, we compare the results of this report with the discussion in Sect. 3 (*Case study 1*). For the second case study, we refer to one of our recent reports, where we used the SHM to enhance the detection of several analytes (ssDNA and *E. coli* bacteria) by a surface plasmon resonance imaging (SPRi) biosensor [50] (*Case study 2*).

Other reports on the topic for the detection of small analytes include that by Foley et al., who used a SPRi biosensor to characterize the spatial distribution of the capture of streptavidin under a channel composed of v-grooves [27]. Recently, Gomez-Aranzadi et al. demonstrated that SHM-based channels improve the capture rate of amine-coated polystyrene nanospheres (252 nm diameter) [64].

Comparison with Experiment: Case Study 1. The mixer used by Golden et al. [28] is shown in Fig. 8a.[12] The performance of this mixer was compared against an unmixed channel in two assay formats: (i) the direct detection of CY5-tagged biotin (with immobilized NeutrAvidin) and (ii) the detection of anti-botulinium toxin in a sandwich assay. The results for the direct detection assay are shown in Fig. 8a; the integrated fluorescent signal is proportional to the total number of analyte molecules captured from the beginning of the channel. They observed an increasing difference in fluorescence intensity between the mixed and unmixed channels along the length of the sensor. The mixed channel was observed to have a 26% higher fluorescence signal over the entire length of the channel ($L = 140$ mm).

The results shown by Golden et al. [28] represent a convenient example for comparison with the scaling behavior discussed in Sect. 3.3. Although their mixer did

[12]Reprinted from *Biosensors and Bioelectronics*, 22, Golden J.P., Floyd-Smith T.M., Mott D.R., and Ligler F.S., Target delivery in a microfluidic immunosensor, 2763-2767, 2007, with permission from Elsevier.

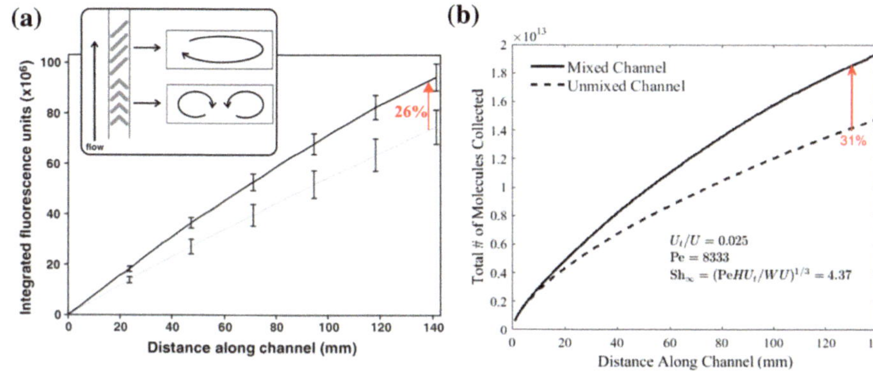

Fig. 8 **a** Results from Golden et al., adapted with permission from [28]: use of a grooved micromixer (shown schematically in the inset) for the direct detection of CY5-labeled biotin. Their mixer consisted of alternating sequences of v- and slanted-grooves. The plot shows the integrated fluorescence intensity as a function of axial distance for both mixed (black) and unmixed (gray) channels. Their channel had dimensions of $W = 200\,\mu\text{m}$, $H = 80\,\mu\text{m}$, and $h_g = 50\,\mu\text{m}$. The flow rate was $Q = 1\,\mu\text{L/min}$, and the assay proceeded for $t_a = 10$ min. **b** Predictions of the total number of molecules bound as a function of axial distance for the results shown in [28]. The total number of molecules captured was calculated as $JWLt_a$, where J was calculated via. Eqs. (20)–(22) ($Sh_\infty = 4.37$). The specific CY5-biotin marker was not given; however, using the respective sizes of CY5 and biotin, we estimated its diffusivity as $D \approx 10^{-10}\text{m}^2\text{s}^{-1}$

not specifically use herringbone grooves, the use of alternating cycles of slanted and v-grooves represents the right "recipe" to create Lagrangian chaos [65]. Hence, we only need an estimate of Sh_∞ to predict the increase in analyte capture due to the inclusion of mixing. For an order of magnitude estimate, we can use their reported groove depth of $h_g/H = 0.625$ to estimate a transverse velocity ratio of $U_t/U \approx 0.025$ (via the results shown in [55]). From their reported flow conditions, we can estimate an order of magnitude estimate of $Sh_\infty \approx (PeU_tH/UW)^{1/3} \approx 4.37$, which leads to a maximum enhancement of $E_{max} \approx 1.34$ (34% enhancement) occurring at a channel length of $L \approx 112$ mm—values very close to those reported by the authors!

Taking our predictions one step further, we can apply our estimate of Sh_∞ to Eqs. (20)–(22) to estimate the total number of molecules captured by their sensor as $JLWt_a$, where t_a is the time of the assay. The results of this approach are shown in Fig. 8b. Comparison of our predicted values match very well with those shown in Fig. 8a. It should be stressed that there are undoubtedly some aspects that are overlooked in this predictive approach (e.g., equilibrium effects were not considered); however, the similarity in the results shown in Fig. 8 is quite impressive, especially considering that no numerical computation was required.

Comparison with Experiment: Case Study 2. In our previous study, we used a SPRi biosensor to monitor the direct detection of both ssDNA and *E. coli* bacteria, where detection was conducted over a range of flow rates and SHM designs. Because of the relatively large channel width, herringbone patterns were arranged in a sym-

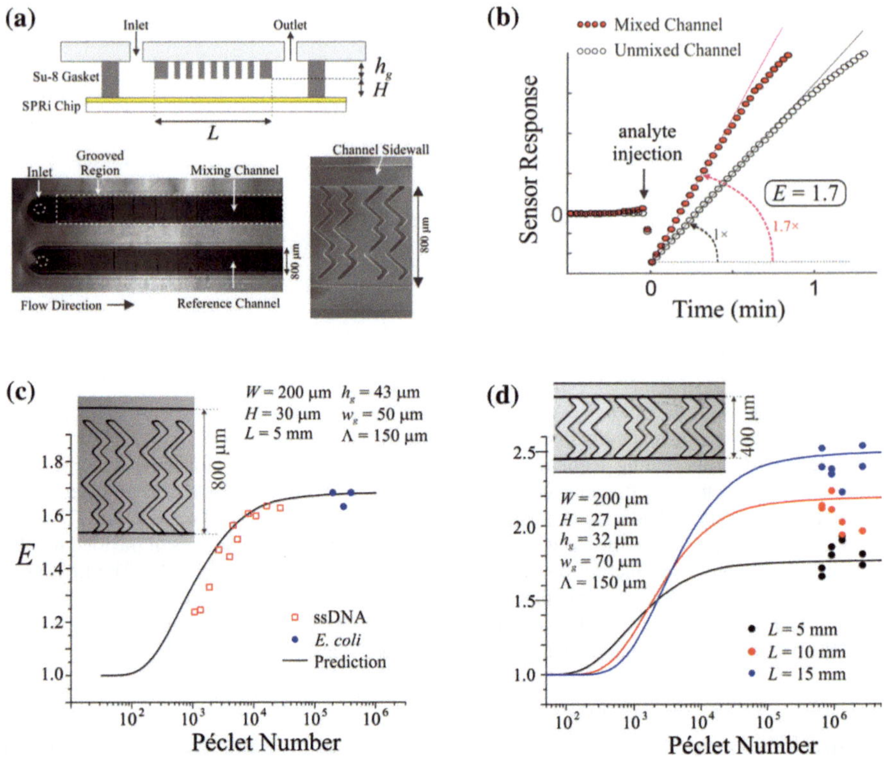

Fig. 9 Adapted with permission from [50]. Copyright 2015 American Chemical Society. **a** Schematic of the biosensor, which consisted of both mixed (SHM) and unmixed channels fabricated from Su8 and sealed to the SPRi chip. **b** Example sensor response in a mixed and unmixed channel for the detection of 20 b.p. ssDNA ($C_o = 20$ nM). **c** and **d** Sensing enhancement E as a function of the Péclet number for the detection of both 20 b.p. ssDNA ($D = 9.9 \times 10^{-11} \mathrm{m^2 s^{-1}}$) and heat-killed *E. coli* ($D = 3.2 \times 10^{-13} \mathrm{m^2 s^{-1}}$). The insets show light images of the SHM used in each experiment

metrical fashion; these patterns were optimized using the numerical methods in [47]. For each detection experiment, the performance increase due to the grooves was calculated by the ratio of initial binding rates between the SHM-mixed channel and an unmixed channel (Fig. 9b). Due to both the large difference in diffusivities between ssDNA and *E. coli* as well as the wide range of flow rates ($5 < Q < 120\,\mu\mathrm{L/min}$), the detection conditions in this case study spanned a wide range of Péclet numbers ($10^3 < \mathrm{Pe} < 10^6$).

The effect of the Péclet number on the increase in sensor response due to mixing is shown in Fig. 9c, d. As expected from the discussion in Sect. 3.2, at sufficiently low Pe, there is no enhancement due to mixing, whereas increases in Pe result in a monotonic increase in E. The magnitude of such increase maintains a complex dependency on both Sh_{∞} and L; as shown in Fig. 9d, increases in the latter result in higher sensing enhancements (for lengths such that $L < L_{opt}$).[13]

The data shown in Fig. 9 highlight the importance of the Péclet number on such mixed sensors. In general, if conditions are such that Pe $\lesssim 100$, mixing is not expected to provide any biosensing enhancement: a condition that remains applicable for sensors under severe mass transfer limitations (i.e., Da $\gg 1$). Mixing will provide the greatest benefit under conditions of very high Péclet number, where Pe $\gtrsim 10^5$. For smaller analytes, the required flow rates and channel sizes to meet these conditions are often undesirable. Conversely, these conditions are regularly encountered in microfluidic sensors for the detection of very large analytes, such as whole cell bacteria (as shown in Fig. 9) or circulating tumor cells.

4.3 Detection of Circulating Tumor Cells Using the SHM

Recently, there has been an increase in the use of the SHM-based biosensors for the detection of circulating tumor cells (CTCs). These cells are shed from cancerous tumors at rate such that their concentration in the bloodstream remains very low. The detection of these cells in a patient's bloodstream is thus an indicator of the presence of a primary tumor. Due to their low concentration, microfluidic approaches often aim toward maximizing the SHM capture efficiency, measured as the number of cells captured divided by the number of cells flown through the channel. There are currently several reviews on the microfluidic-based capture of CTCs, to which we refer the reader [66–71]. The diameters of relevant CTCs are in the order of 10 μm, which correlates to a diffusivity in aqueous media of $D \approx 10^{-14}\,m^2s^{-1}$ [71]. Thus, the detection of CTCs thus often corresponds to conditions in the range of $10^7 <$ Pe $< 10^9$.

Use of all SHM surfaces for cell capture. The first use of the SHM for the detection of CTCs was reported by Stott et al. [29]. As shown in Fig. 10, they used an HB chip (consisting of eight SHM-based channels in series) to capture PC3 cells (a human prostate cancer cell) in whole blood. Unlike the model example shown in Fig. 1, bioreceptors were immobilized to all of the SHM surfaces. Using a scaled down single-channel device, the authors demonstrated that a SHM-mixed channel provided a markable increase in capture efficiency when compared to an unmixed channel developed in the previous study [72]. A decrease in flow rate resulted in an increase in capture efficiency for both mixed and unmixed channels (Fig. 10).

There have been since several reports on the use of the HB chip (either exactly as in [29] or with slight deviations in dimensions) for the detection of CTCs, where the entire SHM surface is used to capture CTCs. Sheng et al. used an HB chip to study the effect of various bioreceptors on the capture of CEM cells (a human leukemia T-cell)

[13]The scaling laws in Sect. 3.3 predict that there will be an indefinite increase in the *maximum* sensor enhancement E_{max} with increases in Pe; however, this scaling law applies only to mixers of length L_{opt}.

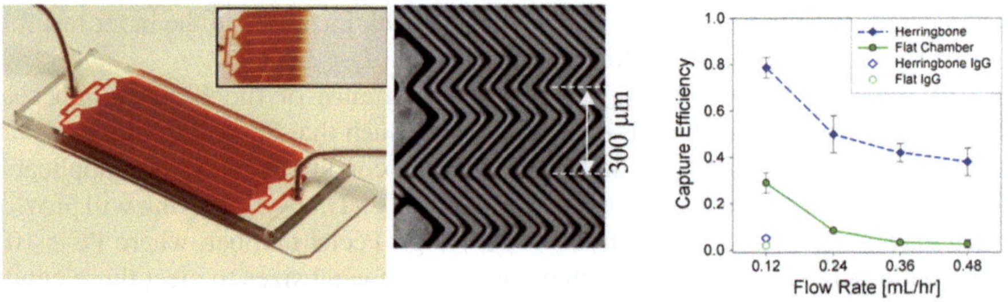

Fig. 10 Adapted with permission from Stott et al. [29]: (**left**) The HB chip consisted of a SHM channel fabricated from PDMS sealed to a glass slide. Bioreceptors (anti-EpCAM) were immobilized to both the glass and PDMS channel walls. Captured cells were detected using fluorescence microscopy. (**middle**) Symmetric placement of herringbone grooves (symmetric width W_s = 300 μm, w_g = 50 μm, Λ = 100 μm). (**right**) Capture efficiency of PC3 cells for both a SHM-based channel (W = 2mm, h_g = 30 μm, H = 70 μm, L = 25 μm) and an unmixed channel (H = 100 μm) as a function of the flow rate through a smaller, single-channel device. Also shown is the capture efficiency of IgG

[73]; they showed that bioreceptors consisting of aptamer-coated gold nanoparticles performed significantly better than both aptamers alone (no nanoparticles) as well as anti-PTK7 antibodies. A pair of studies by Xue et al. used both the original HB chip [74] and a modified HB chip [75] (with posts opposite to the grooves) to capture Hep3B cells (human liver cancer cells, anti-EpCAM bioreceptors). A device by Liu et al. used enrichment channels to separate MCF7 cells from other blood cells (using deterministic lateral displacement [76]), which were subsequently captured by a SHM-based channel [77]. A follow-up study by the same authors demonstrated that changing the dimensions of the herringbone grooves (in accordance with the results of Forbes and Kralj [61]) resulted in both a slight increase in capture efficiency as well as a large increase in the capture purity (measured as the number of targeted versus total cells captured) [73].

In another study using the results of Forbes and Kralj [61] to optimize the HB chip, Hyun et al. studied the effect of several SHM designs on the capture efficiency of Jurkat cells (a human T-cell lymphoblast cell line) [78]. The three different SHM designs used in their study are shown in Fig. 11. In comparison to the design by Stott et al. [29], they showed that increases in capture efficiency can be obtained by increasing the groove width w_g (constant W_s). Further increases were obtained by decreasing W_s, which is in accordance with the results in [47] (albeit for a single surface).

A note regarding transport in "all-surface" SHM channels. An analysis of transport in SHM channels using all surfaces for capture requires a slightly different approach from the discussion in Sect. 3.2.[14] In a follow-up study to their seminal work, Kirtland et al. demonstrated that for a chaotic lid-driven micromixer (Fig. 4), the asymp-

[14]For *unmixed* flow in wide channels, the asymptotic local Sherwood number is $\mathrm{Sh}_\infty \approx 7.5$ [79] and furthermore, the integral in Eq. (21) is multiplied by a factor of 2 into account for the additional reactive surface.

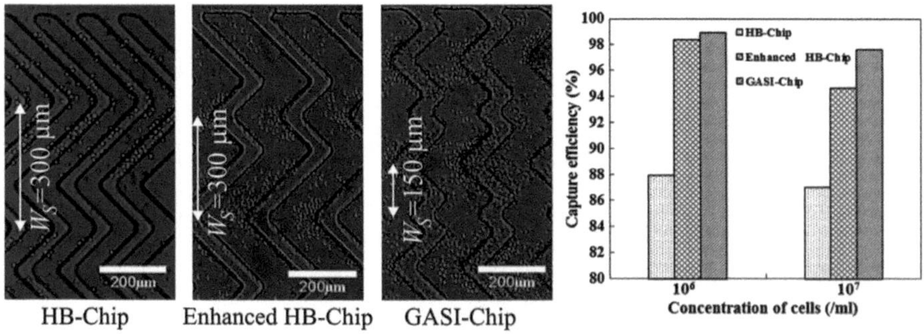

Fig. 11 Adapted with permission from [78]. Copyright 2013 American Chemical Society: capture of Jurkat cells by a SHM. Their HB chip used similar dimensions as [29] ($W = 2.1$ mm, $L = 50$ mm, $H = 50\,\mu$m, $h_g = 45\,\mu$m, $w_g \approx 50\,\mu$m, $\Lambda = 100\,\mu$m). The enhanced HB chip was designed using the results found in [61] and applied to the HB chip keeping W_s constant ($h_g = 50\,\mu$m, $w_g = 125\,\mu$m, and $\Lambda = 200\,\mu$m). The GASI chip used a decreased herringbone width. The capture efficiency was measured for all three chips at two cell concentrations

totic local Sherwood number for the *moving* wall scales as $\mathrm{Sh}_\infty \propto (\mathrm{Pe}U_t/U)^{1/2}$ [48]. As mentioned previously, the lid-driven mixer has previously been shown to emulate flow in the SHM [42]; however, it is unclear if this scaling relationship holds for the SHM geometry, where the grooves themselves are stationary. To our knowledge, there are currently no methods to predict how the capture efficiency of these sensors is dependent on either the SHM architecture or the operating conditions.

Case Study 3: Use of a single SHM surface for cell capture. Another early use of the SHM for CTC detection was shown by Wang et al. [80]. Contrary to the previous studies, where all of the SHM surfaces were active in CTC capture, the device used in [80] had immobilized bioreceptors only on the surface opposite the SHM grooves. Hence, we use this as our third case study for comparison to the predictive approach in Sect. 3.2.

The device in [80] consisted of a wandering SHM used for the capture of MCF7 cells (a human breast cancer cell) onto a nanostructured surface consisting of silicon nanopillars (SiNPs). Their device was shown to have a capture efficiency of > 95% for flow rates up to 2 mL h^{-1}, where similar to [29], they observed a drop in capture efficiency at higher flow rates (down to ≈ 30% at 7 mL h^{-1}). An unmixed channel had a capture efficiency of ≈ 60% (1 mL h^{-1}).

Similar devices have since been used by the same authors for (i) the capture of other CTCs (SKBR3 cells, anti-EpCAM bioreceptors) and subsequent analysis by laser capture microdissection (LCM) [81], (ii) the capture and LCM analysis of circulating melanoma cells (CMCs) (anti-CD146 bioreceptors) [82], and (iii) the capture of A549 cells (human lung cancer) via immobilized aptamers [83]. Reported capture efficiencies for these studies were similar to that in [80].

The previous two case studies pertained to the detection of analytes having relatively well-defined sizes. Conversely, MCF7 cells have been shown to have a large size distribution, with cell radii measured to be $r_c = 9.6 \pm 4.2\,\mu$m [85]. For purposes

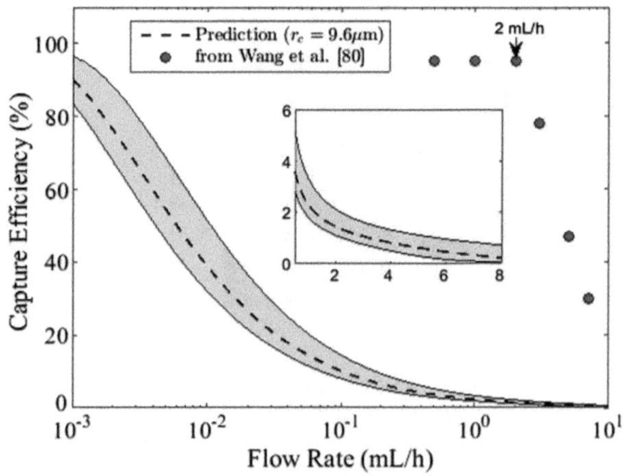

Fig. 12 Comparison of the data in [80]. SHM dimensions were reported as $W = 1$ mm, $H = 100\,\mu$m, $L = 880$ mm, $h_g = 35\,\mu$m, $w_g = 35\,\mu$m, and $\Lambda = 100\,\mu$m. The reported groove depth of $h_g/H = 0.35$ was used to estimate $U_t/U = 0.02$ [55], where we assumed $\text{Sh}_\infty = (\text{Pe}U_tH/UW)^{1/3}$. Predictions of capture efficiency were calculated as $JLW/QCo \cdot 100$, where J was calculated using Eqs. (20)–(22). The dashed line pertains to predictions regarding cells of size $r_c = 9.6\,\mu$m. The upper and lower bounds of the shaded area pertain to cells of $r_c = 5.4$ and $13.8\,\mu$m, respectively. Diffusivities were calculated via the Einstein–Stokes relationship at 25C, where the viscosity of the DMEM medium was taken to be $\mu = 0.94$ cP [84]. Data representative of [80] is shown in the red symbols

of prediction, it is reasonable to assume that the average rate of transport for a distribution of cell sizes is bound by that for fixed distributions of cells having sizes on the upper and lower ends of what is experimentally observed; we take such an approach here.

Figure 12 shows the predicted capture efficiency as a function of the flow rate for the conditions described in [80]. In terms of J, the capture efficiency can be taken as the rate of cells captured by the sensor (JLW) divided by the rate of cells entering the channel (QC_o) times 100.

In contrast to the previous two case studies, predictions via Eqs. (20)–(22) do not match well with the observed experimental behavior in [80] for the capture of CTCs. At a flow rate of $Q = 1\,\text{mL}\,\text{h}^{-1}$, the predicted upper and lower bounds for the capture efficiency are 3.3% and 1.8%, respectively. These low values cannot be attributed to errors in estimating U_t: a much larger (and experimentally unreasonable) value of $U_t/U = 1$ leads to a capture efficiency of 7.7% ($r_c = 9.6\,\mu$m, $Q = 1\,\text{mL}\,\text{h}^{-1}$). Factors contributing to these discrepancies are discussed in the next section.

4.4 Prediction of CTC Capture: Factors to Consider

The discrepancies between the predicted and experimentally observed transport behavior of CTCs in mixed channels (Fig. 12) can primarily be attributed to the large size of the CTCs with respect to the smaller analytes considered in Figs. 8 and 9. The relatively large size of a CTC with respect to that of a microchannel affects transport behavior through a variety of mechanisms.

One source of discrepancy is that under certain conditions a CTC might not follow the fluid streamlines in a microchannel, that is, where inertial forces become greater than the drag force of the fluid. The relative magnitude between these two forces can be estimated (in dimensionless fashion) by the Stokes number (St):

$$\mathrm{St} = \frac{2\rho_c U_o r_c^2}{9\mu H}, \tag{23}$$

where ρ_c is the CTC mass density, and we have taken the channel height to be the characteristic "obstacle size". A cell can be considered to follow a fluid streamline for conditions such that $\mathrm{St} \ll 1$, where drag forces become dominant. Taking the mass density of MCF7 cells to be $\rho_c = 1.04\ \rho$ [86], we can estimate this value for the results shown in Fig. 12 as $\mathrm{St} \approx 0.6$ ($Q = 1\ \mathrm{mL\,h^{-1}}$). Thus, the advection of cells at this flow rate (and higher) cannot be assumed to precisely follow fluid streamlines throughout the chamber.

Aside from inertial effects, the parabolic nature of pressure-driven flow can influence a CTC in several ways. Sufficiently large cells (with respect to H) can be subjected to a large range in fluid velocities across the length of the cell. Cells close to a channel surface experience significant shear forces, which may damage cells and prevent cell capture [87]. Furthermore, cells in the center of a channel can have velocities much smaller with respect to the fluid velocity if no cell were present [88]; this effect would be significant for a CTC flowing within a herringbone groove. It is unclear if either of these effects contribute to the discrepancies shown in Fig. 12.

Perhaps, the most important source of discrepancy is the effect of cell settling caused by the difference in the mass density between cell and fluid. The settling velocity of a spherical cell in a static flow field can be taken from Stokes' law as

$$v_s = \frac{2(\rho_c - \rho)G r_c^2}{9\mu}, \tag{24}$$

where G is the gravitational constant. Applying this to the size distribution found in MCF7 cells ($r_c = 9.6 \pm 4.2\ \mu\mathrm{m}$) leads to estimated settling velocities in the range of $2.7 < v_s < 17.7$ mm/s. In comparison, the experimental conditions in [80] correspond to axial fluid velocities in the range of $1.4 < U < 19.4$ mm/s. Although there are other factors that must be considered (e.g., inertial effects from the helical flow field, deformability of the CTCs), the settling of CTCs is likely to have lead to increased rates of capture when compared to a pure convection and diffusion approach as discussed in Sect. 3.2. This effect was also noted by Stott et al. to have an influence on CTC capture [29]; however, no details were given on the frequency of cells captured by the upper and lower microchannel surfaces.

Although not relevant to Fig. 12, another factor to consider is the viscous nature of the detection media. For example, whole blood is shear thinning, where fluid viscosities at high shear rates are over an order of magnitude lower with respect to viscosities at static conditions [89]. Thus, the diffusivity of CTCs (or any other analyte) cannot be considered to be constant across the channel height,[15] where from a microscopic perspective, fluid shear is zero at the channel center and maximized at the channel walls.

The combined effects from these factors thus serve to muddy the predictive approach discussed herein. It is clear that such an approach has a good match when applied to the detection of smaller analytes; however, as demonstrated in Fig. 12, the same cannot be said regarding the larger CTCs.

In a similar fashion, these additional effects also pertain to the streamline-based approach shown by Forbes and Kralj [61]. Although their approach is certainly useful—as evidenced by the results of several experimental reports [73, 78]—questions remain if those positive results were a result of other factors: increased fluid streamline contact does not always imply increased CTC contact.

Nonetheless, there still remains a large gap in the literature on this topic, regardless of what SHM surface is doing the capture.

5 Final Notes

Although there has been a vast deal of work on the design and implementation of microfluidic mixers, their direct use with a biosensing process has been relatively limited. An obvious reason for this seems to lie in the technical complexity of the task: taken separately, both biosensing and micromixing are complicated processes, and the number of issues to troubleshoot when simultaneously considering both does not scale linearly. On the other hand, one must wonder how many researchers were successful in incorporating the two devices with one another, yet were unable to find any "interesting" results for mixed channels. As per the discussion above, there is a wide range of conditions for when mixing will *not* be beneficial to a biosensor. Many of these conditions are relatively unintuitive, even to those who have backgrounds in transport phenomena.

So when is mixing worth the trouble? A nontechnical answer is that the sensing region needs to be sufficiently long, such that the effect of mixing is realized, and furthermore, flow rates need to be sufficiently fast, such that analyte transport is dominated by convection and detection is far from being limited by interaction kinetics. Referring back to Sect. 2, these conditions are better stated in dimensionless terms,

[15]From the Einstein–Stokes equation, the diffusivity of an analyte is inversely proportional to the fluid viscosity.

where conditions should be such that $Pe \gg 1$, $Pe > \eta$, and $Da \gg 1$. Although far from being precise, another rule of thumb condition should be $\eta > 1$. In general, sensing conditions outside these bounds will remain unaffected by any microfludic mixing process.[16]

And how much benefit can mixing provide? The upper bounds of this question can be estimated as $E_{max} \approx 1/2(PeU_t/U)^{2/9}$, which is valid for a mixer having a length of $L_{opt} \approx H(U/U_t)^{1/3}Pe^{2/3}$. Thus, mixing will be most beneficial for the detection of analytes having low diffusivity in channels that are (from a biosensing perspective) relatively long. The discussion in Sect. 3.2 provides a relatively simple method that one can use to predict the mass transfer response of a biosensor in a mixed channel. As per the results shown in Figs. 8 and 9, this method can be used with confidence for the detection of sufficiently small analytes.

The vast majority of the experimental literature on the topic herein involves the use of the staggered herringbone mixer to increase the rate of CTC capture. In contrast to smaller analytes, there are several transport mechanisms that become dominant due to the large size of the CTC, and the methods discussed in Sect. 3.2 lead to erroneous predictions. Owing to the importance of the problem, further study on the transport of CTCs in mixed channels would certainly be beneficial.

We hope that the reader finds this chapter useful in elucidating the complex behavior involved in coupling a biosensor with a microfluidic mixer. We encourage the reader to explore the preceding works on the topic [15, 46, 48, 49], which, despite their complexity, are useful for researchers in a variety of fields. This chapter represents an attempt to translate those works into the language used by those in the biosensing community.

Acknowledgements The authors would like to acknowledge funding from both the Praemium Academiae of the Academy of Sciences of the Czech Republic as well as the Czech Science Foundation (contract no. P205/12/G118).

References

1. Luka G, Ahmadi A, Najjaran H, Alocilja E, DeRosa M, Wolthers K, Malki A, Aziz H, Althani A, Hoorfar M (2015) Microfluidics integrated biosensors: a leading technology towards lab-on-a-chip and sensing applications. Sensors 15(12):30011–30031
2. Estevez M-C, Alvarez M, Lechuga LM (2012) Integrated optical devices for lab-on-a-chip biosensing applications. Laser Photonics Rev 6(4):463–487
3. Homola J (2008) Surface plasmon resonance sensors for detection of chemical and biological species. Chem Rev 108(2):462–493
4. White IM, Yazdi SH, Wei WY (2012) Optofluidic sers: synergizing photonics and microfluidics for chemical and biological analysis. Microfluid Nanofluid 13(2):205–216
5. Rackus DG, Shamsi MH, Wheeler AR (2015) Electrochemistry, biosensors and microfluidics: a convergence of fields. Chem Soc Rev 44(15):5320–5340

[16]This applies for mixing in the sense as shown in Fig. 1, where there is no selective transport specific to the analyte (e.g., (di)electrophoresis, thermophoresis).

6. Reverté L, Prieto-Simón B, Campàs M (2016) New advances in electrochemical biosensors for the detection of toxins: nanomaterials, magnetic beads and microfluidics systems. a review. Anal Chim Acta 908:8–21
7. Sang S, Zhao Y, Zhang W, Li P, Hu J, Li G (2014) Surface stress-based biosensors. Biosens Bioelectron 51:124–135
8. Skládal P (2016) Piezoelectric biosensors. TrAC Trends Anal Chem 79:127–133
9. Yakovleva M, Bhand S, Danielsson B (2013) The enzyme thermistora realistic biosensor concept. A critical review. Anal Chimica Acta 766:1–12
10. Wang T, Zhou Y, Lei C, Luo J, Xie S, Pu H (2016) Magnetic impedance biosensor: a review. Biosens Bioelectron
11. Faustino V, Catarino SO, Lima R, Minas G (2015) Biomedical microfluidic devices by using low-cost fabrication techniques: a review. J Biomech
12. Whitesides GM (2006) The origins and the future of microfluidics. Nature 442(7101):368–373
13. Squires TM, Quake SR (2005) Microfluidics: fluid physics at the nanoliter scale. Rev Mod Phys 77(3):977
14. Stone HA, Stroock AD, Ajdari A (2004) Engineering flows in small devices: microfluidics toward a lab-on-a-chip. Annu Rev Fluid Mech 36:381–411
15. Squires TM, Messinger RJ, Manalis SR (2008) Making it stick: convection, reaction and diffusion in surface-based biosensors. Nat Biotechnol 26(4):417–426
16. Lynn NS, Šípová H, Adam P, Homola J (2013) Enhancement of affinity-based biosensors: effect of sensing chamber geometry on sensitivity. Lab Chip 13(7):1413–1421
17. Hofmann O, Voirin G, Niedermann P, Manz A (2002) Three-dimensional microfluidic confinement for efficient sample delivery to biosensor surfaces. Application to immunoassays on planar optical waveguides. Anal Chem 74(20):5243–5250
18. Yoon SK, Fichtl GW, Kenis PJ (2006) Active control of the depletion boundary layers in microfluidic electrochemical reactors. Lab Chip 6(12):1516–1524
19. Mokrani A, Castelain C, Peerhossaini H (1997) The effects of chaotic advection on heat transfer. Int J Heat Mass Transf 40(13):3089–3104
20. Jana S, Ottino J (1992) Chaos-enhanced transport in cellular flows. Philosophical transactions of the Royal Society of London a: mathematical. Phys Eng Sci 338(1651):519–532
21. Hessel V, Löwe H, Schönfeld F (2005) Micromixersa review on passive and active mixing principles. Chem Eng Sci 60(8):2479–2501
22. Chang C-C, Yang R-J (2007) Electrokinetic mixing in microfluidic systems. Microfluid Nanofluid 3(5):501–525
23. Ottino JM Wiggins S (2004) Introduction: mixing in microfluidics. Philos Trans Math Phys Eng Sci pp. 923–935,
24. Lee C-Y, Wang W-T, Liu C-C, Fu L-M (2016) Passive mixers in microfluidic systems: a review. Chem Eng J 288:146–160
25. Ward K, Fan ZH (2015) Mixing in microfluidic devices and enhancement methods. J Micromech Microeng 25(9):094001
26. Stroock AD, Dertinger SK, Ajdari A, Mezić I, Stone HA, Whitesides GM (2002) Chaotic mixer for microchannels. Science 295(5555):647–651
27. Foley JO, Mashadi-Hossein A, Fu E, Finlayson BA, Yager P (2008) Experimental and model investigation of the time-dependent 2-dimensional distribution of binding in a herringbone microchannel. Lab Chip 8(4):557–564
28. Golden JP, Floyd-Smith TM, Mott DR, Ligler FS (2007) Target delivery in a microfluidic immunosensor. Biosens Bioelectron 22(11):2763–2767
29. Stott SL, Hsu C-H, Tsukrov DI, Yu M, Miyamoto DT, Waltman BA, Rothenberg SM, Shah AM, Smas ME, Korir GK et al (2010) Isolation of circulating tumor cells using a microvortex-generating herringbone-chip. Proc Natl Acad Sci 107(43):18392–18397
30. McPeak KM, Baxter JB (2009) Zno nanowires grown by chemical bath deposition in a continuous flow microreactor. Cryst Growth Des 9(10):4538–4545
31. Khinast JG, Bauer A, Bolz D, Panarello A (2003) Mass-transfer enhancement by static mixers in a wall-coated catalytic reactor. Chem Eng Sci 58(3):1063–1070

32. Isfahani RN, Bigham S, Mortazavi M, Wei X, Moghaddam S (2015) Impact of micromixing on performance of a membrane-based absorber. Energy 90:997–1004
33. Myszka DG, He X, Dembo M, Morton TA, Goldstein B (1998) Extending the range of rate constants available from biacore: interpreting mass transport-influenced binding data. Biophys J 75(2):583–594
34. Eddowes M (1987) Direct immunochemical sensing: basic chemical principles and fundamental limitations. Biosensors 3(1):1–15
35. Glaser RW (1993) Antigen-antibody binding and mass transport by convection and diffusion to a surface: a two-dimensional computer model of binding and dissociation kinetics. Anal Biochem 213(1):152–161
36. Biacore A (1997) Biaevaluation software handbook, version 3.0. Biacore AB, Uppsala, Sweden
37. Newman J (1973) The fundamental principles of current distribution and mass transport in electrochemical cells. Electroanal Chem 6:279–297
38. Ackerberg R, Patel R, Gupta S (1978) The heat/mass transfer to a finite strip at small péclet numbers. J Fluid Mech 86(01):49–65
39. Zhang W, Stone H, Sherwood J (1996) Mass transfer at a microelectrode in channel flow. J Phys Chem 100(22):9462–9464
40. Lynn NSS Jr, Homola J (2016) (bio) sensing using nanoparticle arrays: on the effect of analyte transport on sensitivity. Anal Chem 88(24):12145–12151
41. Lynn NS, Henry CS, Dandy DS (2008) Microfluidic mixing via transverse electrokinetic effects in a planar microchannel. Microfluid Nanofluid 5(4):493–505
42. Stroock AD, McGraw GJ (2004) Investigation of the staggered herringbone mixer with a simple analytical model. Philosophical transactions of the Royal Society of London a: mathematical. Phys Eng Sci 362(1818):971–986
43. Wiggins S, Ottino JM (2004) Foundations of chaotic mixing. Philosophical transactions of the Royal Society of London a: mathematical. Phys Eng Sci 362(1818):937–970
44. Ottino JM (1989) The kinematics of mixing: stretching, chaos, and transport, vol. 3. Cambridge University Press
45. Ottino J (1990) Mixing, chaotic advection, and turbulence. Annu Rev Fluid Mech 22(1):207–254
46. Kirtland JD, McGraw GJ, Stroock AD (2006) Mass transfer to reactive boundaries from steady three-dimensional flows in microchannels. Phys Fluids 18(7):073602 (1994-present)
47. Lynn NS Jr, Homola J (2015) Biosensor enhancement using grooved micromixers: Part i, numerical studies. Anal Chem 87(11):5516–5523
48. Kirtland JD, Siegel CR, Stroock AD (2009) Interfacial mass transport in steady three-dimensional flows in microchannels. New J Phys 11(7):075028
49. Sundararajan P, Stroock AD (2012) Transport phenomena in chaotic laminar flows. Annu Rev Chem Biomol Eng 3:473–496
50. Lynn NS Jr, Bockova M, Adam P, Homola J (2015) Biosensor enhancement using grooved micromixers: part ii, experimental studies. Anal Chem 87(11):5524–5530
51. Kang TG, Kwon TH (2004) Colored particle tracking method for mixing analysis of chaotic micromixers. J Micromech Microeng 14(7):891
52. Aubin J, Fletcher DF, Xuereb C (2005) Design of micromixers using cfd modelling. Chem Eng Sci 60(8):2503–2516
53. Yang J-T, Huang K-J, Lin Y-C (2005) Geometric effects on fluid mixing in passive grooved micromixers. Lab Chip 5(10):1140–1147
54. Hassell D, Zimmerman W (2006) Investigation of the convective motion through a staggered herringbone micromixer at low reynolds number flow. Chem Eng Sci 61(9):2977–2985
55. Lynn NS, Dandy DS (2007) Geometrical optimization of helical flow in grooved micromixers. Lab Chip 7(5):580–587
56. Yun S, Lim G, Kang KH, Suh YK (2013) Geometric effects on lateral transport induced by slanted grooves in a microchannel at a low reynolds number. Chem Eng Sci 104:82–92
57. Kim DS, Lee SW, Kwon TH, Lee SS (2004) A barrier embedded chaotic micromixer. J Micromech Microeng 14(6):798

58. Sato H, Ito S, Tajima K, Orimoto N, Shoji S (2005) Pdms microchannels with slanted grooves embedded in three walls to realize efficient spiral flow. Sens Actuators A Phys 119(2):365–371
59. Jain M, Rao A, Nandakumar K (2013) Numerical study on shape optimization of groove micromixers. Microfluid Nanofluid 15(5):689–699
60. Liu Y, Deng Y, Zhang P, Liu Z, Wu Y (2013) Experimental investigation of passive micromixers conceptual design using the layout optimization method. J Micromech Microeng 23(7):075002
61. Forbes TP, Kralj JG (2012) Engineering and analysis of surface interactions in a microfluidic herringbone micromixer. Lab Chip 12(15):2634–2637
62. Wesseling P (2009) Principles of computational fluid dynamics, vol 29. Springer Science & Business Media
63. Salamon P, Fernàndez-Garcia D, Gómez-Hernández J (2006) Modeling mass transfer processes using random walk particle tracking. Water Resour Res 42(11)
64. Gomez-Aranzadi M, Arana S, Mujika M, Hansford D (2015) Integrated microstructures to improve surface-sample interaction in planar biosensors. IEEE Sens J 15(2):1216–1223
65. Aref H (1984) Stirring by chaotic advection. J Fluid Mech 143:1–21
66. Dong Y, Skelley AM, Merdek KD, Sprott KM, Jiang C, Pierceall WE, Lin J, Stocum M, Carney WP, Smirnov DA (2013) Microfluidics and circulating tumor cells. J Mol Diagn 15(2):149–157
67. Hajba L, Guttman A (2014) Circulating tumor-cell detection and capture using microfluidic devices. TrAC Trends Anal Chem 59:9–16
68. Miyamoto DT, Sequist LV, Lee RJ (2014) Circulating tumour cells [mdash] monitoring treatment response in prostate cancer. Nat Rev Clin Oncol 11(7):401–412
69. Murlidhar V, Rivera-Báez L, Nagrath S (2016) Affinity versus label-free isolation of circulating tumor cells: who wins?. Small
70. Qian W, Zhang Y, Chen W (2015) Capturing cancer: emerging microfluidic technologies for the capture and characterization of circulating tumor cells. Small 11(32):3850–3872
71. Smith JP, Barbati AC, Santana SM, Gleghorn JP, Kirby BJ (2012) Microfluidic transport in microdevices for rare cell capture. Electrophoresis 33(21):3133–3142
72. Nagrath S, Sequist LV, Maheswaran S, Bell DW, Irimia D, Ulkus L, Smith MR, Kwak EL, Digumarthy S, Muzikansky A et al (2007) Isolation of rare circulating tumour cells in cancer patients by microchip technology. Nature 450(7173):1235–1239
73. Sheng W, Ogunwobi OO, Chen T, Zhang J, George TJ, Liu C, Fan ZH (2014) Capture, release and culture of circulating tumor cells from pancreatic cancer patients using an enhanced mixing chip. Lab Chip 14(1):89–98
74. Xue P, Ye K, Gao J, Wu Y, Guo J, Hui KM, Kang Y (2014) Isolation and elution of hep3b circulating tumor cells using a dual-functional herringbone chip. Microfluid Nanofluid 16(3):605–612
75. Xue P, Wu Y, Guo J, Kang Y (2015) Highly efficient capture and harvest of circulating tumor cells on a microfluidic chip integrated with herringbone and micropost arrays. Biomed Microdevices 17(2):1–8
76. Huang LR, Cox EC, Austin RH, Sturm JC (2004) Continuous particle separation through deterministic lateral displacement. Science 304(5673):987–990
77. Liu Z, Zhang W, Huang F, Feng H, Shu W, Xu X, Chen Y (2013) High throughput capture of circulating tumor cells using an integrated microfluidic system. Biosens Bioelectron 47:113–119
78. Hyun K-A, Lee TY, Jung H-I (2013) Negative enrichment of circulating tumor cells using a geometrically activated surface interaction chip. Anal Chem 85(9):4439–4445
79. Bird RB, Stewart WE, Lightfoot EN (2007) Transport phenomena. Wiley
80. Wang S, Liu K, Liu J, Yu ZT-F, Xu X, Zhao L, Lee T, Lee EK, Reiss J, Lee Y-K et al (2011) Highly efficient capture of circulating tumor cells by using nanostructured silicon substrates with integrated chaotic micromixers. Angew Chem Int Ed 50(13):3084–3088
81. Zhao L, Lu Y-T, Li F, Wu K, Hou S, Yu J, Shen Q, Wu D, Song M, OuYang W-H et al (2013) High-purity prostate circulating tumor cell isolation by a polymer nanofiber-embedded microchip for whole exome sequencing. Adv Mater 25(21):2897–2902

82. Hou S, Zhao L, Shen Q, Yu J, Ng C, Kong X, Wu D, Song M, Shi X, Xu X et al (2013) Polymer nanofiber-embedded microchips for detection, isolation, and molecular analysis of single circulating melanoma cells. Angew Chem Int Ed 52(12):3379–3383
83. Shen Q, Xu L, Zhao L, Wu D, Fan Y, Zhou Y, OuYang W-H, Xu X, Zhang Z, Song M et al (2013) Specific capture and release of circulating tumor cells using aptamer-modified nanosubstrates. Adv Mater 25(16):2368–2373
84. Fröhlich E, Bonstingl G, Höfler A, Meindl C, Leitinger G, Pieber TR, Roblegg E (2013) Comparison of two in vitro systems to assess cellular effects of nanoparticles-containing aerosols. Toxicol Vitro 27(1):409–417
85. Adams DL, Zhu P, Makarova OV, Martin SS, Charpentier M, Chumsri S, Li S, Amstutz P, Tang C-M (2014) The systematic study of circulating tumor cell isolation using lithographic microfilters. RSC Adv 4(9):4334–4342
86. Czerlinski G, Reid D, Apostol A, Bauer K, Scarpelli D (1987) Determination of the density of cells from sedimentation studies at 1g. J Biol Phys 15(2):29–32
87. Dharmasiri U, Balamurugan S, Adams AA, Okagbare PI, Obubuafo A, Soper SA (2009) Highly efficient capture and enumeration of low abundance prostate cancer cells using prostate-specific membrane antigen aptamers immobilized to a polymeric microfluidic device. Electrophoresis 30(18):3289–3300
88. Staben ME, Zinchenko AZ, Davis RH (2003) Motion of a particle between two parallel plane walls in low-reynolds-number poiseuille flow. Phys Fluids 15(6):1711–1733
89. Cho YI, Kensey KR (1991) Effects of the non-newtonian viscosity of blood on flows in a diseased arterial vessel. part 1: steady flows. Biorheology 28(3–4):241–262

Self-directed Transport on Nanostructured Plasmonic Sensors

Shailabh Kumar

Abstract Analytical sensors using varying detection strategies have been widely and successfully employed for advances in areas such as drug discovery, disease diagnosis and study of biological systems. Many of these sensors utilize plasmonic metallic nanostructures which can concentrate electromagnetic fields in nanoscale regions leading to many fold enhancement in optical signal obtained from the molecules. They employ techniques including fluorescence, surface plasmon resonance (SPR)-based refractive index sensing, surface-enhanced Raman spectroscopy (SERS) and other forms of vibrational spectroscopy for molecular characterization. However, the performance of these devices relies on effective transport of the target molecules to these nanoscale detection sites. Guided transport is extremely important for fast detection in cases where the concentration of molecules is really low and for accurate measurements of protein–protein binding kinetics. In this chapter, we discuss nanostructured biosensing substrates which can spontaneously direct the flow of molecules in solution towards the sensing hotspots. These devices demonstrate improved detection sensitivity, while minimizing the limitations and complexity imposed upon the system. Additionally, they can trap biological particles such as organelles and liposomes on the sensor surface, facilitating on-chip analysis of single particles. This chapter discusses a few methods which have been utilized for concentration of molecules on plasmonic sensing surfaces, without the application of external power sources.

1 Plasmonic Sensors and the Diffusion Limit

Plasmonic sensors rely on concentration of electromagnetic fields near their surface resulting in largely amplified optical signal from molecules. This signal can be collected in several forms including fluorescence, luminescence, Raman vibrational

S. Kumar (✉)
Department of Medical Engineering, California Institute of Technology,
Pasadena, CA 91125, USA
e-mail: shailabh@caltech.edu

S.-H. Oh et al. (eds.), *Miniature Fluidic Devices for Rapid Biological Detection*,
Integrated Analytical Systems, https://doi.org/10.1007/978-3-319-64747-0_4

spectra and SPR-based refractive index change. Since the signal is concentrated in nanoscale regions, guided transport of molecules to these sensing locations is needed. In the absence of guided transport random diffusion of molecules can result in extremely slow detection specially at low concentration levels. In a report, Sheehan and Whitman suggested that detection of molecules at sub-picomolar concentrations on nanoscale sensors would take hours to days, if left to diffusion of molecules [1]. This leads to inefficient detection of biomarkers and inaccurate mass-transport limited analysis of target-receptor interactions being studied. This critical role played by mass-transport and its interplay with receptor-target molecule binding reaction for micro- and nanoscale sensors has been extensively discussed by Squires et al. [2].

Several methods to promote the transport of molecules to the nanoscale sensing sites have been tried. Microfluidic channels with pressure-driven convective flow were utilized to maintain the supply of analytes to receptors on the sensor surface. However it was realized that convective flow over the sensors has limited benefits, due to the formation of a depletion zone near the sensing site [2, 3]. Furthermore since most of the analytes in this case flow over the substrate and do not interact with the receptors, this method is very wasteful towards expensive and difficult to extract analyte molecules. More direct means of controlling molecular transport were attempted. Pressure-driven flow was used to drive aqueous solution through plasmonic nanochannels [3–5]. Electric fields have also been employed to generate flow using electric field gradient focusing or dielectrophoresis [6–8]. However these methods suffer from key drawbacks in terms of portability and ease-of-application. Pressure-driven flow through nanochannels deals with problems derived from a high-pressure environment including leakages, damage to the nanochannels in fragile systems and fluctuation of optical signal. Similarly, electric field derived methods either rely on the charge on the molecules, or are limited by factors such as solution conductivity making it difficult to use them with biological fluids containing various molecules and ions. The requirement for external power sources also limit their portability.

Therefore, miniature fluidic devices which can generate fluidic flow spontaneously towards the sensing spots are a much desired option for diagnostic devices. In this chapter we will discuss generation of flow on nanostructured optical sensing chips using phenomena such as evaporation, surface-tension, and magnetic forces. These methods do not require any external power supplies, can be used for biological liquids such as blood and do not generate any local heating effects. Their application for efficient on-chip molecular sensing as well as trapping and analysis of biological particles will be discussed. Trapping of complex biological particles such as cells and organelles on chip can be used in combination with sensing techniques to understand cell-membrane interactions as well as their functionality.

2 Evaporation-Driven Flow

An example of evaporation-driven aggregation of particles is the famous coffee-stain effect, where the edge of the stain is much darker than the central regions. This effect can be explained by generation of localized flow towards the pinned edges of a drop of liquid on a surface [9]. The loss of water from the drop through evaporation is replenished by generation of flow from other areas towards the pinned edges, since the drop maintains its contact area on the surface. As a result, particles are transported towards this pinned edge and are concentrated there. This idea can be utilized at any pinned liquid–air interface.

Metallic nanohole arrays were chosen as the plasmonic substrate of choice for application of this technique. Metal-coated sub-wavelength nanoholes have been extensively utilized as an extremely efficient and popular plasmonic sensing devices [10–13]. They have enabled measurement of SPR-based real-time binding kinetics of molecules and proteins [14–16], SERS [14, 17, 18], and plasmon-enhanced fluorescence [19, 20]. These nanoholes have been fabricated on substrates such as glass or silicon wafers as dead-ended holes as well as in thin (hundreds of nanometers) freely hanging silicon nitride membranes which are open on both sides (open-ended nanoholes also called nanopores). Solid-state nanopores fabricated in freely hanging membranes have also been applied for DNA sequencing and single-molecule sensing. These open-ended holes have a great potential for spontaneous concentration of biomolecules.

Figure 1 shows one of the standard methods used to fabricate these holes. Photolithography and KOH wet etching were used on silicon wafers with layers of low-stress nitride to create the freely hanging nitride membranes [21]. A layer of gold was deposited on top of the membranes and focused ion beam milling was used to fabricate these structures. Nanoholes of sizes ranging from 200 to 600 nm were milled. When a drop of aqueous solution was added to these substrates, the solution was wicked into the holes due to capillary effect. In the absence of other external forces, the capillary effect does not allow the solution to flow out of the nanoholes. Hence, a water–air interface inside each of the nanohole is created, where evaporation can take place. For the solution inside the nanoholes, loss of volume due to evaporation is replenished by generation of a localized flow towards the holes, similar to pinned edges discussed earlier [9]. This results in transport of any molecules or particles in solution towards the nanoholes.

Molecules in solution can be captured by receptors present at the nanoholes and the continuous flow helps maintain the supply of the molecules to these sites. The performance of this platform was tested for protein concentration and on-chip biosensing. The chip was coated with a conformal layer of silica using atomic layer deposition (ALD). Silane-PEG-biotin was then added to the chip, forming a monolayer of biotin on the surface due to silane-silica binding. A drop of streptavidin, R-phycoerythrin conjugate (SAPE) solution was added to the chip and the effect of evaporation-driven transport was studied. As shown in Fig. 2, compared to a control region with dead-ended holes, significant concentration of the molecules

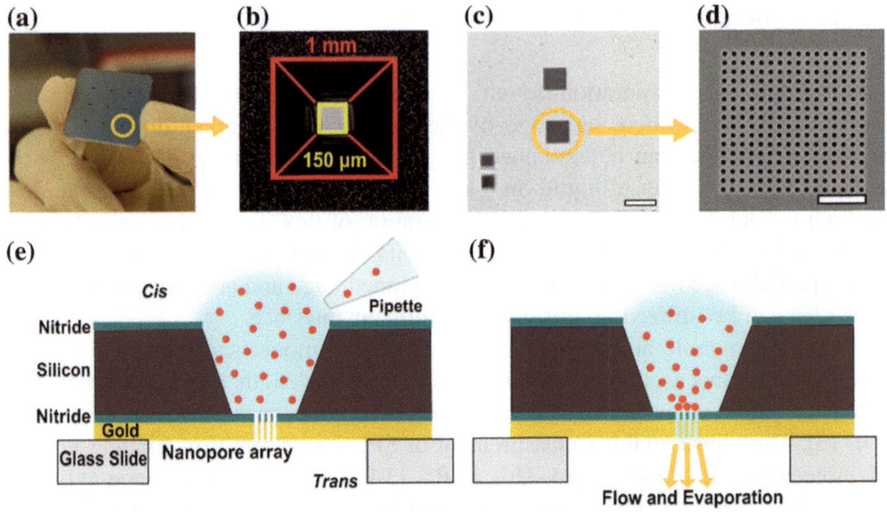

Fig. 1 **a** Chip with nine etched reservoirs. **b** Bright-field image with outlines showing a pyramidal reservoir leading to freely hanging nitride membrane. **c** Bright-field image of the membrane after milling 4 nanohole arrays. Scale bar is 20 µm. **d** Scanning electron micrograph of a nanohole array (hole diameter: 600 nm, periodicity: 1 µm) milled using focused ion beam. Scale bar is 5 µm. **e**, **f** Schematics showing the evaporation-driven mechanism of guided transport towards the holes. Reprinted with permission from [21]. Copyright 2013 American Chemical Society

can be observed at the open-ended nanoholes. This highlights the difference between diffusion-limited transport at the dead-ended holes and directed transport at the open-ended nanohole region. The results show detection of 100 pM streptavidin over the nanoholes using this method.

As mentioned earlier, gold-coated nanoholes are efficient plasmonic sensing substrates. To demonstrate the potential of label-free plasmonic sensing together with evaporation-driven concentration on these chips, SPR-based refractive index measurements were performed. These measurements are performed by recording the transmission spectra of light through the nanoholes, and monitoring the shift in peaks or dips in the spectra. The peak-shift is directly related to the local refractive index close to the nanohole hotspots. Hence binding or unbinding of molecules causes a proportional increase or decrease in the local refractive index, reflected in the peak-shift of the transmission spectra. Figure 3 shows the detection of streptavidin binding on chip at various concentrations using spectra-shift. As seen by these results, the on-chip concentration mechanism helps lower the limit of detection of the sensor by transporting molecules to nanoscale sensing locations. It works within a time-span of minutes, does not require external appendages, power sources and does not place any restrictions on the setup.

This platform and evaporation-driven transport was also utilized for capture of single organelles into the nanoholes [22]. The holes can act as reservoirs where the properties of the organelles can be analyzed individually, while they are arranged in

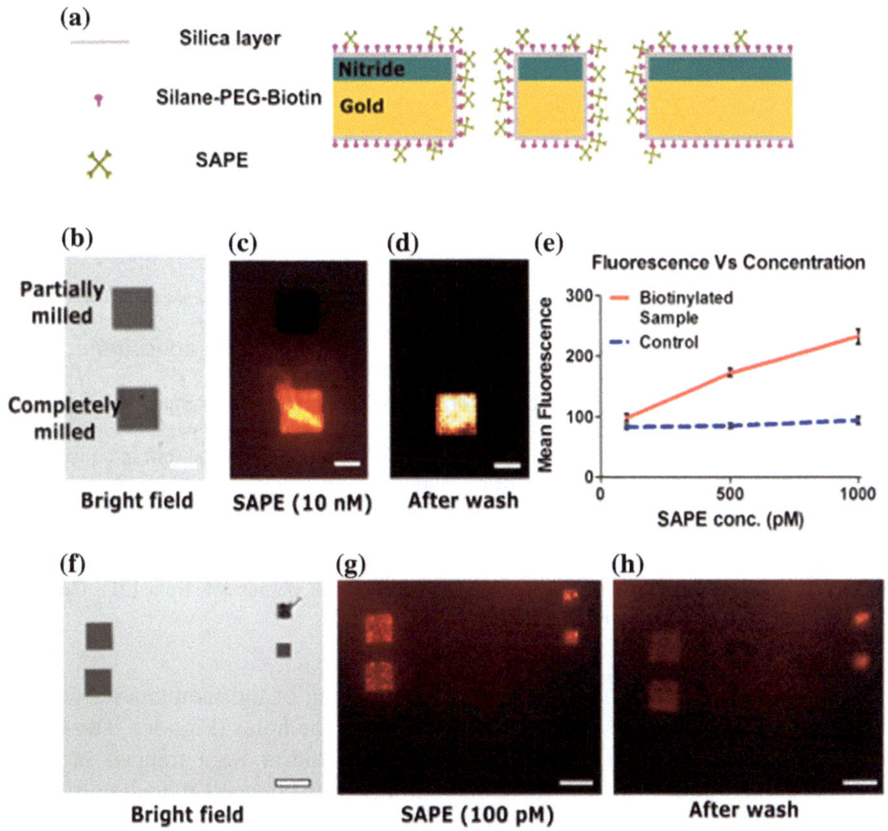

Fig. 2 **a** The chip was coated with a layer of silica followed by silane-PEG-biotin. A solution of SAPE (Streptavidin, R-Phycoerythrin conjugate), which binds to biotin, was added on to the chip. **b** A region with a partially milled nanohole array (dead-ended, control) and an array with completely milled holes (open-ended) were fabricated. **c** Fluorescence image of the sample after addition of 10 nM SAPE solution demonstrating evaporation-driven concentration. **d** Fluorescence image of the sample after rinsing the chip with DI water. **e** A graph showing the mean fluorescence over the holes as a function of SAPE concentration. **f** A suspended nitride region with 4 nanohole arrays. **g** Fluorescence image of the sample 10 mins after addition of 100 pM SAPE solution. **h** Fluorescence image after rinsing the chip with DI water. Scale bar is 10 μm for (**b**, **c**, **d**) and 20 μm for (**f, g**, **h**). Reprinted with permission from [21]. Copyright 2013 American Chemical Society

a large array. Mitochondria samples were obtained, where one batch had the membranes labeled with MitoTracker Green fluorophore and the other half expressed a fluorescent protein, DsRed2. A drop of solution with mitochondria was added to the substrate and allowed to sit for 10 mins. The sample was then washed with buffer and images were taken as shown in Fig. 4. Due to the evaporation-driven flow, organelles were transported towards the holes where they were trapped. Equal volumes of the two differently labeled mitochondria samples

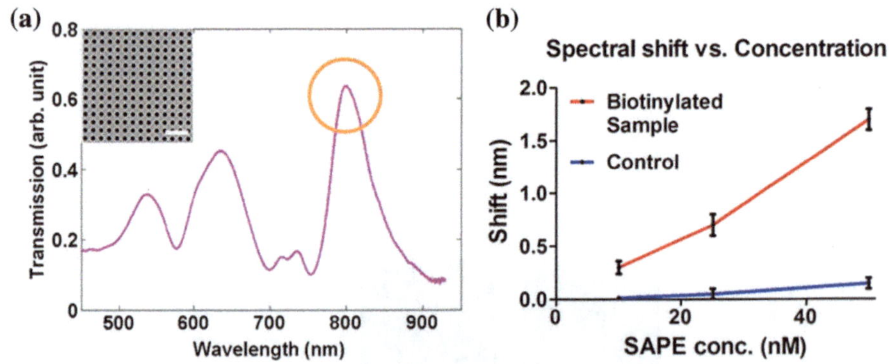

Fig. 3 Surface plasmon resonance detection of SAPE-biotin binding. **a** Normalized transmission spectra as recorded from an array of 200 nm nanopores. The transmission peak at approximately 800 nm corresponds to the gold–water interface and was used to measure the shift in transmission. (inset) SEM image of nanopores (diameter: 200 nm; periodicity: 500 nm) used for spectral measurements. Scale bar is 1.5 µm. Four arrays were milled on the suspended nitride area for measuring the transmission spectra, two large (16 µm × 16 µm) and two smaller (8 µm × 8 µm). **b** A graph showing shift of transmission spectra through the nanopore arrays upon SAPE binding as a function of SAPE concentration. Reprinted with permission from [21]. Copyright 2013 American Chemical Society

were mixed together and added to the chips. Imaging of the sample after washing indicated the presence of single color from most of the holes (Fig. 4c). The lack of color colocalization meant that individual mitochondria were trapped in single nanoholes predominantly. The membrane properties of trapped mitochondria were studied on chip using fluorescent markers of mitochondria function [22].

These results reveal the potential of an evaporation-based delivery scheme on nanoplasmonic sensors. This method increases device performance by performing analyte concentration and lowering the limit of detection. Biological particles can be trapped at nanostructured sites, without the need for any chemical modification of the surface, enabling their on-chip analysis. The platform itself is very simple to use as it does not require any assemblies, external power sources or training. Simply addition of a droplet on to the chip is sufficient to enable concentration of molecules. The limitations of this method include reliance on rate of evaporation which can be dependent on many factors such as area of the liquid–air interface and ambient conditions.

3 Surface-Tension-Based Concentration

The rate of evaporation for aqueous solutions can be slow, dependent on the fluidic volume and ambient conditions. Hence other strategies which can work with larger volumes and promote faster concentration of analytes would be advantageous.

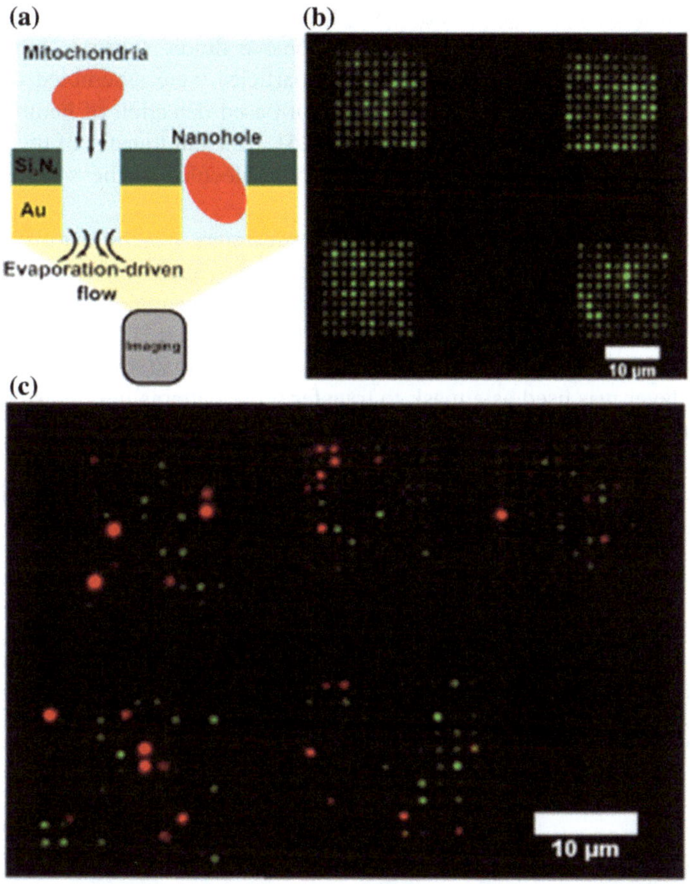

Fig. 4 **a** Mitochondria were allowed to assemble on to the nanoholes as a result of evaporation-driven transport. The samples were washed and imaged from the gold-coated end of the nanoholes. **b** Image shows 4 arrays containing trapped mitochondria tagged with MitoTracker Green dye. **c** Two samples containing mitochondria tagged with a fluorescent marker (DsRed2 in one sample and MitoTracker Green in the other) were diluted and mixed. The mixed mitochondria suspension was then added to the chip over the nanohole arrays for the capture of mitochondria. Image shows 5 arrays containing captured mitochondria tagged with either DsRed2 or MitoTracker Green. Adapted with permission from [22]. Copyright 2015 American Chemical Society

The difference in surface interaction of fluids to various surfaces can be used to drive their flow. This method takes advantage of difference in affinity of surfaces towards water. This method was employed for a microfluidic platform, demonstrating that hydrophobic patterning of microfluidic channels in combination with pressure-driven flow can control the direction of flow as well as split the flow into nanoliter-sized droplets [23]. Another application was shown by Beebe et al. using self-assembled monolayers inside microfluidic channels and creating hydrophilic pathways in order to control fluid flow [24]. Moving towards biosensing applications, another report demonstrated an inexpensive fluidic substrate prepared with hydrophobic paper. Hydrophilic silica nanoparticles were assembled on the substrate defining the path of fluid flow. Luminol-based detection of hemoglobin was demonstrated using the paper–silica hybrid [25]. Efficient transfer of this method to nanoplasmonic sensors directing transport of molecules to the sensing hotspots would enable nanoscale flow control.

In order to transfer this idea to plasmonic substrates, nanohole arrays fabricated in freely hanging silicon nitride membranes were again utilized. Nanoimprint lithography was utilized to fabricate a 1 mm × 1 mm region of freely hanging nanohole arrays. A silicon mold with nanopillars, diameter 200 nm, height 300 nm and periodicity 500 nm was used to imprint nanohole patterns into a resist layer. The resist layer was used as a mask to transfer these patterns into the silicon nitride layer on the chips. A layer of noble metal (gold or silver) was deposited on the substrates. As shown in Fig. 5, nanoimprinting allows reproducible fabrication of nanostructures over a large region. The same mold can be used several times, thus making the process scalable, reproducible, and reducing expenses.

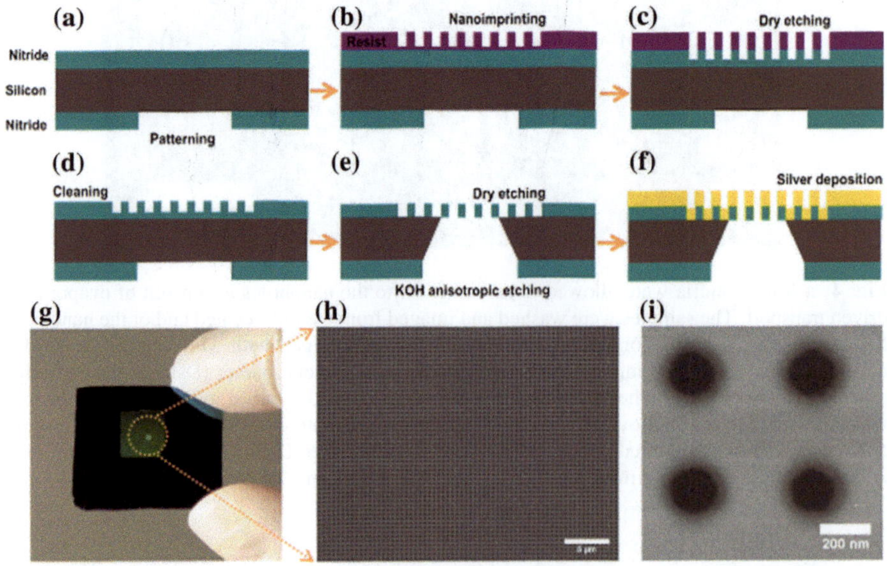

Fig. 5 **a** Deposition of 200 nm low-stress nitride and photolithography on a silicon chip. **b**, **c**, **d** Nanoimprinting, dry etching and cleaning to transfer the nanohole array pattern to the top nitride membrane. **e** Anisotropic KOH etching of silicon. A final dry-etch to remove the remaining nitride from the bottom of the holes. **f** Silver evaporated from the top to obtain freely hanging metallic nanohole arrays. **g** A 1 inch × 1 inch chip with a 1 cm × 1 cm nanoimprinted region in the center. The nanoimprinted region further has a 1 mm × 1 mm nitride membrane in the center. (Circled) **h** SEM image of a nitride membrane with nanoholes. **i** SEM shows magnified image of the individual nitride nanoholes. Adapted from [26]

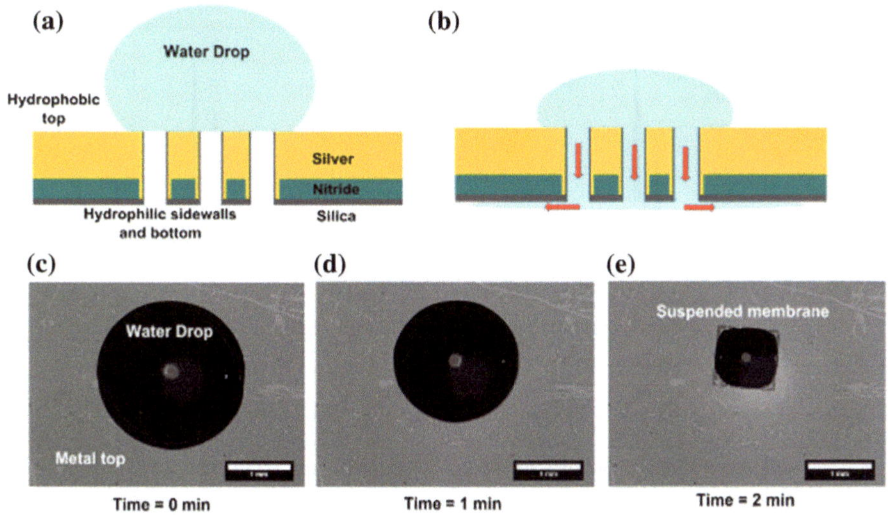

Fig. 6 **a** Addition of an aqueous droplet on the chip. The top metal surface is relatively hydrophobic as compared to the silica layer within the holes and on the backside of the chip. **b** Solution driven into the nanoholes when they come into contact with the hydrophilic sidewalls. **c**, **d**, **e** Flow of a 2 μL drop of water through the nanohole array within 2 min. The water drop shrinks as it is sucked in through the nanoholes. Adapted from [26]

The layer of metal on top of the nanostructures makes the substrate plasmonically active. A layer of silica was deposited on the backside of the chip, such that the nanohole sidewalls were also coated (Fig. 6). The noble metal layer is relatively hydrophobic compared to the silica layer. As a result, when a drop of aqueous solution was placed on the top metal surface, it was driven into the hydrophilic nanopores and flowed onto the backside of the chips. A flow rate of about 1 μL/min was generated for membranes of size 1 mm^2, which compares well to standard on-chip microfluidics. Fluid volume on the order of tens of microliters was easily transported through the holes within minutes. When compared to evaporation-driven concentration on the same chips, surface-tension-driven flow was found to be 10 times faster and the degree of concentration was about an order of magnitude higher as well [26, 27]. The plasmonic hotspots for the nanoholes are located along the edge of the nanoholes at the metal–water interface and the sidewalls. The flow of analyte molecules through the holes forces them to travel close (<100 nm) to the plasmonic hotspots. This distance is small enough to be overcome by diffusion of molecules, such that molecules can bind at or close to the plasmonic hotspots.

In order to test the sensing performance of this device, 10 μL droplets of 4-mercaptopyridine (4-MP) were added to the chips and allowed to flow through the nanoholes. SERS signal from the chips was measured using a 785 nm laser. The signal was compared with control chips which had diffusion-limited binding only. A signal enhancement of 50 times over diffusion-limited substrates was observed for these samples. A low concentration sample containing 50 pM 4-MP was

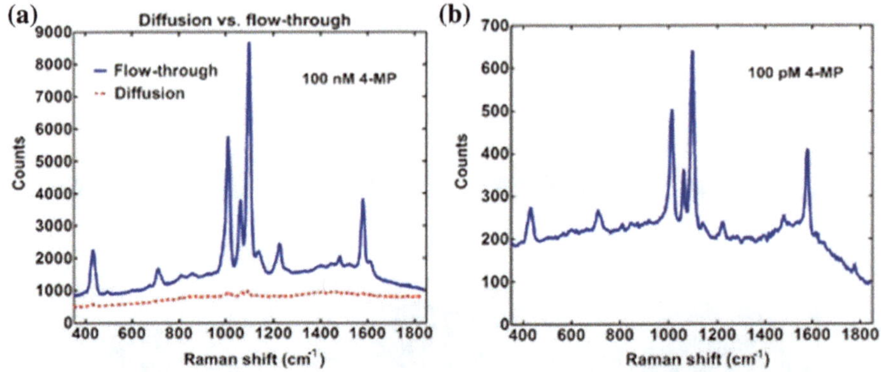

Fig. 7 **a** Comparison of SERS spectra obtained from directed transport of 100 nM 4-MP (blue spectrum) as compared to diffusion-limited transport (red). **b** SERS spectra obtained after flow of 100 pM 4-MP through the nanoholes for an hour. Adapted from [26]

detected after flow through the nanoholes for an hour, demonstrating the ability of the substrates to detect low concentration analytes. The improved sensing efficiency of these plasmonic substrates was due to the excellent plasmonic enhancement and the surface-tension directed delivery of molecules to the hotspots (Fig. 7).

As discussed, the surface-tension directed method was reported to be an order of magnitude faster and a much more efficient concentrator than the evaporation-driven method. It also has applications towards labeled or label-free sensing as well as trapping of particles at plasmonic nanostructured locations. The limitations of this method lie with the need for surface-patterning to create hydrophilic pathways. This can pose some restrictions on attachment of receptors or molecules on the fluidic pathway, which can alter the wettability of the surface.

Some other highly efficient methodologies have been used, which combine surface-patterning with evaporation-driven concentration [28, 29]. These techniques have tried to eliminate the pinning of evaporating aqueous droplets on the substrate surface, and forcing the shrinking droplets to be confined to the sensing region (Fig. 8). The evaporation of the droplet without any surface pinning forced the molecules into a smaller volume, concentrating the molecules and depositing them in an area on the order of 10 μm^2. A substrate with super-hydrophobic surface and plasmonic nanostructures was used [28], enabling molecules to be detected at attomolar concentrations.

Another example allowed molecules and plasmonic gold nanoparticles to evaporate together on a surface, while avoiding pinning of the droplet [29]. Drying droplet concentrated the molecules and gold nanoparticles in a very small region, where SERS measurement was performed. Molecules were detected at sub-femtomolar concentrations using this technique [29].

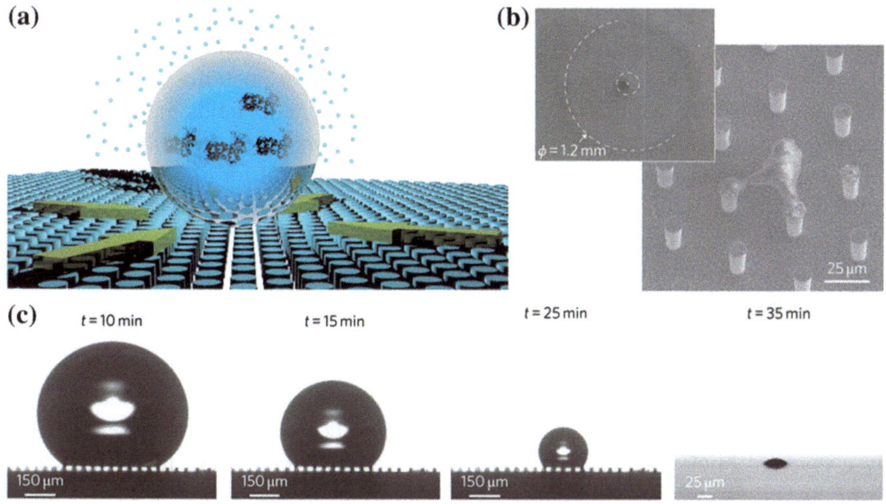

Fig. 8 **a** A schematic representing evaporation process on a super-hydrophobic surface with no pinning of the drop and no solute left on the substrate during drop concentration. **b** SEM images showing the footprint diameter of the drop and deposition of the solute on pillars. The whole content of the drop, with an initial contact area of 1.2 mm (original diameter, 2 mm), is localized on a triangle with lateral sides of ~25 µm after evaporation. **c** Images showing the evaporating droplet on the super-hydrophobic surface. The drop slips on the surface, keeping the contact angle and the shape of the drop constant. The final image shows the collapsing condition of the droplet. Reprinted by permission from Macmillan Publishers Ltd: Nature Photonics (De Angelis et al.) [28], Copyright (2011)

4 Magnetic Concentration in Fluidic Systems

Magnetic forces are well suited to work in complex biological media, including conductive and non-transparent solution like blood. Small magnets can be easily integrated with on-chip fluidic systems. They require no additional power sources, and can create extremely strong magnetic field gradients on the chip-scale. They are generally used in conjugation with magnetic particles with sizes ranging from microns to tens of nanometers. These particles offer a high surface area to volume ratio and can be chemically functionalized with different receptors for various bioassays [31]. Magnetic nanoparticles have been used for aptamer selection against specific antigens [32]. Metal-coated magnetic nanoparticles have been demonstrated to be useful for measurements such as SERS [33, 34]. They have also been utilized for amplification of signal for SPR-based refractive index measurements [35].

Integration of ferromagnetic metal layers with plasmonic substrates can allow more precise control over on-chip capture of these magnetic particles and their subsequent application towards plasmonic or non-plasmonic sensing. Specially for nanostructures, the shape of the structures can have profound effect on the electromagnetic field [36, 37]. Electrostatic fields exhibit singular behavior near sharp tips due to the lightning rod effect [38]. Electromagnetic fields can also be highly localized near sharp tips made of noble metals by nanofocusing of surface plasmons [39, 40]. Similarly concentration of magnetic fields near sharp tips, leading to singular behavior can be theoretically predicted [41]. Intense highly localized magnetic field and field gradients are very useful for directed capture and concentration of molecules and cells on chip [42–44]. Fabrication of ultrasharp ferromagnetic tips with noble metals such gold and silver can be used to combine their plasmonic and magnetic advantages.

A facile way of fabricating ferromagnetic plasmonic substrates over large regions reproducibly is template stripping [30, 45, 46]. This technique relies on the poor adhesion between noble metals (Au, Ag) and silicon surface. The inverse stencil of desired structures is fabricated on to silicon wafers using standard silicon processing techniques. A noble metal layer is first deposited on the wafer followed by a layer of Nickel. The nanostructures can then be peeled off the silicon mold using an adhesive layer (thermal or UV-curable epoxy) on a substrate of choice. Figure 9 shows wedges and pyramids fabricated on planar glass slides using this method. These structures had a tip with very sharp radius of curvature, close to 10 nm.

Analytical equations derived for these structures demonstrate that the magnetic field close to the tips diverge if the tip radius is close to zero. Computer simulations further show the increase in magnetic field and field gradient close to the tips for

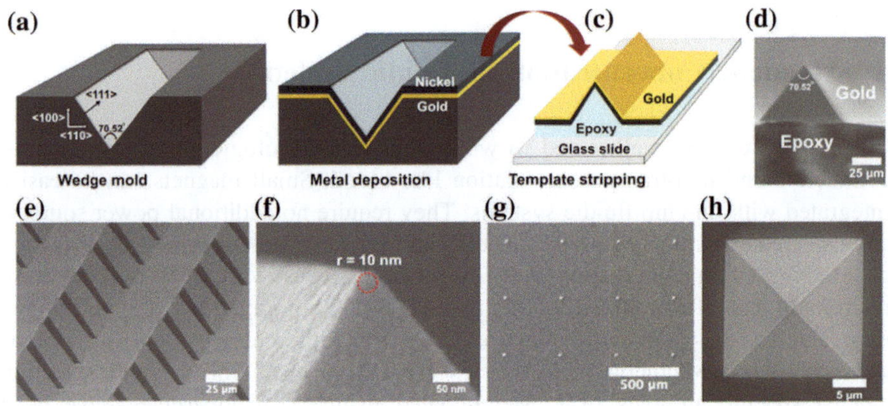

Fig. 9 **a** Wedge-shaped trenches in a silicon mold formed using anisotropic etching of <100> silicon wafer in KOH. **b** Deposition of thin layer of gold followed by thin layer of nickel. **c** UV-curable epoxy was added to the template and a glass slide was placed on top. The sample was then placed under ultraviolet light for curing. The metal film attached to epoxy and glass slide was template-stripped. SEM images showing **d** cross-sectional view of a template-stripped wedge. **e** Bird's eye view of arrays of wedges. **f** Side-view of the tip of a wedge with 10 nm radius of curvature. This wedge had 50 nm gold deposited on top of 125 nm nickel film. **g** Widely separated array of pyramids fabricated using a pyramidal silicon template. **h** Top-down view of a template-stripped pyramid. Adapted from [30]

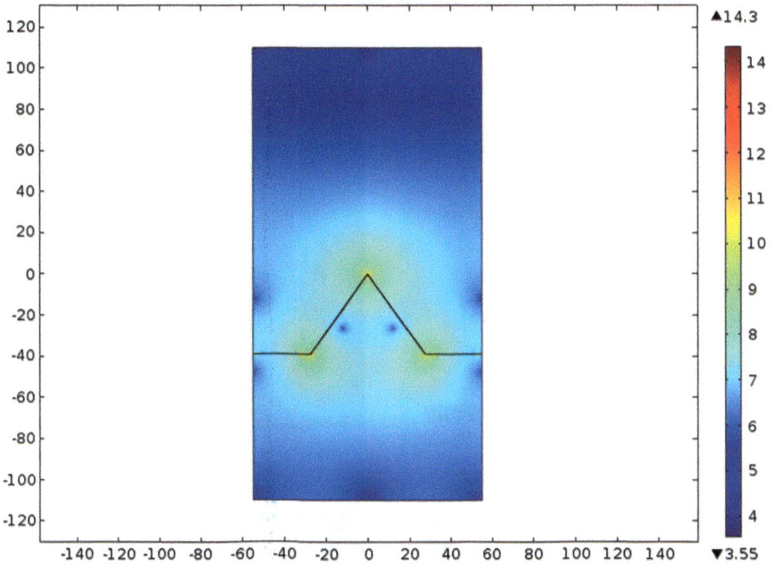

Fig. 10 COMSOL simulation showing order of magnitude plot of gradient of magnetic field (∇H) for a wedge with 10 nm tip radius showing the entire modeled area. Adapted from [30]

various tip radii of curvature. [27, 30] Magnetic particles of various sizes (1.6 µm, 300, and 20 nm) suspended in an aqueous environment were loaded on these substrates and an external magnetic field was introduced. The force experienced by a magnetic particle under the influence of an external magnetic field is directly proportional to the gradient of field. The strong magnetic field close to the sharp tips combined with strong spatial gradient resulted in generation of extremely strong magnetic field gradients near them, directed the particles to be trapped there (Figs. 10, 11). Similar results were obtained for the pyramids. Trapping of single nanoparticles at the pyramidal tips was observed in several cases (Fig. 11d) (Fig. 12).

In order to reveal the plasmonic contribution from the substrates, magnetic nanoparticles labeled with 4-MP were trapped on these substrates. Raman signal was obtained from these molecules and compared with control samples which should not have any contribution towards the enhancement of Raman signal from the molecules. Increased SERS signal from the molecules was observed on the noble metal-coated wedges and pyramids. Hence, these substrates were able to show directed capture of magnetic nanoparticles in a fluidic environment over large areas as well as applications towards plasmonic biosensing. This method can be used for rapid capture of analyte molecules in solution, tagged to magnetic

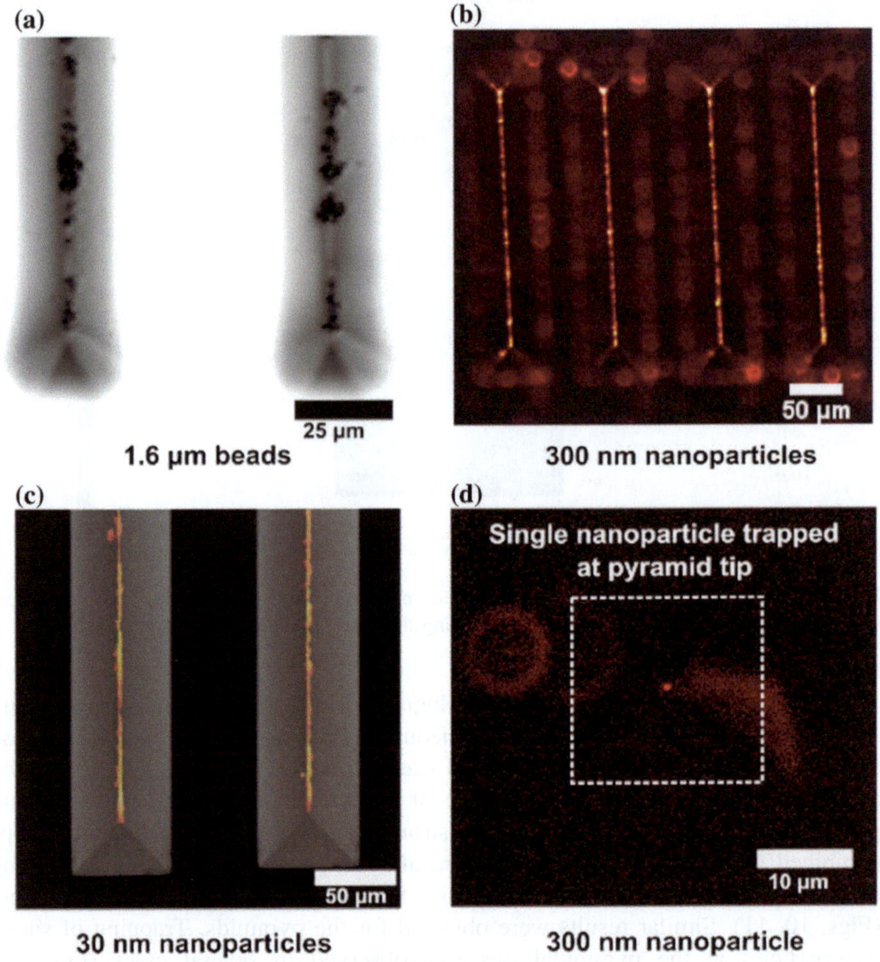

Fig. 11 Images showing the capture of magnetic beads and nanoparticles of various sizes at the tips of magnetic nanostructures. Adapted from [30]

nanoparticles. The molecules are delivered close to the sharp metallic tips, which also serve as plasmonic hotspots. Sharp noble metal tips have been widely used for measurement of tip-enhanced Raman signal [47]. In this case, a high-density of these tips is distributed throughout the substrate and target molecules can be delivered directly to them.

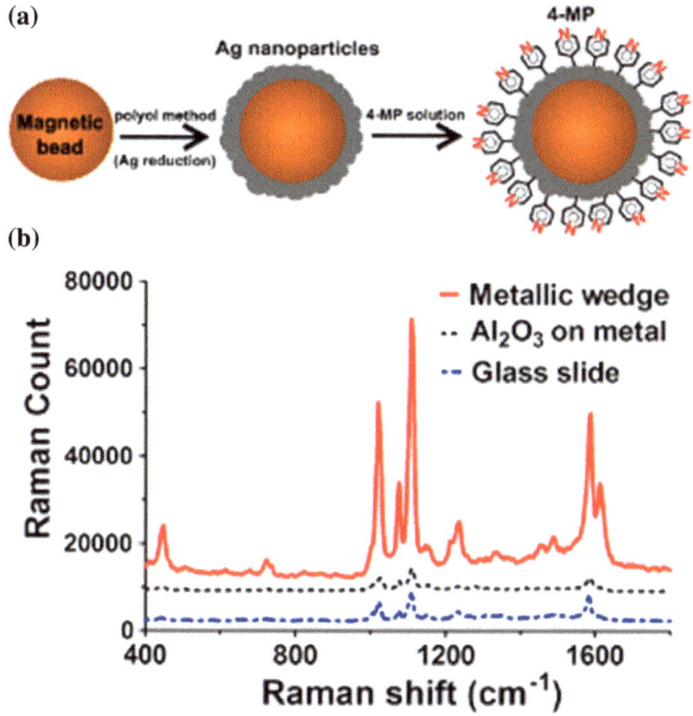

Fig. 12 **a** Cross-sectional schematic showing the chemical functionalization of magnetic beads with 4-mercaptopyrdine (4-MP) after coating them with Ag nanoparticles. **b** Raman signal obtained from the beads captured on SERS-active metallic wedges as compared to on metallic wedges coated with 10 nm of Al_2O_3 and standard glass slide. Glass slides as well as Al_2O_3 coated wedges should not have any contribution towards the SERS signal obtained from the molecules. Adapted from [30]

5 Conclusion

In summary, nanostructured plasmonic substrates which can direct the flow of molecules in solution towards sensing sites can result in drastic improvements in device performance. The limit of detection as well as time taken for sensing can be significantly improved. Three mechanisms for analyte concentration and biosensing were discussed in this chapter. These were based on evaporation, surface-tension and magnetic forces. Localized concentration of molecules near the plasmonic sensing sites was reported for all the cases. Among these techniques, evaporation-dependent directed flow is the simplest as it does not have any preset requirements for surface functionalization of receptors on the sensor surface and does not require integration with external elements such as magnetic beads. It is likely also the slowest among the three concentration mechanisms. Magnetic particle based analyte capture can be extremely fast and has other unique advantages, such as analyte capture independent of the direction of fluid flow. None of these methods needed external power sources, bulky apparatus, or assemblies. No heating artifacts or solution conductance limitations were encountered, which is often the case with electric field-driven concentration methods.

These methods can be easily used in combination with standard microfluidics or just by addition of a droplet on the chip-surface. Droplet-based methods are advantageous as they are simple to use and have zero dead-volume lacking any fluidic interconnects and hence do not waste a lot of analytes. These substrates can also be used for electrochemical biosensing using the metal layers as electrodes. Furthermore, they have shown potential towards trapping of biological particles such as organelles on chip and studying their properties. Moving further these techniques can lead to the development of point-of-care or commercial nanostructured sensing devices with significantly better performance and improved usability as compared to current state of the art.

References

1. Sheehan PE, Whitman LJ (2005) Detection limits for nanoscale biosensors. Nano Lett 5(4): 803–807
2. Squires TM, Messinger RJ, Manalis SR (2008) Making it stick: convection, reaction and diffusion in surface-based biosensors. Nat Biotechnol 26(4):417–426
3. Escobedo C, Brolo AG, Gordon R, Sinton D (2010) Flow-through vs flow-over: analysis of transport and binding in nanohole array plasmonic biosensors. Anal Chem 82(24): 10015–10020
4. Eftekhari F, Escobedo C, Ferreira J, Duan X, Girotto EM, Brolo AG, Gordon R, Sinton D (2009) Nanoholes as nanochannels: flow-through plasmonic sensing. Anal Chem 81(11): 4308–4311
5. Yanik AA, Huang M, Artar A, Chang T-Y, Altug H (2010) Integrated nanoplasmonic-nanofluidic biosensors with targeted delivery of analytes. Appl Phys Lett 96(2):021101
6. Escobedo C, Brolo AG, Gordon R, Sinton D (2012) Optofluidic concentration: plasmonic nanostructure as concentrator and sensor. Nano Lett 12(3):1592–1596
7. Barik A, Chen X, Oh S-H (2016) Ultralow-power electronic trapping of nanoparticles with sub-10 nm gold nanogap electrodes. Nano Lett 16(10):6317–6324
8. Barik A, Otto LM, Yoo D, Jose J, Johnson TW, Oh S-H (2014) Dielectrophoresis-enhanced plasmonic sensing with gold nanohole arrays. Nano Lett 14(4):2006–2012
9. Deegan RD, Bakajin O, Dupont TF, Huber G, Nagel SR, Witten TA (1997) Capillary flow as the cause of ring stains from dried liquid drops. Nature 389(6653):827–829
10. Ebbesen TW, Lezec HJ, Ghaemi HF, Thio T, Wolff PA (1998) Extraordinary optical transmission through sub-wavelength hole arrays. Nature 391(6668):667–669
11. Barnes WL, Murray WA, Dintinger J, Devaux E, Ebbesen TW (2004) Surface plasmon polaritons and their role in the enhanced transmission of light through periodic arrays of subwavelength holes in a metal film. Phys Rev Lett 92(10)
12. Gao HW, Henzie J, Odom TW (2006) Direct evidence for surface plasmon-mediated enhanced light transmission through metallic nanohole arrays. Nano Lett 6(9):2104–2108
13. Homola J, Yee SS, Gauglitz G (1999) Surface plasmon resonance sensors: review. Sensors Actuators B Chem 54(1–2):3–15
14. Brolo AG, Arctander E, Gordon R, Leathem B, Kavanagh KL (2004) Nanohole-enhanced Raman scattering. Nano Lett 4(10):2015–2018

15. Lesuffleur A, Im H, Lindquist N, Oh S (2007) Periodic nanohole arrays with shape-enhanced plasmon resonance as real-time biosensors. Appl Phys Lett 90(24): 243110
16. Im H, Lee SH, Wittenberg NJ, Johnson TW, Lindquist NC, Nagpal P, Norris DJ, Oh S-H (2011) Template-stripped smooth Ag nanohole arrays with silica shells for surface plasmon resonance biosensing. ACS Nano 5(8):6244–6253
17. Yu QM, Guan P, Qin D, Golden G, Wallace PM (2008) Inverted size-dependence of surface-enhanced Raman scattering on gold nanohole and nanodisk arrays. Nano Lett 8(7): 1923–1928
18. Lee S, Bantz K, Lindquist N, Oh S, Haynes C (2009) Self-assembled plasmonic nanohole arrays. Langmuir 25(23):13685–13693
19. Brolo AG, Kwok SC, Moffitt MG, Gordon R, Riordon J, Kavanagh KL (2005) Enhanced fluorescence from arrays of nanoholes in a gold film. J Am Chem Soc 127(42):14936–14941
20. Saboktakin M, Ye X, Chettiar U, Engheta N, Murray C, Kagan C (2013) Plasmonic enhancement of nanophosphor upconversion luminescence in Au nanohole arrays. ACS Nano 7(8):7186–7192
21. Kumar S, Wittenberg NJ, Oh S-H (2013) Nanopore-induced spontaneous concentration for optofluidic sensing and particle assembly. Anal Chem 85(2):971–977
22. Kumar S, Wolken GG, Wittenberg NJ, Arriaga EA, Oh S-H (2015) Nanohole array-directed trapping of mammalian mitochondria enabling single organelle analysis. Anal Chem 87(24): 11973–11977
23. Handique K, Gogoi BP, Burke DT, Mastrangelo CH, Burns MA (1997) Microfluidic flow control using selective hydrophobic patterning. In: Micromachining and microfabrication. International Society for Optics and Photonics pp 185–195
24. Zhao B, Moore JS, Beebe DJ (2001) Surface-directed liquid flow inside microchannels. Science 291(5506):1023–1026
25. Chitnis G, Ding Z, Chang C-L, Savran CA, Ziaie B (2011) Laser-treated hydrophobic paper: an inexpensive microfluidic platform. Lab Chip 11(6):1161–1165
26. Kumar S, Cherukulappurath S, Johnson TW, Oh S-H (2014) Millimeter-sized suspended plasmonic nanohole arrays for surface-tension-driven flow-through SERS. Chem Mater 26(22):6523–6530
27. Kumar S (2015) Directed transport-enabled improved biosensing and bioanalysis on plasmonic nanostructured substrates. University of Minnesota
28. De Angelis F, Gentile F, Mecarini F, Das G, Moretti M, Candeloro P, Coluccio M, Cojoc G, Accardo A, Liberale C (2011) Breaking the diffusion limit with super-hydrophobic delivery of molecules to plasmonic nanofocusing SERS structures. Nat Photonics 5(11):682–687
29. Yang S, Dai X, Stogin BB, Wong T-S (2016) Ultrasensitive surface-enhanced Raman scattering detection in common fluids. Proc Natl Acad Sci 113(2):268–273
30. Kumar S, Johnson TW, Wood CK, Qu T, Wittenberg NJ, Otto LM, Shaver J, Long NJ, Victora RH, Edel JB, Oh S-H (2016) Template-stripped multifunctional wedge and pyramid arrays for magnetic nanofocusing and optical sensing. ACS Appl Mater Interfaces 8(14): 9319–9326
31. Grancharov S, Zeng H, Sun S, Wang S, O'Brien S, Murray C, Kirtley J, Held G (2005) Bio-functionalization of monodisperse magnetic nanoparticles and their use as biomolecular labels in a magnetic tunnel junction based sensor. J Phys Chem B 109(26):13030–13035
32. Lou X, Qian J, Xiao Y, Viel L, Gerdon AE, Lagally ET, Atzberger P, Tarasow TM, Heeger AJ, Soh HT (2009) Micromagnetic selection of aptamers in microfluidic channels. Proc Natl Acad Sci USA 106(9):2989–2994
33. Jun BH, Noh MS, Kim J, Kim G, Kang H, Kim MS, Seo YT, Baek J, Kim JH, Park J (2010) Multifunctional silver-embedded magnetic nanoparticles as SERS nanoprobes and their applications. Small 6(1):119–125
34. Jun BH, Noh MS, Kim G, Kang H, Kim JH, Chung WJ, Kim MS, Kim YK, Cho MH, Jeong DH, Lee YS (2009) Protein separation and identification using magnetic beads encoded with surface-enhanced Raman spectroscopy. Anal Biochem 391(1):24–30

35. Soelberg SD, Stevens RC, Limaye AP, Furlong CE (2009) Surface plasmon resonance detection using antibody-linked magnetic nanoparticles for analyte capture, purification, concentration, and signal amplification. Anal Chem 81(6):2357–2363
36. Maier SA (2007) Plasmonics: fundamentals and applications. Springer
37. Halas NJ, Lal S, Chang WS, Link S, Nordlander P (2011) Plasmons in strongly coupled metallic nanostructures. Chem Rev 111(6):3913–3961
38. Jackson JD (1998) Classical electrodynamics, 3rd edn. Wiley
39. Stockman M (2004) Nanofocusing of optical energy in tapered plasmonic waveguides. Phys Rev Lett 93(13):137404
40. Gramotnev D, Bozhevolnyi S (2010) Plasmonics beyond the diffraction limit. Nat Photonics 4:83–91
41. Van Bladel J (1983) Field singularities at the tip of a cone. Proc IEEE 71(7):901–902
42. Tanase M, Felton EJ, Gray DS, Hultgren A, Chen CS, Reich DH (2005) Assembly of multicellular constructs and microarrays of cells using magnetic nanowires. Lab Chip 5(6): 598–605
43. Pamme N, Wilhelm C (2006) Continuous sorting of magnetic cells via on-chip free-flow magnetophoresis. Lab Chip 6(8):974–980
44. Chen G, Alberts C, Rodriguez W, Toner M (2010) Concentration and purification of human immunodeficiency virus type 1 virions by microfluidic separation of superparamagnetic nanoparticles. Anal Chem 82(2):723–728
45. Nagpal P, Lindquist NC, Oh S-H, Norris DJ (2009) Ultrasmooth patterned metals for plasmonics and metamaterials. Science 325(5940):594–597
46. Park JH, Nagpal P, McPeak KM, Lindquist NC, Oh S-H, Norris DJ (2013) Fabrication of smooth patterned structures of refractory metals, semiconductors, and oxides via template stripping. ACS Appl Mater Interfaces 5(19):9701–9708
47. Pettinger B (2006) Tip-enhanced Raman spectroscopy (TERS). Surf Enhanc Raman Scatt Phys Appl 103:217–240

Enhanced Plasmonic Detection with Dielectrophoretic Concentration

Avijit Barik and Sang-Hyun Oh

Abstract Performance of surface-based plasmonic sensors is often plagued by diffusion-limited transport, which complicates detection from low-concentration analytes. By harnessing gradient forces available from the sharp metallic edges, tips, or gaps that are often found in the plasmonic sensors, it is possible to combine a dielectrophoretic concentration approach to overcome mass transport limitations. A transparent electrode is combined with the plasmonic substrates that allow dielectrophoresis without interfering with the optical detection. Detection from pM-level protein solution is expedited by more than 1000 times as compared to the case of diffusion. Also, enhanced Raman spectroscopic detection is demonstrated using carbon nanotubes and biological particles. Finally, to improve the performance of dielectrophoresis, the gap between the electrodes is reduced to sub-10 nm and ultralow voltage trapping experiments are shown. The ultralow power electronic operation combined with plasmonic detection can enable high-density on-chip integration and portable biosensing.

1 Introduction

Surface plasmon resonance (SPR) is widely used in biosensors to study molecular binding kinetics and affinity [1–3]. SPR is defined as the charge density oscillations of conduction electrons present at the metal dielectric interface. Analyte binding on sensor surface changes the local refractive index (RI) that is measured by monitoring the SPR angle or wavelength in real time [3]. However detection of analytes at low concentration is still challenging because of the fundamental mass transport limitations due to diffusion [4, 5]. For instance, even with a minimum detectable surface coverage of 0.1 ng/cm^2 (commercial SPR instruments), it takes several hours to detect from pM-level analytes [6]. Previous efforts to overcome diffusion

A. Barik · S.-H. Oh (✉)
Department of Electrical and Computer Engineering, Department of Biomedical Engineering, University of Minnesota, Minneapolis, MN 55455, USA
e-mail: sang@umn.edu

S.-H. Oh et al. (eds.), *Miniature Fluidic Devices for Rapid Biological Detection*, Integrated Analytical Systems, https://doi.org/10.1007/978-3-319-64747-0_5

limits include various schemes such as dispersing nanoparticles for rapid sampling, [4] electrokinetic preconcentration of charged particles, [7] evaporation-driven concentration [8] or flowing solutions through nanopores to reduce the diffusion distance [9–11]. Although it is desirable to combine these electrical and optofluidic schemes with SPR sensing, it is difficult to use flat gold films for altering electrical fields. Nanoplasmonic substrates such as nanohole array, [12, 13] metallic nanogaps [14, 15] or tips [16–18] provide options to enhance SPR sensing as well as surface-enhanced spectroscopies. By leveraging the unique geometry of these structures another alternative preconcentration scheme that can be employed is dielectrophoresis (DEP).

DEP is widely used as a particle manipulation technique that is independent of the particle's inherent charge but depends on its induced charge in presence of an external electric field [19]. It is often used to trap polarizable objects such as cells, vesicles, nanoparticles, DNA molecules, or proteins. The strength of the DEP force depends on the gradient of the electric field intensity, which is the greatest near the metallic edges, tips, or micro-/nanoscale gaps. The plasmonic field is also generally the strongest at the sharpest features of the sensors. Thus by combining DEP on nanoplasmonic sensors, it is possible to drive analytes towards the region of highest sensitivity. This directed transport in one hand helps to maximize the signal from a specific quantity of analytes, on the other reduces the time of detection.

2 Overview of Dielectrophoresis

DEP is a phenomenon experienced by a polarizable object in presence of a non-uniform electric field. The principle is illustrated in Fig. 1. When a polarizable object is placed in an electric field, surface charges are induced—positive on one side and negative on the other of same magnitude. Because of coulomb interaction between induced charge and the external electric field, there will be a pulling force on the positive side and a pushing force on the negative side. In the presence of a uniform electric field, the force on either side of the particle cancels each other and

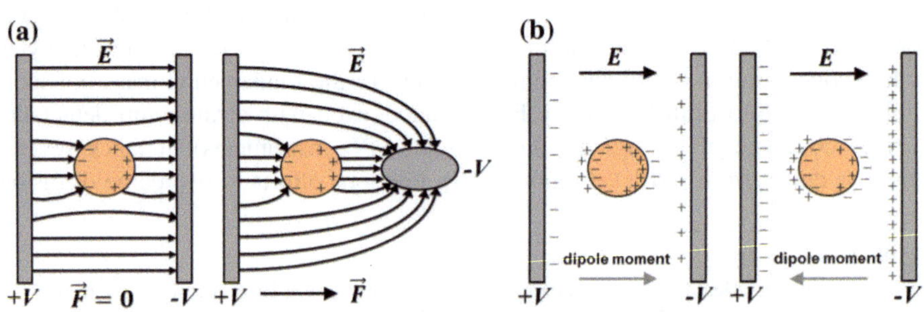

Fig. 1 **a** Schematic of dielectrophoresis. **b** Induced charge for two different conditions where a particle has higher or lower polarizability than the surrounding medium

the particle does not move. However, in the presence of a non-uniform electric field, the force on either side is unequal—resulting in a net force in the direction of the higher field strength (right side of the object in Fig. 1a). However, net polarization of a particle does not only depend on the particle but also is a function of the polarizability of the surrounding medium. The effect of surrounding medium is illustrated in Fig. 1b. There are three possible cases to consider. When the polarizability of the particle is higher than the surrounding medium, there are more charges just inside the interface than outside. In this situation, the net dipole moment of the particle is aligned with the applied field and the particle will move toward the region of stronger electric field strength, also known as positive DEP or pDEP. However, in case of the converse situation, where the medium has higher polarizability, the net dipole moment is in the opposite direction. As a result the particle is pushed toward the region of weaker field strength, and is called negative DEP or nDEP. The third case is when the particle's polarizability matches that of the surrounding medium. In this situation, there is no net dipole and even in the presence of a non-uniform electric field, the particle does not move. Also once the electric field is removed, the charges disappear, hence the term "induced charge" is used to describe this system. The strength of the dipole moment depends on the amount of induced charge and the size of the particle.

The time-averaged DEP force acting on a spherical particle of radius R and calculated from the electric field amplitude, E, is given by

$$\vec{F}_{DEP}(\omega) = \pi \varepsilon_m R^3 \cdot Re(f_{CM}(\omega)) \nabla |E|^2, \tag{1}$$

where $|E|$ is the magnitude of the electric field, ε_m is the permittivity of the surrounding medium and $Re(f_{CM}(\omega))$ is the real part of the Clausius–Mossotti (CM) factor. The CM factor, $f_{CM}(\omega)$, determines the polarity of the DEP force, which for a spherical particle is given by

$$f_{CM}(\omega) = \frac{\varepsilon_p^*(\omega) - \varepsilon_m^*(\omega)}{\varepsilon_p^*(\omega) + 2\varepsilon_m^*(\omega)}, \tag{2}$$

where $\varepsilon_p^*(\omega)$ and $\varepsilon_m^*(\omega)$ are the complex permittivities of the particle and the medium, respectively. The complex permittivities are related to the conductivity σ and the angular frequency ω of the applied electric field by

$$\varepsilon^* = \varepsilon - i(\sigma/\omega) \tag{3}$$

The value of the CM factor varies between −0.5 and +1.0. Particles experience pDEP when the value is positive or nDEP if it is negative. Equation 1 shows that the DEP force grows smaller as the particle volume, which has major implications while manipulating nanoparticles. Equation 1 also points out that the DEP force depends on the electric field intensity gradient, $\nabla |E|^2$. One avenue to boost DEP forces is by making miniaturized geometries such as sharp electrode edges, metallic tips, or micro/nanoscale gap electrodes that can enhance $\nabla|E|^2$. Finally, the direction of DEP force does not depend on the direction of the electric field itself. Thus, both alternating current (AC) and direct current (DC) signal can be used. However, AC has advantages over DC in terms of reducing electrophoretic effects on charged particles, avoiding electrochemical surface reactions and reducing electroosmosis and other electrohydrodynamic effects.

3 Nanohole Array

Nanohole arrays have been extensively used as a SPR sensor [20–24] after the discovery of extraordinary optical transmission (EOT) effect in 1998 [25]. When illuminated with a broadband light source, it shows a series of peaks and dips in the transmitted light collected from the other side. The positions of these resonances are associated with SPR in the gold film and are very sensitive to the local RI of the nanohole surface. By monitoring the shift in the resonances on analyte binding, one can measure molecular binding kinetics and affinity. Also, as the holes itself provide sharp metallic edges, they can act as DEP trap on application of an AC voltage with respect to a ground terminal. Instead of using suspended flow-through nanoholes, [26] a dead-ended structure is used because of the simplicity in the fabrication process as well as equally strong DEP trap. Transport of analytes towards the nanohole surface is significantly accelerated by DEP, which is further quantified and compared to diffusion-based detection (Fig. 2).

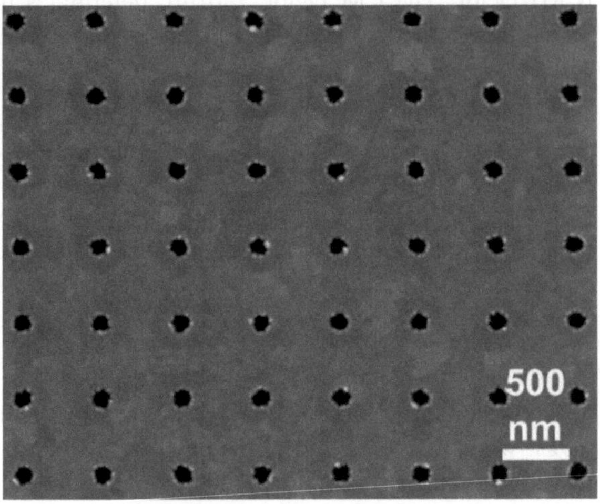

Fig. 2 Scanning electron micrograph (SEM) of the nanohole array showing a hole diameter of 140 nm and a periodicity of 600 nm

3.1 Colocalization of Electrostatic Field with Optical Field

The rims of each hole in a metallic nanohole array concentrate charge and thereby create strong local field gradients, thus nanohole films can concurrently act as a DEP electrode and SPR sensing substrate, which is not possible with flat gold films used in conventional SPR. The electrostatic field around a single hole (140 nm in diameter and 20 nm hole edge radius of curvature) is calculated using a 2D axisymmetric FEM model in COMSOL Multiphysics (Fig. 3). From Eq. 1, for DEP force calculation, we need to know the gradient of electric field intensity, $\nabla|E|^2$. The edge of the nanohole as being the sharpest feature in the geometry provides the maximum $\nabla|E|^2$—enabling particle transport towards it. The effectiveness of this DEP trap depends on the distance between the nanoholes and the top electrode. The dependence is determined by observing $\nabla|E|^2$ at a fixed data collection point at the edge of the hole for electrode gaps, d, varying between 1 and 500 μm and a $1/d^2$ dependence is found, which matches theoretical expectations for relatively long waves interacting with a small hole in a thin sheet of metal [27]. Additionally, it is

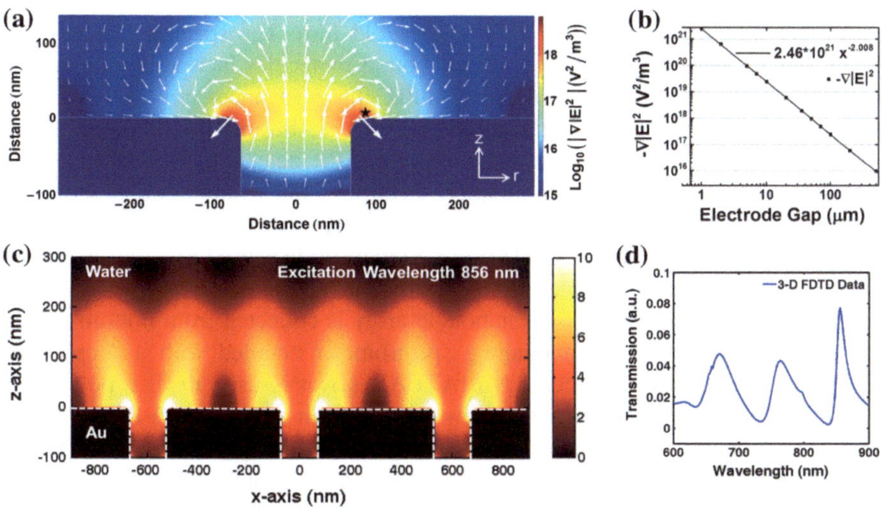

Fig. 3 Colocalization of the electrostatic field with the optical field. **a** FEM simulation using COMSOL shows the gradient of electric field intensity ($\nabla|E|^2$) near a single 140 nm diameter nanohole with a 20 nm hole edge radius of curvature. A 6 V DC signal is applied between the electrodes that are separated by 40 μm. The maximum $\nabla|E|^2$ obtained from this geometry is 6×10^{18} V^2/m^3. **b** Dependence of $\nabla|E|^2$ is plotted as a function of the distance between the electrodes. The $\nabla|E|^2$ value corresponding to different gap widths are collected from the data collection point at r = 86 and 5 nm above the gold surface (starred). **c** 3D FDTD simulation showing time-averaged electric field intensity distribution around gold nanohole array in water at an excitation wavelength of 856 nm. **d** The simulated transmission spectrum from the nanohole due to EOT effect when illuminated with a broadband light source

possible to generate an order of magnitude higher DEP force by reducing the gap between the electrodes by three times.

A 3D finite-difference time-domain (FDTD) method is used to calculate the optical field distribution around a 140 nm diameter hole with periodic boundary conditions (periodicity 600 nm) using a commercial software package (Fullwave, RSoft Design Group). The location of maximum plasmonic fields, i.e., the edge of each hole, coincides with the location of maximum electric field gradients, thereby trapping analytes in the region of maximum detection sensitivity. Metallic apertures had been previously used to trap particles via optical forces generated from laser illumination, which also depends on the gradient of field intensity. DEP, a similar physical principle but typically occurs at radiofrequencies, is easy to implement, and can be used to trap or repel particles by positive and negative DEP, respectively —providing another attractive option for particle manipulation.

3.2 Combining Plasmonic Sensing with DEP Manipulation

The transmission spectrum through the nanohole array reveals a series of resonant peaks and dips associated with SPR in the gold film. The spectral features shift toward longer wavelengths as the surface-bound molecules increase the interfacial RI. Whereas unbinding of molecules shifts the resonances toward the lower wavelengths. Large-area gold nanohole array (8 × 8 mm) is integrated with a transparent top electrode to combine DEP with SPR detection. Nanohole array is prepared using nanoimprint lithography, electron-beam evaporation of gold and template stripping that transfers gold nanohole array from silicon template onto a glass slide. An indium tin oxide (ITO)-coated glass slide is used as a transparent top electrode that allows illumination from top with a tungsten-halogen lamp. The transmitted light through the gold nanohole array is collected from bottom using a 2× objective and sent to an imaging spectrometer coupled with a deep-cooled 1340 × 400 pixel CCD camera. A MATLAB script is used to fit a polynomial function to the appropriate resonance in the spectrum and measure the resonance shift automatically. Sample solution is injected into the gap and an AC bias is applied between the ITO electrode and nanohole array to attract analyte molecules toward the nanohole array surface. The gap between the electrodes is 40 μm. Transport of molecules from the bulk solution to the sensor surface is normally governed by diffusion. But in presence of DEP forces, it is significantly accelerated (Fig. 4).

Polystyrene beads (190 nm in diameter) are used as a model system because of their well-known DEP behavior in water medium. The frequency response of a particle can be understood from the CM factor plot. For frequencies lower than the crossover frequency ($\omega = 1.67$ MHz where $f_{CM} = 0$, for 190 nm polystyrene beads in water of conductivity 0.28 mS/m), particles are attracted toward the nanoholes by pDEP and for frequencies larger than the crossover frequency particles are repelled

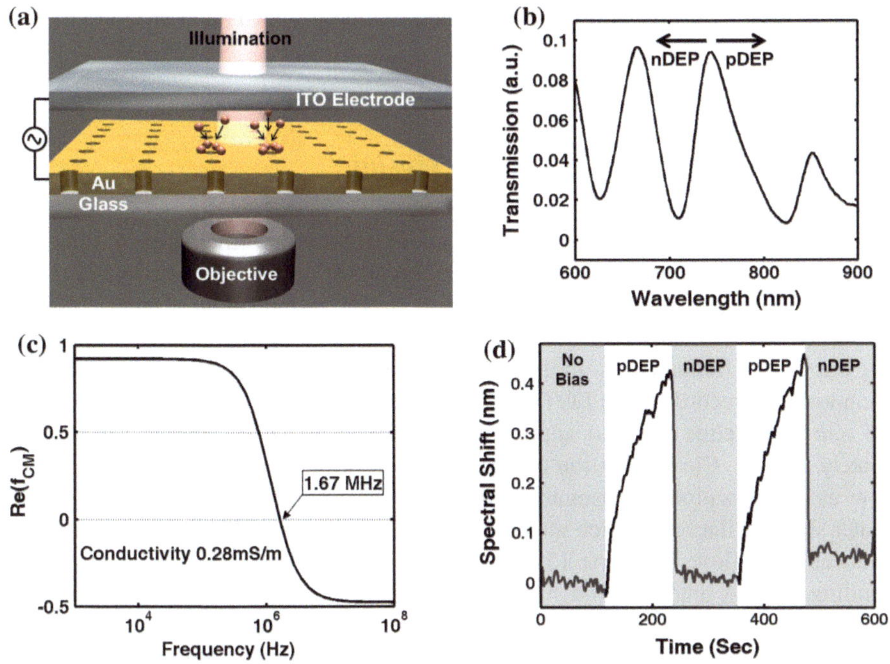

Fig. 4 Combining plasmonic sensing with DEP manipulation. **a** Gold nanohole array is integrated with a top ITO electrode and the whole setup is illuminated from the bottom side while collecting the transmitted light from the top side, which is sent to an imaging spectrometer for analysis. **b** Measured transmission spectrum. The peak positions match well with the simulated spectrum. **c** The CM factor plot of polystyrene beads in water of conductivity 0.28 mS/m. The crossover frequency is 1.67 MHz. **d** DEP manipulations of polystyrene beads on the gold nanohole array while monitoring the event in real time by measuring the spectral shift. A frequency of 1 kHz is used for pDEP and 10 MHz for nDEP at a voltage of 10 $V_{p\text{-}p}$

due to nDEP. So, pDEP shifts the resonances toward the longer wavelengths (red shift) and nDEP reverts it back (blue shift) to the original position. A bias of 10 $V_{p\text{-}p}$ (peak to peak voltage) between the ITO electrode and the gold nanohole array creates an electric field strength of 350 kV/m at the edge of the nanohole that overcomes the thermal motion of the polystyrene particles and trap them at the region of highest SPR sensitivity. A frequency of 1 kHz is used for pDEP and 10 MHz for nDEP. The resonance dip close to 830 nm is used to monitor the DEP manipulation in real time. The choice of resonance peak or dip position is made empirically by comparing the sensitivity. During the positive DEP cycle, the resonance red shifts as the beads move toward and adsorb around the nanoholes. Beads are rapidly repelled from the surface during the negative DEP cycle, as evidenced by the sudden reversal of the spectrum to the baseline level. We show multiple cycles of pDEP and nDEP during a single experiment, demonstrating a robust reversibility. The slight offsets in the measured spectral shift after each negative DEP cycle are attributed to a small number of beads that are immobilized on the surface by van der Waals forces.

3.3 DEP-Enhanced Plasmonic Sensing

The utility of this approach is further characterized by detecting protein molecules at different concentrations, especially in the regime where mass transport limitations become important. Bovine serum albumin or BSA (molecular weight: 65 kDa) dissolved in water is used at a concentration of 1 nM, 10 pM and 1 pM. A bias of 6 $V_{p\text{-}p}$ at 1 kHz frequency is applied between the electrodes and binding of BSA is monitored by recording the EOT spectrum every 2.5 s. Each experiment consists of 15 min of baseline with no applied bias followed by DEP trapping for approximately 40 min. Clear detection of BSA trapping is observed from concentration as low as 1 pM within a reasonable time frame. Negative control experiment with water shows a flat line, which means the spectral shift is not due to any electric field induced fluid flow. The first 15 min of baseline also shows that diffusion-limited binding of BSA onto the gold surface is negligible for the concentrations chosen in this work. Furthermore water being the suspending medium used in this work, it reduces any heat generation as well as unwanted surface reactions. Irreversible spectral shifts are observed while using high-conductivity buffers such as phosphate-buffered saline (PBS) solution, possibly due to the dissolution of the gold surface [28]. In such cases, perhaps an alternative scheme could be used—trapping receptor molecules first in a low-conductivity buffer followed by injecting high-conductivity solution for subsequent detection (Fig. 5).

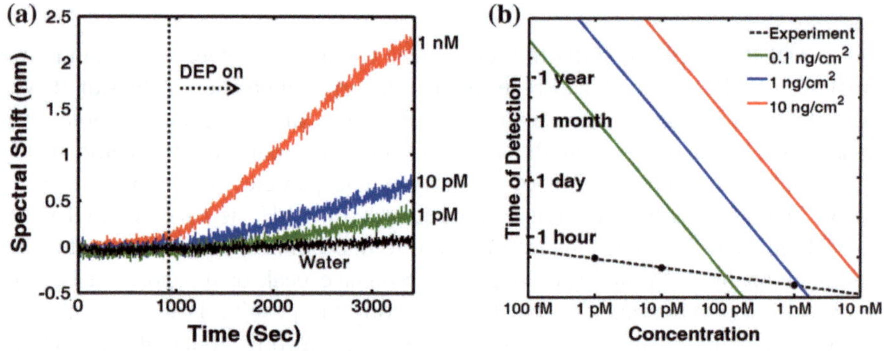

Fig. 5 DEP-enhanced plasmonic sensing. **a** Time-resolved spectral shift observed from various concentrations of BSA and control experiment (water). **b** Comparing t_{DEP} as a function of the BSA concentration to $t_{\mathrm{diffusion}}$. The dependence is drastically changed in presence of DEP—enabling ~1000 times faster detection

The trapping volume of a single BSA molecule is estimated by calculating the threshold force, which is a function of its thermal energy

$$F_{th} = \sqrt{\frac{2}{D\Delta t}} \cdot k_B T, \tag{4}$$

where D is the diffusion constant, Δt is the experimental time, k_B is the Boltzmann constant, and T is the ambient temperature [29, 30]. The threshold force for BSA molecule is calculated to be 12.8 aN ($\Delta t = 3300$ s and $D = 63.8$ $\mu m^2/s$). From Eq. 1 the trapping volume of BSA molecule (volume 163 nm^3) [31] is estimated to extend 30–40 nm above the nanohole edge. Furthermore the performance of the sensor is characterized by estimating the time of detection (t_{DEP})—time to achieve limit of detection (LOD) that corresponds to a signal three times the noise level of the sensor. The t_{DEP} for different concentrations is compared to the time of detection for purely diffusive capture ($t_{diffusion}$). For one-dimensional diffusion in a stagnant solution and assuming an irreversible binding condition, the time-dependent surface coverage can be expressed as [32, 33]

$$\Gamma(t) = 2C_{bulk}\sqrt{\frac{Dt}{\pi}}, \tag{5}$$

where C_{bulk} is the bulk concentration. We can estimate the $t_{diffusion}$ for diffusion-based SPR biosensors from Eq. 5—time it takes to reach a surface coverage value equaling its LOD. The range of typical LOD for SPR sensors lies between 0.1 and 1 ng/cm^2. The dependence of $t_{diffusion}$ is observed as a function of BSA concentrations and is compared with the t_{DEP} values. For example, a sensor with LOD = 0.1 ng/cm^2 will require 8 h to reach the detection limit from a 10 pM BSA solution in a stagnant condition, whereas it only takes 4 min for 100 pM BSA. This drastic difference in $t_{diffusion}$ is a result of the inverse square dependence ($t_{diffusion} \propto C_{bulk}^{-2}$), which complicates detection at ultralow concentrations. Even with convection-aided transport, detection of analytes at sub-10 pM is very challenging [6]. The t_{DEP} for each BSA concentration equals to the time required to achieve a signal to noise ratio (S/N) of 3. Accelerating particle transport by DEP allows detection of analytes at ultralow concentrations as also evidenced from the improved dependence, $t_{DEP} \propto C_{bulk}^{-0.22}$. For example, t_{DEP} for 10 pM BSA is 8 min as compared to 8 h for $t_{diffusion}$, which is 60 times faster. This effect is even more pronounced at lower concentrations—more than 1000 times for 1 pM BSA ($t_{DEP} = 14$ min; $t_{diffusion} = 800$ h).

A novel approach to combine DEP concentration with SPR sensing is shown by simply adding a top transparent electrode that does not interfere with the optics but achieve fast detection of analytes. Colocalization of the electrostatic field with the optical field utilizes the most sensitive regions of the sensor, thus achieving high S/N. The DEP-enhanced detection limit of 1 pM is a significant advancement over previously reported detection limit of 100 nM for BSA using flow-through nanohole-based electrokinetic concentration [11]. This scheme can also be combined with other nanoplasmonic sensors, such as noble metal nanoparticles, [34] single isolated or random nanohole arrays, [21] or flow-through nanoholes [9]. Furthermore engineering designs to reduce the electrode gap from 40 μm will improve the DEP performance further, especially to reduce the working voltages that work well with high-conductivity buffers.

4 Sharp Pyramidal Tips

Gradient forces generated by sharp metallic tips have been used to trap sub-micron particles that are polarizable with respect to the surrounding medium by DEP [35–38]. In addition to the ability to generate strong electric field gradients, sharp tips made with noble metals can also localize surface plasmons [39, 40]. This could be useful for plasmonic trapping, [41] near-field scanning optical microscopy, [42] nonlinear spectroscopy [43]. Considerable effort has been invested to fabricate sharp metallic tips, but it is not trivial to produce high-quality reproducible tips. Conventional method to make sharp tips employs electrochemical etching technique, [42] which is low-throughput and has reproducibility issues. Furthermore these tips are prone to nanoscale roughness that causes unwanted trapping sites [44]. To overcome these hurdles, a template-stripping method has been implemented that produces wafer-scale nanometrically smooth sharp pyramid tips—useful for NSOM and tip-enhanced Raman spectroscopy [45–47].

4.1 Creating a DEP Trap at the Tip of the Pyramid

To create a DEP hotspot at the tip of the pyramid tip a second electrode is added to apply an AC bias between them. This can be achieved either by using an array of pyramid tips and adding a transparent ITO electrode on top or by approaching a single pyramid tip from top onto a bottom ITO electrode (Fig. 6). The transparency of the second electrode is important to combine optical detection such as Raman spectroscopy. To better understand the electrostatic field distribution around a pyramid tip, FEM simulation is performed using COMSOL. A pyramid tip is simulated using a 2D axisymmetric model and assuming a conical geometry with a half angle of 35° and a base radius of 10 μm. The distance between the electrodes is set to be 66 μm. Applying a voltage of 8 V between the electrodes creates an electrostatic hot spot at the tip of the pyramid, which depends on the distance between the electrodes (*d*). Jose et al. has presented a more detailed theoretical model for the same geometry including the dependence of the electric field intensity gradient, $\nabla|E|^2$ on d [17]. For d values less than 1 μm, there is a $1/d^{0.88}$ dependence. Whereas for large d values (> 100 μm), the dependence changes to $1/d^{1.95}$. Clearly, reducing d will increase $\nabla|E|^2$ and the dependence is more prominent at

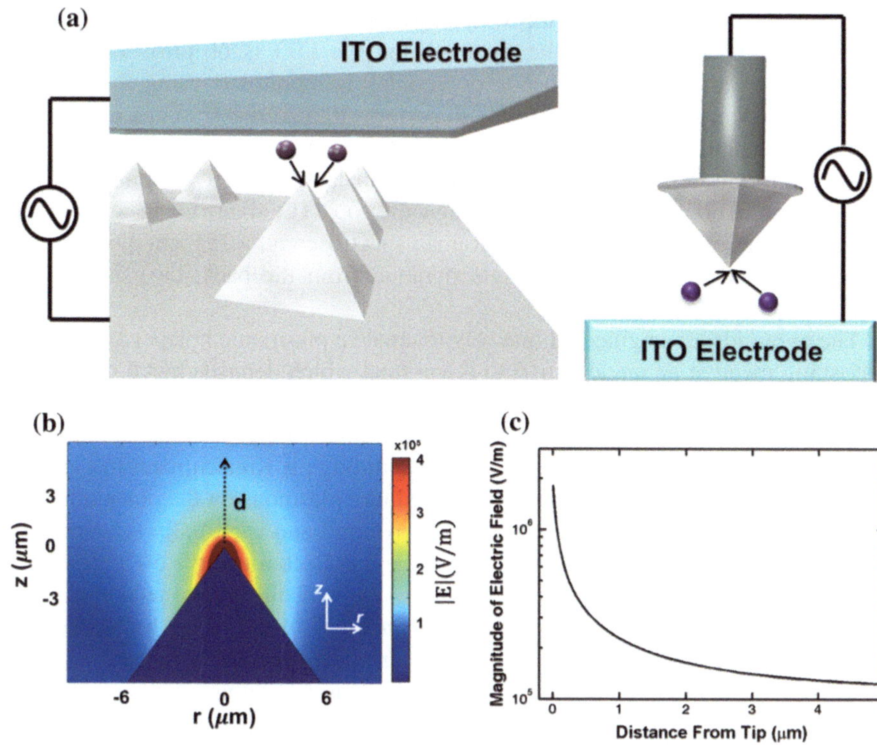

Fig. 6 DEP trap on the tip of the pyramids. **a** Sharp metallic tips can concentrate charges due to lightning rod effect that can induce a DEP trap at the tip of the pyramid. An array of pyramids can be combined with a top ITO electrode or a single pyramid tip can be gradually lowered to a bottom ITO electrode to perform DEP experiments. As the pyramids are made out of noble metals such as gold or silver, it enables plasmonic detection capabilities. **b** FEM simulation of a single pyramid tip using COMSOL shows the presence of strong electric field at the tip of the pyramid. An 8 V DC signal is applied between the electrodes that are separated by 66 μm. **c** The magnitude of the electric field goes down away from the tip. Thus a particle is trapped at the tip only when it reaches within a threshold volume—defined as the trapping volume

larger values of *d*. Reducing the electrode gap to sub-0.5 μm length scale will increase the DEP force by one or even two orders of magnitude.

4.2 Array of Pyramids

A combination of photolithography, anisotropic etching, metal evaporation, and template-stripping (transferring metallic patterns from silicon wafer to glass slide by using an adhesive) is used to fabricate wafer-scale array of pyramid tips made out of gold or silver. A small volume of the sample solution (5 μL) is injected and a top

ITO electrode is assembled on top with a spacer in between, which is 80 μm thick (distance between the tip of the pyramid and the ITO is 66 μm). Trapping of particles such as polystyrene beads or vesicles is monitored using fluorescence imaging. To demonstrate the capability to perform Raman spectroscopy at the tip of the pyramid, Raman-active particles are made by encapsulating 4-mercaptopyridine (4MP) molecules within the vesicles. A 532 nm laser is used in epifluorescence mode (illumination and collection through the top ITO slide) using a 50 × objective for fluorescence imaging. For Raman spectroscopy, a 785 nm diode laser is loosely focused at the tip of a single pyramid from the backside using a 20× objective (NA 0.45).

The pyramids are engineered in a way to create a plasmonic hotspot at the tip by depositing metal at an angle of 10° to the normal which deposits metal of different thickness on the opposite facets of the pyramid. A 135-nm-thick metal deposition will result in 40 nm one side and 120 nm on the other. When illuminated internally, free-space light is coupled from the backside through a Kretschmann-like configuration across the thin side of the pyramid and into plasmons on the outer side [48]. The plasmons then travel up the face of the pyramid and is partially scattered into free-space light when it reaches the tip, which is collected using a 50× objective. Most of the incident light is reflected back by the optically thick metal elsewhere. The power of the laser is maintained at less than 5 mW, which is not high enough in this setup to induce optical trapping or any unwanted thermal effects. This spectroscopy scheme offers certain advantage over confocal method such as collection of Raman signal is achieved from a highly localized point (single pyramid tip). Also the readout of the spectrum is faster here as the region of interest is small. Finally this setup is simpler, compact, and provides real time in situ measurement capabilities. This platform is used to detect analytes encapsulated within vesicles that represent a large class of biological particles. In cells many important molecules or biomarkers are packaged into membrane-bound containers such as nucleus, golgi apparatus, mitochondria, lysozomes, endoplasmic reticuli along with secretory vesicles containing neurotransmitters. Fluorescence imaging of these structures requires labeling each species of interest with distinct fluorophores [49, 50]. However vibrational spectroscopic methods such as Raman spectroscopy offers chemical fingerprinting ability without the need for any external labeling step [51, 52] (Fig. 7).

Vesicle solutions are prepared by mixing 1,2-dimyristoyl-*sn*-glycero-3-phosphocholine (DMPC) and cholesterol in 80:20 molar ratios followed by drying and rehydration in water. For fluorescence microscopy the vesicles are labeled with 1,2-dimyristoyl-s*n*-glycero-3-phosphoethanolamine-N-lissamine Rhodamine B-sulfonyl, ammonium salt (Rhodamine-DMPE). For Raman spectroscopy the hydration step is done in 5 mM 4MP solution followed by dialysis to get rid of the 4MP molecules outside the vesicles. This step encapsulated 4MP molecules within the vesicles. The presence of cholesterol molecules in the lipid membrane ensures tight packing that prevents any unwanted leakage of 4MP. The vesicles are extruded through a 200 nm filter for uniform size distribution. On applying an AC

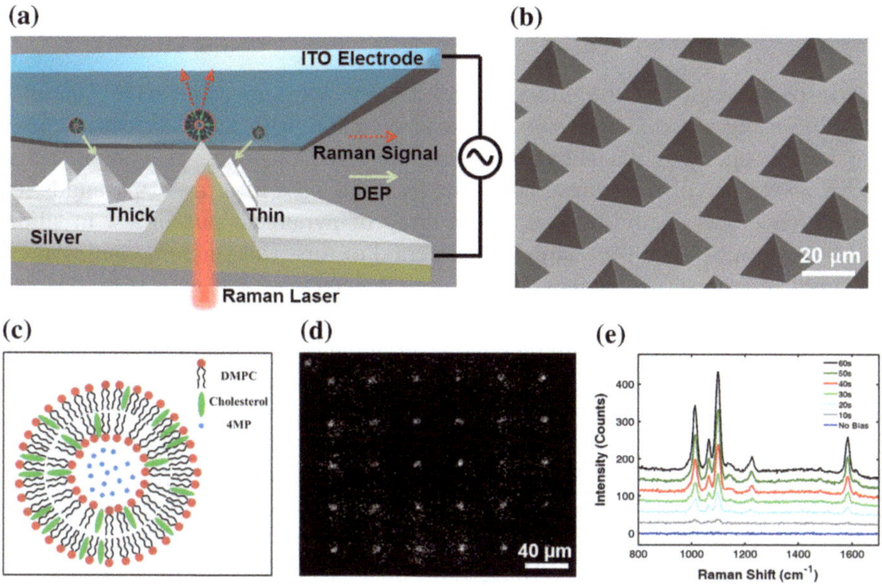

Fig. 7 DEP trapping and concurrent Raman detection. **a** An ITO electrode is assembled on top of an array of asymmetric pyramids. Applying an AC signal between the ITO and the pyramid surface creates a DEP trap at the tip of the pyramid, which is used trap vesicles. A monochromatic laser light is loosely focused at the tip of the pyramid and the far field scattered light is collected through the ITO electrode and sent to a spectrometer. **b** SEM of an array of pyramids with a base length of 20 μm. **c** Schematic of a vesicle encapsulating 4MP molecules. **d** Fluorescence image showing trapped vesicles at the tip of the pyramid across an array of 5 × 6 pyramids. **e** Evolution of Raman signal over time showing the signature spectrum from the 4MP molecules inside the vesicles

signal (10 $V_{p\text{-}p}$, 1 kHz) the vesicles are attracted towards the tip of the pyramid due to pDEP. Electric field induces charge separation at the interface of the lipid membrane and the surrounding medium, enabling DEP trapping. Trapping occurs across an array of pyramids, which is observed by imaging a 5 × 6 pyramid array using fluorescence microscopy. Next Raman-active vesicles are trapped at the tip of the pyramids and Raman signal is obtained from a single pyramid tip in real time. Signature spectrum corresponding to 4MP molecules start to appear within seconds. The peaks at 1013, 1065, 1099, 1226, and 1584 cm^{-1} are assigned to the ring-breathing mode, the in-plane C-H bend, the ring-breathing mode coupled with the ν(C-S) stretching mode, the in-plane C-H bend and the ring-stretching mode respectively [53–55]. This platform will open up several avenues to fingerprint molecules in cellular environments such as detecting neurotransmitters in secretory vesicles [56, 57]. Furthermore it could be used to identify and quantify encapsulated drugs in nanoparticle or liposomal drug delivery preparations.

4.3 Single Pyramid Tip

Single pyramid tips at the end of a conductive wire are achieved by template stripping using a conductive epoxy instead of traditionally used dielectric epoxy. Conductive epoxy is made by dispersing silver nanoparticles in nonconductive epoxy above percolation threshold. Same as in the case of array of pyramids, the fabrication process starts by making an array of pyramidal pits in a 4 in silicon wafer that can be stored for an extended period of time and single pyramid tip can be template-stripped on demand. A 50-μm-diameter tungsten wire is dipped in a droplet of electrically conductive epoxy mixed with two-component fast-curing epoxy. It is then contacted with the gold layer at the base of the pyramidal pit using a micrometer stage and cured. The wire is gradually pulled from the pit and a gold pyramid is detached from the silicon template. The conductive epoxy is then thermally cured at 100 °C for 1 h. The tungsten wire is then soldered to a copper wire and positioned onto a XYZ micro-positioning stage on an inverted microscope for DEP experiments. A drop of sample solution is placed on an ITO electrode and the pyramid tip is gradually lowered down to a distance of 50–70 μm. Due to practical limitation such as crashing the tip, it is challenging to lower the gap further. Also transparent ITO electrode is used to perform fluorescence imaging at the tip as well as for spectroscopic applications (Fig. 8).

Fluorescent polystyrene beads of diameters 190 nm and 2 μm are used to demonstrate single particle manipulation by positive and negative DEP. Based on the frequency of the applied bias single bead can be trapped at the tip of the pyramid—100 kHz for pDEP and 10 MHz for nDEP that corroborates well with the theoretically observed frequency dependence of the polystyrene beads in water. Finally, this platform is used to trap single-walled carbon nanotubes (SWCNTs) at the tip and perform simultaneous spectral detection. A 785 nm laser source is focused at the tip and the far field scattered light is collected through the ITO slide, filtered and sent to the spectrometer for detection. As DEP concentrates SWCNTs at the tip, the Raman G-band (1595 cm^{-1}) and the photoluminescence (PL) signal intensities (758, 1411 and 1898 cm^{-1}) start to rise. The G-band [58] is a characteristic of SWCNTs and the three PL peaks are attributed to (5,4), (6,4) and (9,1) [59, 60]. The Raman and PL signal can be enhanced further by creating a plasmonic hotspot at the tip by using a radially polarized light source [61]. These conductive gold pyramid tips have many applications such as electrochemistry, [62] scanning probe microscopy [63], and photoemission studies [64]. Furthermore a 3D mobile DEP trap is useful for studies involving single molecules or nanoparticles.

Sharp tips and edges made out of noble metals such as silver or gold is useful for DEP concentration of nanoparticles and concurrent plasmonic detection by EOT (nanohole array) or Raman spectroscopy (pyramid tips or array). These sharp features concentrate charges due to lightning rod effect and create electrostatic hotspots for DEP manipulation. As it turns out, the optical field also is the strongest at the sharpest features of the metal patterns—making a case for colocalization of both fields. This can lead to a class of novel nanoplasmonic sensors with

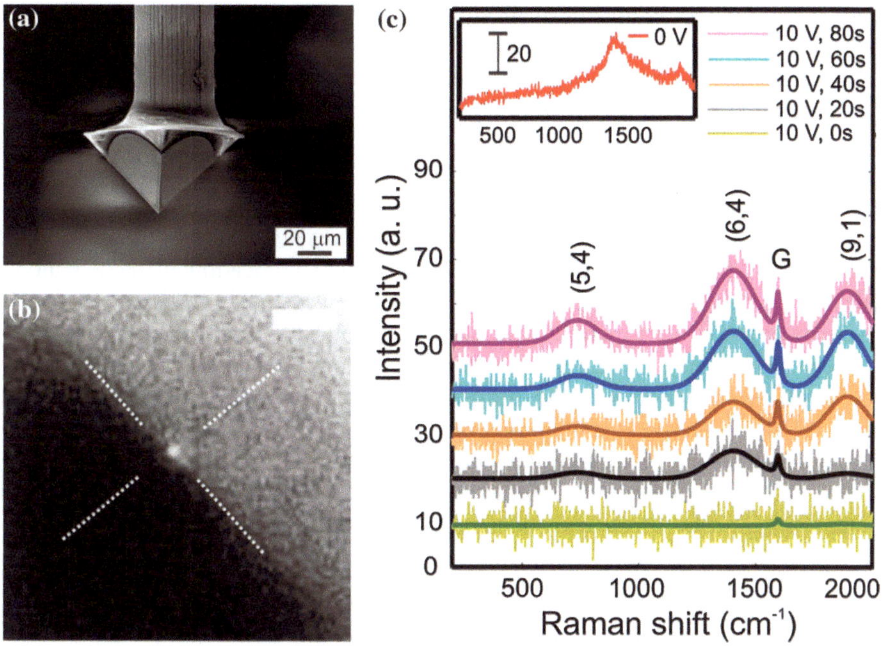

Fig. 8 Single pyramid tip for DEP and Raman detection. **a** A single pyramid tip pulled out using a 50 μm tungsten wire and conductive epoxy. **b** A 190 nm diameter polystyrene bead is trapped at the tip of the pyramid by pDEP (10 $V_{p\text{-}p}$, 100 kHz). The scale bar is 5 μm. **c** Background subtracted Raman spectra (2 s acquisition time) at different time points show cleat emergence of the Raman G band at 1595 cm^{-1} and the three different photoluminescence peaks. Adapted from Jose et al. ACS Photonics (2014) 1, 464–470 [17]

capabilities to concentrate as well as detect analytes in a label-free manner with high sensitivity and low detection limits. The performance can be further improved if the gap between the electrodes is reduced to sub-micron length scale. The setups described here use an ITO-coated glass slide as the second electrode, which is great for optical measurements but limits the ability to reduce the gap down to nanometer scale. In the next section, another class of nanoplasmonic sensor based on nanoscale gap between gold electrodes is discussed, which will improve the DEP performance significantly and also allow plasmonic detection.

5 Nanogap Electrodes

DEP is a widely used technique for biomolecule manipulation including cells, vesicles, DNA molecules or proteins using conventional microelectrode structures [65]. However hurdles exist such as heat/bubble generation and unwanted surface reaction due to the need to apply high voltages [66]. Optical trapping is an alternative but it often suffers from photobleaching, refractive index contrast between the object and its medium, and smaller trapping volume [67–69]. Plasmonic trapping can reduce photobleaching by decreasing the laser power requirement, but is often limited by trapping range because of the evanescent plasmonic field [41, 70–72]. On the other hand, DEP experiments are done with simple and inexpensive setup in a highly parallel fashion [73]. Barik et al. has showed that reducing the electrode gap to sub-10 nm can induce nanoparticle manipulation by applying 100 mV-range biases that overcomes the disadvantages of traditional DEP experiments. [15] The most challenging part is the fabrication of these nanogap electrodes. Previous work relied on slow and expensive electron-beam lithography to create bow-tie like structures [29, 74]. These point-like nanogap junctions in one had can probe only a small amount of sample volume and in other hand require high voltages to overcome the diffusion limitations. Recently, Chen et al. has reported an alternative method based on atomic layer lithography to create gaps as narrow as 1 nm in width with the side length as large as a centimeter, producing strong enhancement of optical fields [75]. Barik et al. has adopted this method to make wafer-scale electronic nanogap electrodes for DEP manipulation of nanoparticles using ultralow voltages [15].

5.1 Nanoscale DEP

Equation 1 shows that it is possible to enhance the DEP force either by (1) applying higher voltage between the electrodes (thus $|E|$), which is undesirable because of heat/bubble generation and electrochemical surface reactions, or (2) by reducing the gap between electrodes to boost $\nabla|E|^2$. Also, from the unit of $\nabla|E|^2$, i.e., V^2m^{-3}, it is clear that reducing the characteristic length of the system can significantly increase the DEP force. To elaborate this effect further, a comparison of electrostatic field distribution is made for different gap widths (1 μm and 10 nm) using COMSOL Multiphysics simulation. The DEP force calculated from the simulation is then compared to the random thermal motion of a particle, which for a particle of radius R is given by [65].

$$F_T = \frac{k_B T}{2R} \tag{6}$$

A particle is trapped if the DEP force is high enough to overcome its thermal motion. While applying 1 V, the maximum DEP force generated from a 1-μm-gap electrode is 0.2 pN, about two orders of magnitude lower than the maximum DEP force (37 pN) obtained from a 10-nm-gap electrode. Thus by using a 1-μm-gap electrode, it is impossible to generate enough force that can overcome the thermal motion of a 10 nm diameter particle (0.4 pN from Eq. 6). Another interesting parameter to look at is the minimum trapping voltage, V_{min}, which is defined as the minimum voltage you need to apply to trap a particle of certain diameter and is

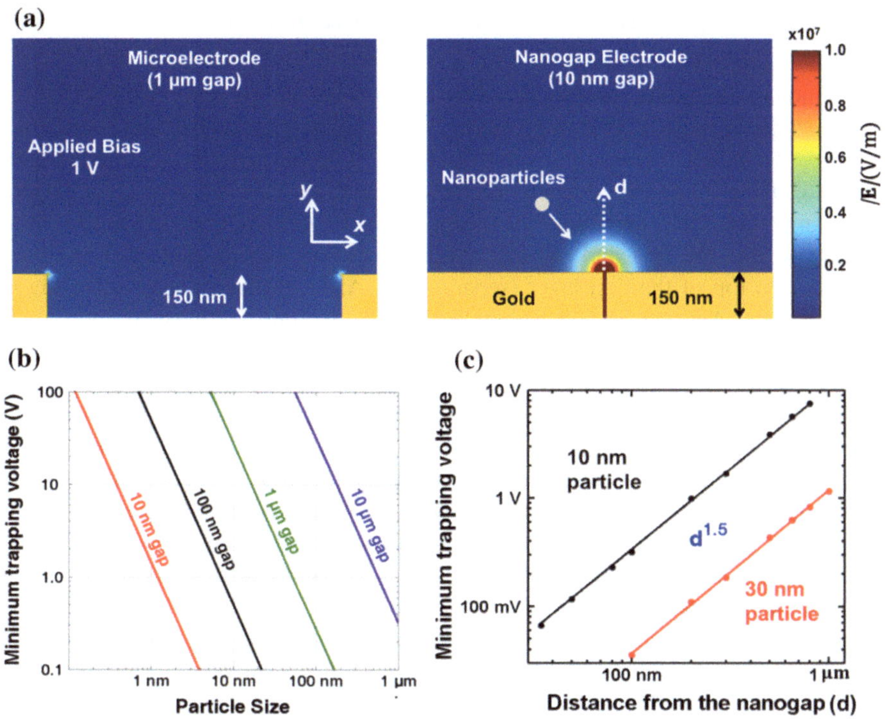

Fig. 9 Nanogap electrodes for DEP. **a** Comparison of electrostatic field maps for a gap width of 1 μm against 10 nm. At an applied bias of 1 V, nanogaps produce enhanced electric fields useful for DEP trapping. **b** Scaling analysis between V_{min}, particle size and the gap between the electrodes. **c** Dependence of V_{min} on the distance away from the gap. This is important to overcome the diffusion limitations

calculated by equating the DEP force with the thermal force obtained from Eq. 6. V_{min} scales with the gap between the electrodes as well the particle diameter. This scaling analysis further shows the importance of nanogap electrodes in trapping small particles (<100 nm) while using ultralow voltages. For example, a 10 nm particle is trapped by a 1-μm-gap electrode by applying ~30 V; whereas ~500 mV is needed for 100-nm-gap electrode. However, the scaling analysis only measures the voltage required to hold a particle on top of the electrode. In actual experiments to achieve trapping within a reasonable time frame, it is important to overcome the diffusion limitations, which is possible if the DEP force can be exerted far from the electrode. Hence the dependence of V_{min} is tested as a function of the distance from the nanogap electrode (d) for two different particle diameters (10 and 30 nm) and it is observed to be $V_{min} \propto d^{1.5}$. Putting it into perspective, a 10-nm-gap electrode can capture a 30 nm particle from ~1 μm away by applying 1 V, whereas the trapping range only extends up to ~70 nm while using a 1 μm gap (Fig. 9).

5.2 DEP Experiments with Nanogap Electrodes

The nanogap electrodes are fabricated by atomic layer lithography, [75] which is a combination of atomic layer deposition (ALD), photolithography, and scotch tape peeling. An Al_2O_3 layer of desired thickness is deposited on a patterned gold substrate using ALD that deposits dielectric materials in a layer-by-layer fashion with Ångstrom scale precision. A second layer of gold is then deposited directionally so that the first and second layers of gold are not in contact. The excess gold film on top of the first layer (separated by Al_2O_3) is then peeled off using an adhesive tape. Another round of photolithography is done followed by ion milling to create electrically disconnected array of gold nanogap electrodes. The gap width is fixed by the ALD step and the side length can be varied on demand by tweaking the photolithography step. Barik et al. has made 32 nanogaps (9 nm wide) per chip with side lengths varied between 20 and 800 μm [15]. The quality of the gaps is tested using a current-voltage (IV) characterization setup. For a working gap, no current is expected while sweeping the voltage. An added benefit of atomic layer lithography is the presence of Al_2O_3 layer in the gap, which prevents any unwanted migration of gold atoms across the gap. Finally, a fluidic cell is assembled on top to hold the sample solution and DEP experiments are carried out. To demonstrate pDEP and nDEP of nanoparticles, simple fluorescence imaging is used. It is also possible to combine label-free SPR sensing by adding metallic sidewalls (Fig. 10).

Fluorescently labeled nanoparticles such as polystyrene beads, nanodiamonds, and quantum dots are used to demonstrate nanoparticle manipulation using ultralow bias voltages. 190 nm diameter polystyrene particles are trapped and released from a 9-nm-wide gap by applying 600 $mV_{p\text{-}p}$ at a frequency of 1 MHz for positive and 10 MHz for negative DEP. The crossover frequency of a polystyrene bead in water (conductivity 4 μS/cm) is 4 MHz—justifying the choice of the frequencies used for pDEP and nDEP. Furthermore, particle movements in solution are monitored by using particle-tracking software that helps us gauge the effective trapping range [76]. Before applying any bias randomly diffusing particles are observed that move in a deterministic fashion in presence of an electric field. For 190 nm particles a range of approximately 5–10 μm is observed which demonstrates the long-range particle manipulation capability offered by the nanogap electrodes. Next, 30 nm polystyrene particles are used to perform a voltage dependence study and to calculate the V_{min}. For 9-nm-wide gap the V_{min} for 30 nm particles is observed to be 400 $mV_{p\text{-}p}$, which is an unprecedented improvement over previously reported DEP studies for nanoparticle manipulation. Reducing the operating voltage helps to manage heat generation within the solution. The rise in solution temperature (T_s) is predicted from the Poisson's equation with Joule heating as the heat source $\left(k\nabla^2 T_s = -\sigma\langle E^2\rangle\right)$ and is approximated as [77]

$$\Delta T_S \sim \frac{\sigma V^2}{2k}, \tag{7}$$

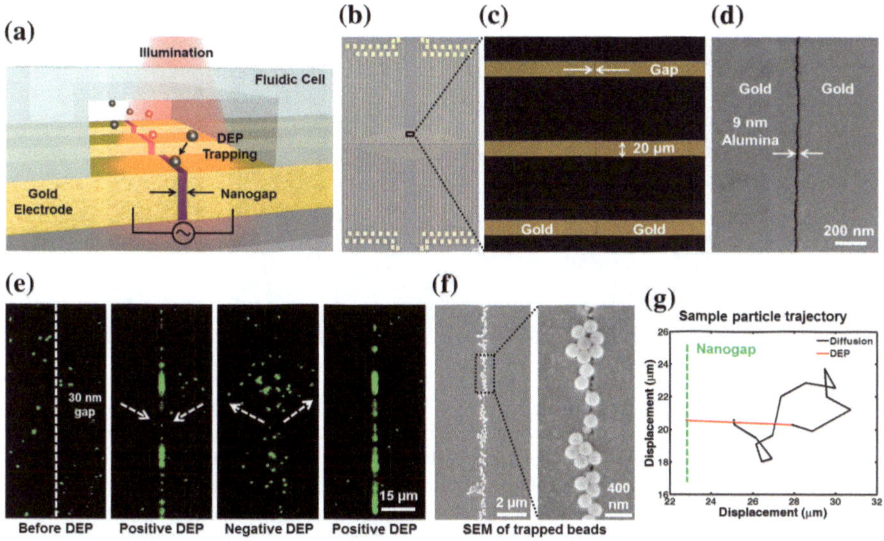

Fig. 10 Long-range DEP manipulation of polystyrene beads. **a** Schematic showing an array of nanogap electrodes, each of which is individually addressable for DEP trapping. **b** A microscopic image of a 32-electrode device. **c** Magnified image of three electrodes with the gaps at the center of the lines. **d** SEM image showing gold electrodes separated by 9-nm-thick Al_2O_3 layer. **e** Fluorescent images showing floating beads before DEP, followed by trapping and releasing due to pDEP and nDEP, respectively, in a reversible fashion. **f** SEMs of trapped polystyrene beads along the nanogap at different magnifications. **g** Particle-tracking analysis shows a sample trajectory with random thermal motion in black and deterministic movement due to DEP in red. 190 nm beads can be trapped from 5–10 μm away from the gap

where V is the amplitude of the AC signal and k is the thermal conductivity of the solution. The V_{min} reported here is about an order of magnitude lower than typical voltage values used for DEP experiments involving nanoparticles. Lowering the operating voltage by 10-fold reduces the temperature rise as well as the power consumption by 100-fold. Additionally, this nanogap structure provides a larger trapping volume because of its elongated geometry as compared to the bow-tie structures. For instance, a 30-nm particle experiences a trapping force from 1 μm away from the nanogap by applying 1 V. For point-like nanogap junctions this corresponds to a trapping volume of 2.1 fL. However for an 800 μm long gap, the trapping volume is about 1000 times bigger, 1.26 pL. Thus by having a millimeter-long trapping zone, it is possible to overcome diffusion limitations and perform DEP experiments at ultralow voltages (Fig. 11).

Positioning of quantum emitters on precise locations in a rapid and reproducible fashion is of great interest in the nanooptics community [68, 78, 79]. But because of their ultrasmall dimensions and aggressive Brownian motion, it is very difficult to achieve. Using the nanogaps (9 nm), it is possible to trap quantum dots of core/shell size 8 nm (λ_{em}: 625 nm) while using a reasonably low voltage of 1.5 V.

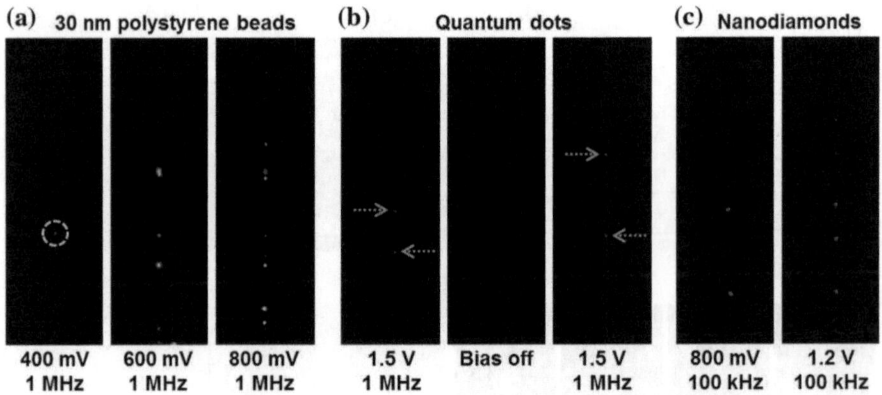

Fig. 11 Ultralow voltage trapping of sub-100 nm particles. **a** The V_{min} observed for 30 nm polystyrene beads using 9-nm-gap electrode is 400 $mV_{p\text{-}p}$ (circled in the first frame). Applying higher voltage traps more particles. **b** Quantum dots with a core/shell size of 8 nm are trapped on the 9-nm-gap electrode by applying 1.5 $V_{p\text{-}p}$. Reversible trapping is shown simply by turning off the bias. **c** For 40 nm nanodiamond particles, a V_{min} of 800 $mV_{p\text{-}p}$ is observed

The particles are also released from the gap simply by turning off the voltage. Furthermore, 40 nm nanodiamond (ND) particles with 10–15 nitrogen-vacancy (NV) centers (λ_{em}: 637 nm) are trapped on the gap while using sub-1 V bias. Trapping quantum emitters by ultralow voltages is an important finding in for building solid-state quantum optics devices [80].

A new approach to make nanogap electrodes for ultralow voltage DEP experiments is shown that can overcome hurdles of conventional DEP measurements such as heat generation, bubble formation, and electrochemical surface reactions. Using elongated gap helps to probe a much larger trapping volume as compared to bow-tie structures that enables fast detection. Furthermore, these nanogaps are made of gold because of its capability to confine optical energy, which can be used for concurrent plasmonic biosensing and on-chip spectroscopy. Thus, this ultralow power nanogap platform can be integrated as nanoplasmonic sensors, [81] nanophotonic circuits [79] and single-photon source arrays [79].

6 Conclusion

A dielectrophoretic concentration approach is implemented to improve the performances of nanoplasmonic sensors. Nanohole arrays, a widely used SPR sensor based on the EOT effect, is combined with DEP by adding an ITO electrode. This simple addition to the experimental setup has expedited detection from ultralow concentrations such as 1 pM by more than 1000 times than the diffusion-based detection. A similar approach has also been implemented with sharp metallic pyramids to trap nanoparticles or vesicles at the tip and perform enhanced Raman spectroscopy. To improve the DEP performance further, the distance between the electrodes is reduced to sub-10 nm that allows ultralow voltage DEP of nanoparticles. The electrodes are made of gold to combine plasmonic detection to it. Future work will involve adding a metallic sidewall on either side of the nanogap to perform real-time monitoring of nanoparticle manipulation. Also, as the nanogap electrodes allow long-range analyte trapping, it is possible to combine a microfluidic setup on top and capture all the analytes present within the channel—creating a "perfect sink". The ultralow power electronic nanogaps have potential for high-density on-chip integration that allows portable biosensing applications.

References

1. Liedberg B, Nylander C, Lundstrom, I (1983) Surface plasmon resonance for gas detection and biosensing. Elsevier
2. Homola J, Yee SS, Gauglitz G (1999) Surface plasmon resonance sensors: review. Sens Actuat B-Chem 54(1–2):3–15
3. Homola J (2008) Surface plasmon resonance sensors for detection of chemical and biological species. Chem Rev 108(2):462–493
4. Sheehan P, Whitman L (2005) Detection limits for nanoscale biosensors. Nano Lett 5(4): 803–807
5. Squires T, Messinger R, Manalis S (2008) Making it stick: convection, reaction and diffusion in surface-based biosensors. Nat Biotechnol 26(4):417–426
6. Feuz L, Höök F, Reimhult E (2012) Design of intelligent surface modifications and optimal liquid handling for nanoscale bioanalytical sensors. intelligent surfaces in biotechnology: scientific and engineering concepts, enabling technologies, and translation to bio-oriented applications, pp 71–122
7. Cho HS et al (2009) Label-free and highly sensitive biomolecular detection using SERS and electrokinetic preconcentration. Lab Chip 9(23):3360–3363
8. De Angelis F et al (2011) Breaking the diffusion limit with super-hydrophobic delivery of molecules to plasmonic nanofocusing SERS structures. Nat Photon 5(11):682–687
9. Eftekhari F et al (2009) Nanoholes as nanochannels: flow-through plasmonic sensing. Anal Chem 81(11):4308–4311
10. Jonsson MP et al (2010) Locally functionalized short-range ordered nanoplasmonic pores for bioanalytical sensing. Anal Chem 82(5):2087–2094
11. Escobedo C et al (2012) Optofluidic concentration: plasmonic nanostructure as concentrator and sensor. Nano Lett 12(3):1592–1596
12. Brolo AG (2012) Plasmonics for future biosensors. Nat Photon 709
13. Barik A et al (2014) Dielectrophoresis-enhanced plasmonic sensing with gold nanohole arrays. Nano Lett 14(4):2006–2012
14. Im H et al (2010) Vertically oriented sub-10-nm plasmonic nanogap arrays. Nano Lett 10(6):2231–2236
15. Barik A, Chen X, Oh S-H (2016) Ultralow-power electronic trapping of nanoparticles with sub-10 nm gold nanogap electrodes. Nano Lett 16(10):6317–6324
16. Cherukulappurath S et al (2013) Template-stripped asymmetric metallic pyramids for tunable plasmonic nanofocusing. Nano Lett 13(11):5635–5641

17. Jose J et al (2014) Individual template-stripped conductive gold pyramids for tip-enhanced dielectrophoresis. ACS Photon 1(5):464–470
18. Barik A et al (2016) Dielectrophoresis-assisted raman spectroscopy of intravesicular analytes on metallic pyramids. Anal Chem 88(3):1704–1710
19. Pohl HA, Pohl H (1978) Dielectrophoresis: the behavior of neutral matter in nonuniform electric fields. Cambridge University Press Cambridge
20. Brolo AG et al (2004) Surface plasmon sensor based on the enhanced light transmission through arrays of nanoholes in gold films. Langmuir 20(12):4813–4815
21. Dahlin A et al (2005) Localized surface plasmon resonance sensing of lipid-membrane-mediated biorecognition events. J Am Chem Soc 5043–5048
22. Tetz KA, Pang L, Fainman Y (2006) High-resolution surface plasmon resonance sensor based on linewidth-optimized nanohole array transmittance. Opt Lett 31(10):1528–1530
23. Lesuffleur A et al (2007) Periodic nanohole arrays with shape-enhanced plasmon resonance as real-time biosensors. Appl Phys Lett
24. Yang JC et al (2008) Metallic nanohole arrays on fluoropolymer substrates as small label-free real-time bioorobes. Nano Lett 8(9):2718–2724
25. Ebbesen TW et al (1998) Extraordinary optical transmission through sub-wavelength hole arrays. Nature 391(6668):667–669
26. Sinton D et al (2009) Microfluidic and nanofluidic integration of plasmonic substrates for biosensing. In: Proceedings of SPIE
27. Jackson JD (1998) Classical electrodynamics. 3rd ed. Wiley
28. Dahlin AB, Zahn R, Voros J (2012) Nanoplasmonic sensing of metal-halide complex formation and the electric double layer capacitor. Nanoscale 4(7):2339–2351
29. Holzel R et al (2005) Trapping single molecules by dielectrophoresis. Phys Rev Lett. 95(12)
30. Morgan H, Green NG (2003) AC Electrokinetics: colloids and particles. Research Studies Press, Baldock
31. González Flecha FL, Levi V (2003) Determination of the molecular size of BSA by fluorescence anisotropy. Biochem Mol Biol Educat. 31(5): 319–322
32. Hibbert DB, Gooding JJ, Erokhin P (2002) Kinetics of irreversible adsorption with diffusion: application to biomolecule immobilization. Langmuir 18(5):1770–1776
33. Dahlin AB (2012) Plasmonic biosensors: an integrated view of refractometric detection. vol. 4, Ios Press
34. Anker JN et al (2008) Biosensing with plasmonic nanosensors. Nat Mater 7(6):442–453
35. Hulman M, Tajmar M (2007) The dielectrophoretic attachment of nanotube fibres on tungsten needles. Nanotechnology 18(14)
36. Freedman KJ et al (2016) Nanopore sensing at ultra-low concentrations using single-molecule dielectrophoretic trapping. Nat Commun 7:10217
37. Freedman KJ et al (2016) On-demand surface- and tip-enhanced raman spectroscopy using dielectrophoretic trapping and nanopore sensing. ACS Photon 3(6):1036–1044
38. Yeo W et al (2012) Dielectrophoretic concentration of low-abundance nanoparticles using a nanostructured tip. Nanotechnology 23(48)
39. Maier SA et al (2003) Local detection of electromagnetic energy transport below the diffraction limit in metal nanoparticle plasmon waveguides. Nat Mater 2(4):229–232
40. Barnes WL, Dereux A, Ebbesen TW (2003) Surface plasmon subwavelength optics. Nature 424(6950):824–830
41. Pang Y, Gordon R (2012) Optical trapping of a single protein. Nano Lett 12(1):402–406
42. Novotny L, Hecht B (2006) Principles of nano-optics
43. Bouhelier A et al (2003) Near-field second-harmonic generation induced by local field enhancement. Phys Rev Lett 90(1)
44. Yeo BS et al (2009) Tip-enhanced Raman spectroscopyIts status, challenges and future directions. Chem Phys Lett 472(1–3):1–13
45. Lindquist N et al (2010) Three-dimensional plasmonic nanofocusing. Nano Lett, 1369–1373
46. Johnson T et al (2012) Highly reproducible near-field optical imaging with sub-20-nm resolution based on template-stripped gold pyramids. ACS Nano 6(10):9168–9174

47. Henzie J et al (2009) Nanofabrication of plasmonic structures. Ann Rev Phys Chem 147–165
48. Cherukulappurath S et al (2013) Template-stripped asymmetric metallic pyramids for tunable plasmonic nanofocusing. Nano Lett 13(11):5635–5641
49. Grynkiewicz G, Poenie M, Tsien RY (1985) A new generation of Ca_2^+ indicators with greatly improved fluorescence properties. J Biol Chem 260(6):3440–3450
50. Belousov VV et al (2006) Genetically encoded fluorescent indicator for intracellular hydrogen peroxide. Nat Methods 3(4):281–286
51. Schaefer JJ, Ma C, Harris JM (2012) Confocal Raman microscopy probing of temperature-controlled release from individual. Optical Trapped Phosphol Ves Anal Chem 84(21): 9505–9512
52. Klein K et al (2012) Label-free live-cell imaging with confocal Raman microscopy. Biophys J 102(2):360–368
53. Wang Y et al (2007) Raman scattering study of molecules adsorbed on ZnS nanocrystals. J Raman Spectrosc 38(1):34–38
54. Wang Z, Rothberg LJ (2005) Origins of blinking in single-molecule raman spectroscopy. J Phys Chem B 109(8):3387–3391
55. Baldwin J et al (1996) Integrated optics evanescent wave surface enhanced raman scattering (IO-EWSERS) of mercaptopyridines on a planar optical chemical bench: binding of hydrogen and copper ion. Langmuir 12(26):6389–6398
56. Finnegan JM et al (1996) Vesicular quantal size measured by amperometry at chromaffin, mast, pheochromocytoma, and pancreatic β-cells. J Neurochem 66(5):1914–1923
57. Colliver TL et al (2000) VMAT-Mediated changes in quantal size and vesicular volume. J Neurosci 20(14):5276–5282
58. Saito R et al (2011) Raman spectroscopy of graphene and carbon nanotubes. Adv Phys 60 (3):413–550
59. Nugraha ART et al (2010) Dielectric constant model for environmental effects on the exciton energies of single wall carbon nanotubes. Appl Phys Lett 97(9):091905
60. Weisman RB, Bachilo SM (2003) Dependence of Optical Transition Energies on Structure for Single-Walled Carbon Nanotubes in Aqueous Suspension: an empirical kataura plot. Nano Lett 3(9):1235–1238
61. Cançado LG et al (2009) Mechanism of near-field Raman enhancement in one-dimensional systems. Phys Rev Lett 103(18):186101
62. Fan F, Bard A (1995) Electrochemical detection of single molecules. Science 267(5199): 871–874
63. Odom TW et al (1998) Atomic structure and electronic properties of single-walled carbon nanotubes. Nature 391(6662):62–64
64. Bharadwaj P, Bouhelier A, Novotny L (2011) Electrical excitation of surface plasmons. Phys Rev Lett 106(22):226802
65. Pohl HA (1978) Dielectrophoresis. Cambridge University Press, Cambridge, England
66. Squires TM (2009) Induced-charge electrokinetics: fundamental challenges and opportunities. Lab Chip 9(17):2477–2483
67. Grigorenko AN et al (2008) Nanometric optical tweezers based on nanostructured substrates. Nat Phot 2(6):365–370
68. Geiselmann M et al (2013) Three-dimensional optical manipulation of a single electron spin. Nat Nano 8(3):175–179
69. Yang AHJ et al (2009) Optical manipulation of nanoparticles and biomolecules in sub-wavelength slot waveguides. Nature 457(7225):71–75
70. Novotny L, Bian R, Xie X (1997) Theory of nanometric optical tweezers. Phys Rev Lett 79(4):645–648
71. Juan ML et al (2009) Self-induced back-action optical trapping of dielectric nanoparticles. Nat Phys 5(12):915–919
72. Ndukaife JC et al (2016) Long-range and rapid transport of individual nano-objects by a hybrid electrothermoplasmonic nanotweezer. Nat Nano 11(1):53–59

73. Barik A et al (2014) Dielectrophoresis-enhanced plasmonic sensing with gold nanohole arrays. Nano Lett 14(4):2006–2012
74. Bezryadin A, Dekker C, Schmid G (1997) Electrostatic trapping of single conducting nanoparticles between nanoelectrodes. Appl Phys Lett 71(9):1273–1275
75. Chen X et al (2013) Atomic layer lithography of wafer-scale nanogap arrays for extreme confinement of electromagnetic waves. Nat Commun 4:2361
76. Sbalzarini IF, Koumoutsakos P (2005) Feature point tracking and trajectory analysis for video imaging in cell biology. J Struct Biol 151(2):182–195
77. Castellanos A et al (2003) Electrohydrodynamics and dielectrophoresis in microsystems: scaling laws. J Phys D Appl Phys 36(20):2584
78. Curto AG et al (2010) Unidirectional emission of a quantum dot coupled to a nanoantenna. Science 329(5994):930–933
79. Pelton M (2015) Modified spontaneous emission in nanophotonic structures. Nat Phot 9(7):427–435
80. Kress SJP et al (2015) Wedge waveguides and resonators for quantum plasmonics. Nano Lett 15(9):6267–6275
81. Lal S, Link S, Halas NJ (2007) Nano-optics from sensing to waveguiding. Nat Photonics 1(11):641–648

An Introduction to Lensless Digital Holographic Microscopy

Nathan C. Lindquist

Abstract Digital Holographic Microscopy (DHM) is a technique that uses optical interference patterns to record a three-dimensional optical field for imaging, sensing, and microscopy applications. “Lensless” in-line DHM is the simplest arrangement, requiring no lenses, no mirrors, and typically only a light source, sample, and a digital imager chip such as a CCD or CMOS pixel array. Despite this simplicity, lensless in-line DHM is capable of producing high-resolution images over a wide field of view and allows researchers to record the amplitude and phase of a light field, and to digitally reconstruct the shape, thickness, 3D position, velocity, refractive index, and other parameters of cells or small particles. There are therefore many potential opportunities for combining in-line DHM with microfluidics, optical flow velocimetry, low-cost imaging, point-of-care diagnostics, single cell tracking, cell cytometry, counting, sorting, and lab-on-a-chip technologies.

Keywords Digital holographic microscopy • Holography • Microscopy
Optical imaging

1 Introduction

When Dennis Gabor was working to improve the resolution of electron microscopy in the 1940s, he coined the term “hologram” by developing a technique to record (“gram”) the whole (“holo”) amplitude and phase of a wavefield [1]. After the invention of the laser, his technique saw dramatic improvements in concept and applicability, eventually winning him the Nobel Prize in physics in 1971 “for his invention and development of the holographic method” [2]. The critical concept of holography is that both the amplitude and phase of a wavefield can be recorded into a medium that only responds to intensity (e.g., a photographic film or, for the purposes of this chapter, an electronic imager such as a CMOS or CCD camera).

N.C. Lindquist (✉)
Physics Department, Bethel University, St. Paul, MN 55112, USA
e-mail: n-lindquist@bethel.edu

S.-H. Oh et al. (eds.), *Miniature Fluidic Devices for Rapid Biological Detection*,
Integrated Analytical Systems, https://doi.org/10.1007/978-3-319-64747-0_6

This is possible if light scattered from an object interferes coherently with a reference wave that has not scattered from the object. With such a setup, reconstruction of an entire three-dimensional optical field is possible, as is extracting both the amplitude as well as the phase information. This has traditionally been done by recording the hologram (the interference pattern) into a photographic plate [3]. Then, illuminating the hologram with another reference beam would reconstruct or "play back" the original wavefield. Recently, however, "digital" recording media have become more popular. Indeed, "digital holography" and "analog holography" differ mainly in their recording medium—instead of a photographic plate, a CCD or CMOS electronic imager pixel array is used. This concept emerged [4, 5] in the 1960s and has seen rapid development since [6–18]. In this case, a numerical playback beam is used to digitally reconstruct the original wavefield at any point in three-dimensional space. The advantages are significant in that a digital hologram does not require chemical photo-development and darkroom facilities and is therefore orders of magnitude faster, allowing real-time holographic imaging and processing. Modern digital CMOS and CCD imager chips are also more sensitive than photographic plates, capable of detecting even only a few photons, meaning that exposure times can be orders of magnitude shorter (milliseconds instead of seconds) and that standard vibration isolation techniques and bulky optical benches become less necessary. Most significantly, however, is that the data is easily recorded and shared while further numerical filtering and processing can be done after-the-fact such as compensating for aberrations or numerically interfering several holograms together. While giving all these benefits, the most significant drawback of digital holography also lies in the recording medium itself: most advanced imager pixels are still a few microns wide, limiting the spatial frequency of the recorded holographic interference fringes and hence the angular size of the object to only a few degrees. Analog photographic plates provide much higher spatial resolutions and a full 180-degree view, giving those holograms a remarkable three-dimensional parallax when viewed.

When applied to microscopy, for example to view living cells, digital holography is often called "Digital Holographic Microscopy" or DHM [16]. This chapter will highlight some of the basic features and theory DHM, numerical reconstruction techniques, and outline the wide range of potential applications. Due to its simplicity and ease of use, e.g., that no lenses are required, this chapter will focus especially on lensless in-line "Gabor" holograms for DHM [6], a subset of an emerging research push towards low-cost lensless imaging and diagnostics tools [19–27] that provide both high-resolution imaging and a wide field of view [28] with significant potential for biological imaging [29–31] and biomedical sensing [32].

The chapter is organized as follows. First, a basic outline of holography and the physical diffraction and interference theory will be discussed in Sect. 2. The discussion is applicable to holography in general, but will be tailored towards in-line DHM. Section 3 will outline a widely used and powerful numerical reconstruction technique, the propagation of the angular plane wave spectrum [3]. The particular advantages and challenges of in-line DHM will be analyzed in detail, including resolution limits and experimental concerns such as pixel density and source, object, and imager distances. Finally, Sect. 4 will briefly outline some of the most promising recent applications of in-line DHM, including digital reconstruction of the shape, thickness, 3D position, topography, velocity, or refractive index of cells and small particles, combining DHM with microfluidics, optical flow velocimetry, or single cell tracking, counting, sorting, and other lab-on-a-chip diagnostics [13].

It is the aim of this chapter to present an accessible, straightforward introduction to in-line DHM for those not familiar with the technique or the theoretical background. The chapter will provide a description of a minimal set of in-line DHM equipment and experimental protocols and outline a simple yet robust reconstruction algorithm. The chapter includes example code written in MATLAB™ at the end in Sect. 5 to provide new researchers with a quick starting point. This will introduce researchers to lensless in-line DHM and provide easy access to its powerful optical imaging and diagnostic potential.

2 Holography

The basic problem of recording both the amplitude and phase of a light field is that the recording medium—be it a piece of photosensitive film or the pixels of a scientific camera—is only sensitive to the intensity of the light. All phase information is lost. However, with the introduction of a known "reference" beam of light that interferes coherently with the "object" beam, phase information can be recorded [3]. This is particularly relevant in biological imaging where many objects of interest, e.g., living cells, are essentially transparent phase-only objects, modifying the amplitude of a beam of light only slightly. But since holography is inherently an interference technique, very precise phase measurements are possible. To begin with, consider an unknown object wavefront $O(x, y)$ that we wish to measure. For simplicity, the (x, y) directions are in the plane of the recording medium, in this case a CMOS or CCD imager chip. The object is located some distance z_0 from the imager in the z direction. Since the wavefront is a complex optical field with an amplitude and a phase, in the plane of the imager pixels it takes the form:

$$O(x, y) = |O(x, y)| \exp[i\phi(x, y)], \tag{1}$$

where $\phi(x, y)$ is the phase and $|O(x, y)|$ is the amplitude. Interference with a reference beam $R(x, y)$ as follows:

$$R(x, y) = |R(x, y)| \exp[i\psi(x, y)] \tag{2}$$

with phase $\psi(x, y)$ results in the following hologram intensity $I(x, y)$

$$I(x, y) = |O(x, y) + R(x, y)|^2 \tag{3}$$

Expanding gives

$$\begin{aligned} I(x,y) &= |O(x,y)|^2 + |R(x,y)|^2 + O^*(x,y)R(x,y) + O(x,y)R^*(x,y) \\ &= |O(x,y)|^2 + |R(x,y)|^2 + 2|O(x,y)||R(x,y)|\cos[\psi(x,y) - \phi(x,y)] \end{aligned} \quad (4)$$

where it is now apparent that the phase $\phi(x,y)$ and amplitude $|O(x,y)|$ of the object wavefront have been recorded on the camera imager chip as interference fringes in the intensity. This interference process is shown graphically in Fig. 1. The first two terms represent the intensities of the reference and scattered object beams separately. The scattered object beam is the subject of classical diffraction theory and occurs when the object blocks most of the reference beam and interferes only with itself. Classical double-slit diffraction is an example. The third term, however, contains the information to reconstruct the phase and amplitude of the object wavefield. Of course, the holographic interference patterns that are recorded on the imager chip must be sampled in space at a high enough spatial frequency to avoid aliasing, e.g., the pixel density of the imager array must satisfy the Shannon–Nyquist sampling criterion to record the highest frequency interference pattern. This issue and others such as magnification, resolution, and pixel size are discussed in detail further below.

2.1 In-Line Digital Holographic Microscopy

While many experimental setups are possible, the lensless "in-line" digital hologram is perhaps the simplest, requiring only a single illumination beam. It is often called a "Gabor" hologram since that is the arrangement originally proposed by

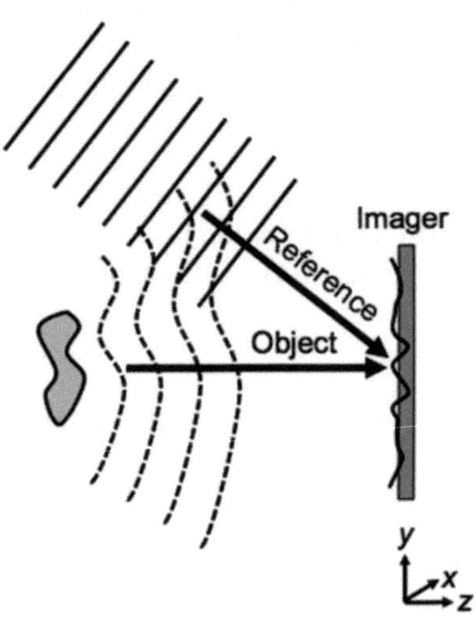

Fig. 1 Basic digital holography setup. An object scatters an "object beam" that interferes with a known "reference beam." For digital holography, an imager chip (e.g. CCD or CMOS) acts as a recording medium of the resulting interference pattern

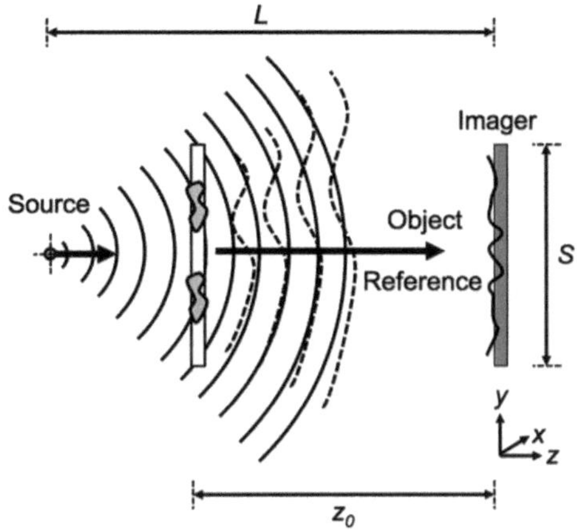

Fig. 2 In-line digital holography. The reference beam is a point source, typically a laser focused through a pinhole. The object is a mostly transparent screen. Weak scattering from the object creates the object beam whereas the reference beam remains an expanding spherical wave. The reference beam source is located some distance L in front of the imager chip surface and the object is at a distance of z_0. The physical size of the imager is S

Gabor with an expanding spherical wave [1] and will be the focus of this chapter. A typical arrangement is shown in Fig. 2. A diverging point source is used as the reference beam, typically a laser focused through a pinhole or emerging from a single-mode optical fiber, although LED illumination has been used [22, 23, 33, 34] often with larger pinholes [35] and relatively larger source-to-object and smaller object-to-imager distances to maintain coherent interference effects. Transmission through a mostly transparent object creates the object beam. The objects themselves don't need to be transparent (the hologram will then reconstruct the outline of the objects) but they should be small and dispersed. These two beams then interfere as before on the imager chip. Typical sizes of the pinhole, for visible light, are on the order of several μm while the distance from the pinhole to the object is typically on the order of millimeters to centimeters. The object–imager distance z_0 or the source–imager distance L is adjusted to capture the entire emission cone of the pinhole. These distances also affect the magnification and resolution of the hologram, as discussed further below.

Assuming the object beam $|O(x,y)|^2$ has a relatively small intensity (typical for imaging cells, small transparent particles, or dispersed opaque particles), the DC terms $|O(x,y)|^2 + |R(x,y)|^2 \approx |R(x,y)|^2$ in Eq. (4) can often be removed by recording a separate image with just the reference beam $|R(x,y)|^2$ and subtracting, thereby leaving only the interference terms. This is sometimes called the "contrast hologram" [6]. If this is not possible, e.g., by recording a moving object or an object that cannot be easily removed, the average DC value of the hologram can be subtracted or the entire hologram digitally high-pass-filtered to simulate the removal of the reference background. In any case, the hologram recording now becomes

$$I'(x,y) = O(x,y)R^*(x,y) + O^*(x,y)R(x,y) \tag{5}$$

Here $I'(x,y) = I(x,y) - |R(x,y)|^2$ is the contrast hologram and $|O(x,y)|^2$ is simply dropped. Again, if the reference wave cannot be recorded separately, subtracting the average value or high-pass-filtering the hologram is also a possibility. However, calculating the contrast image with a separate recording of $|R(x,y)|^2$ is preferred since it can easily remove unwanted imperfections in the illumination profile or in the camera itself. An example hologram is shown in Fig. 3. The source

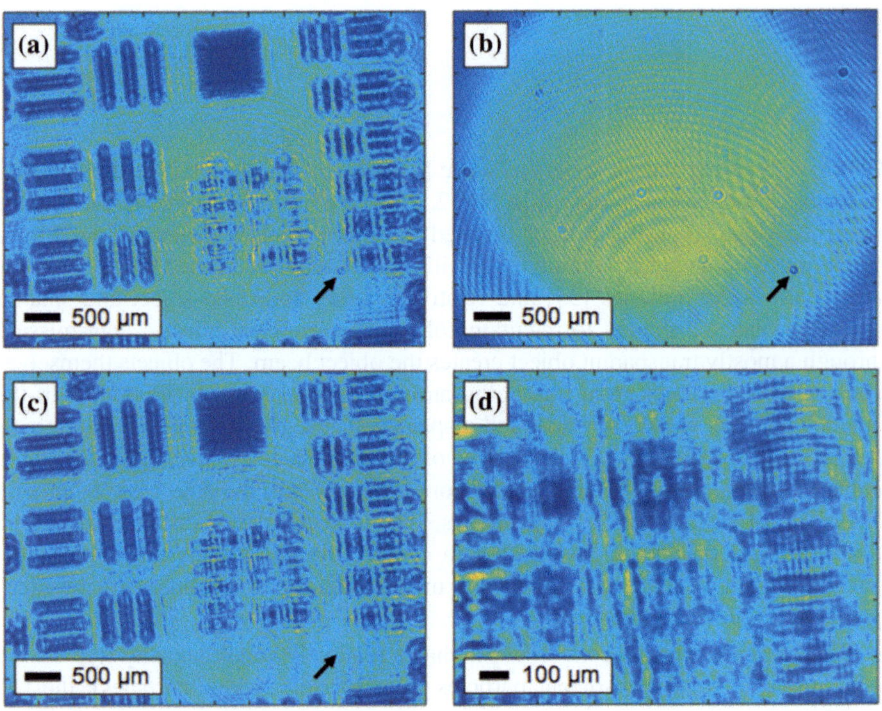

Fig. 3 In-line digital hologram of a USAF resolution test target as the object. **a** The raw recorded hologram from the CCD imager. The small arrow points to a defect. **b** The reference image taken without the object. The defect appears here, indicating that it is a dust particle or scatterer on the CCD window and not part of the object itself. **c** The contrast hologram. Notice that the defect has disappeared. **d** Zoomed in image of the contrast hologram showing some of the interference fringes. In this hologram the source to imager distance $L = 70$ mm and the object-to-imager distance $z_0 = 23.3$ mm. The scale bar is calculated simply by knowing the physical size of the imager array (e.g. the pixel size and the number of pixels)

was 633 nm laser light emerging from a single-mode fiber. The assumption of weak scattering from the object itself in Eq. (5) will be held throughout the rest of this chapter since it is valid for imaging small, disperse, or transparent samples such as cells or particles in solution. Given this single intensity image, constructed without a lens or the need to focus, a reconstruction algorithm will need to determine the two unknown values of the object phase $\phi(x,y)$ and amplitude $|O(x,y)|$. Once this is done, the wavefront can be reconstructed in any xy plane at any z coordinate, or even in three-dimensional sections. This is in stark contrast to traditional light microscopy that can only provide an intensity image in a single plane with a limited depth of focus. Unfortunately, with in-line holography, it is not possible to further separate the two terms $O(x,y)R^*(x,y)$ and $O^*(x,y)R(x,y)$ from each other. These are the "real" and "virtual" images—or the "twin images"—and can provide significant interference during the reconstruction process [36]. Several techniques have been used to separate these terms [37], both physically by using an angled reference wave [3] to spatially separate the twin images or phase-shifting the reference beam [38], recording the holograms at multiple planes [39], using multiple apertures [40], or numerical elimination via various iterative algorithms [41]. Fortunately, the interference from the twin images for in-line digital holographic microscopy will be minimal, provided a few conditions are met as discussed further below.

Consider now the extraction, or reconstruction, of the original object wavefront $O(x,y)$ from $I'(x,y)$. For analog hologram playback, another reference beam illuminates the developed holographic plate [3]. In a digital hologram, this is done numerically, and the full complex wavefield can be numerically back-propagated to the object. In our example, $I'(x,y)$ in Eq. 5 can be simply multiplied by a numerical copy of the original reference beam $R(x,y)$ giving:

$$\begin{aligned} I'(x,y) &= O(x,y)R^*(x,y)R(x,y) + O^*(x,y)R(x,y)R(x,y) \\ &= O(x,y)|R(x,y)|^2 + O^*(x,y)R(x,y)R(x,y) \end{aligned} \tag{6}$$

In this case, the first term is an exact replica of the original object wavefront $O(x,y)$ at the imager surface, multiplied by the known (recorded) intensity of the reference beam. The second term is the twin image. To further understand these two terms, it is helpful to consider a model where the original object wavefront is a simple, unit valued point source. In the plane of the imager

$$O(x,y) = \exp\left[+ik\sqrt{x^2+y^2+z_0^2}\right], \tag{7}$$

where $k = \frac{2\pi}{\lambda}$ is the wavenumber and z_0 is the distance from the object to the imager surface as shown earlier in Fig. 1. The positive sign in the exponent indicates a diverging spherical wavefront. Inserting these into Eq. (6) and dividing by $|R(x,y)|^2$ gives

$$I'(x,y) = \exp\left[+ik\sqrt{x^2+y^2+z_0^2}\right] + \exp\left[-ik\sqrt{x^2+y^2+z_0^2}\right]\frac{R(x,y)R(x,y)}{|R(x,y)|^2} \quad (8)$$

The first term is now seen to represent the original diverging spherical wavefront $O(x,y)$ whereas the second term is a converging wavefront that represents the "twin image." If this were a real wavefield instead of a numerical reconstruction, looking one would see a virtual image of the original object [3] and a real image would form in front of the hologram. If we had used the conjugate $R^*(x,y)$ of the original reference wavefront instead, then the reconstruction would contain the converging wavefront $O^*(x,y)$ instead.

Various techniques have been used to eliminate this "twin image" such as phase-shifting holography [38], off-axis holography [7], spatial filtering [42], and various numerical or iterative algorithms [41, 43]. Fortunately, for "in-line" digital holographic microscopy, the twin image term is often completely negligible [6]. When $I'(x,y)$ is back-propagated to the original location of the object, hence reconstructing the wavefield at the object, the first term in Eq. (8) will become our original object (a point source) and the second term will have spread out or diverged over the entire reconstruction plane, becoming part of a small background. The twin image will appear in focus mirrored on the other side of the source. Therefore, if the source–object distance is relatively large compared to the wavelength and a big enough recording screen is used to achieve the desired resolution (discussed below), when the hologram is reconstructed the twin image will be spread out over the reconstruction and is therefore typically negligible and can be ignored [6].

The reference beam is represented by an ideal point source located a distance L from the imager surface:

$$R(r) = R_0\frac{\exp[ikr]}{r}, \quad (9)$$

where R_0 is some amplitude of the point source and r is the distance from the point source origin to locations on the imager surface. This can be written in terms of the xy coordinates on the imager at a distance L from the point source origin:

$$R(x,y;L) = R_0\exp\left[ik\sqrt{x^2+y^2+L^2}\right]/\sqrt{x^2+y^2+L^2} \quad (10)$$

The emission cone of the pinhole in an actual experiment will modify this ideal point source. However, this can be largely corrected, as stated above, by calculating the contrast image $I'(x,y)$.

3 Wavefront Reconstruction

Once the hologram has been recorded and we have recovered the original object wavefront (plus the twin term) in the plane of the imager by multiplying by a numerical reference beam, the next step is to back-propagate the wavefront to the object plane, thereby reconstructing the object. We will assume plane geometry in the following. If Eq. (6) is then divided by the known reference beam and the twin image effect is ignored, the wavefield at the imager becomes simply that of the original object $O(x, y)$. Scalar diffraction theory can now propagate that wavefield using propagation of the angular spectrum [3, 44]. While other propagation techniques exist such as a so-called direct Fresnel method or a Huygen's Convolution method [7], the Angular Spectrum method has several advantages [45]. In particular, the Fresnel method only provides valid reconstructions at relatively large distances under the paraxial approximation. When the entire path from the source to the object to the imager is only on the order of a few mm, the paraxial approximation may not be valid. Its advantage, however, is that numerical reconstruction can be very fast, requiring only a single Fourier transform to propagate the field from the detector plane to the object plane. While the short distance behavior of the Huygen's Convolution technique is significantly better, it requires three Fourier transforms. The angular spectrum technique does not use any approximations and is therefore valid over short distance, even to zero [7]. Furthermore, it requires only two Fourier transforms and maintains a constant pixel size with distance. While the constant pixel size at any reconstruction distance gives a constant field of view that may limit imaging microscopic samples [45], consideration of the geometric magnification provided with a point-source reference wave alleviates this concern. This is described in detail in the next section.

3.1 Angular Spectrum Technique

Diffraction of a wavefield can be understood as the propagation of a series of plane waves that make up that wavefield [3]. An object field $O(x, y; z)$ has an angular spectrum $A(k_x, k_y; z)$ directly related to its Fourier transform $\mathfrak{F}[\,]$

$$A(k_x, k_y; z) = \mathfrak{F}[O(x, y; z)], \tag{11}$$

where $k = (k_x, k_y, k_z)$ is the wavevector of each plane wave component of the angular spectrum. The object wavefront and the angular spectrum are explicitly evaluated at a particular z location. To describe a propagating wavefront, e.g. non-evanescent, all components of the wavevector must be real. Therefore, since

$$k_x^2 + k_y^2 + k_z^2 = k^2 = \left(\frac{2\pi}{\lambda}\right)^2 \tag{12}$$

for propagation in the z direction we require that $k_z = \sqrt{k^2 - k_x^2 - k_y^2}$ be real so that:

$$k^2 \geq k_x^2 + k_y^2 \tag{13}$$

The angular spectrum component plane waves, e.g., the k_x and k_y components of the Fourier transform, are then propagated through space by simply accumulating phase in the z direction. Unlike the spherical wavelets of Huygens–Fresnel diffraction, the amplitudes of the plane waves do not change. Therefore, the angular spectrum propagates from $z = 0$ to $z = z_0$ through the following expression:

$$A(k_x, k_y; z_0) = A(k_x, k_y; 0)\exp[iz_0 k_z], \tag{14}$$

where $A(k_x, k_y; z_0)$ represents the angular spectrum at a given z_0 location. Since the original wavefront was the inverse Fourier transform of its original angular spectrum, the new wavefront—propagated to a new z position—is simply the inverse Fourier transform $\Im^{-1}[]$ of the new angular spectrum:

$$O(x, y; z_0) = \Im^{-1}[A(k_x, k_y; z_0)] \tag{15}$$

Putting everything together

$$O(x, y; z_0) = \Im^{-1}\left[\Im[O(x, y; 0)]\exp\left[iz_0\sqrt{k^2 - k_x^2 - k_y^2}\right]\right] \tag{16}$$

This is the angular spectrum reconstruction method. Again, the condition $k^2 \geq k_x^2 + k_y^2$ must be satisfied. If it is not, then the angular spectrum is simply set to zero for all $k^2 < k_x^2 + k_y^2$. However, as long as the imaging is done within a diffraction-limited system, as it will be unless evanescent wave components are somehow recorded via a near-field interaction, this condition will always be met. However, with geometric magnification or if lenses are used, this condition may need to be revisited [7]. As an example, Fig. 4 shows the reconstructed image using Eq. 16 at several distances from the hologram data shown in Fig. 3. Because the reconstruction distance can be chosen freely, the optical field can be reconstructed at any plane in 3D space. This is also shown by reconstructing the object image to be both in-focus and out-of-focus. With this method, the field of view is constant, i.e. the physical size of the CCD or CMOS array. With microscopic objects, this may not be desirable. However, as discussed below, this drawback is fixed by taking into account the magnification factor introduced by the expanding point source [45].

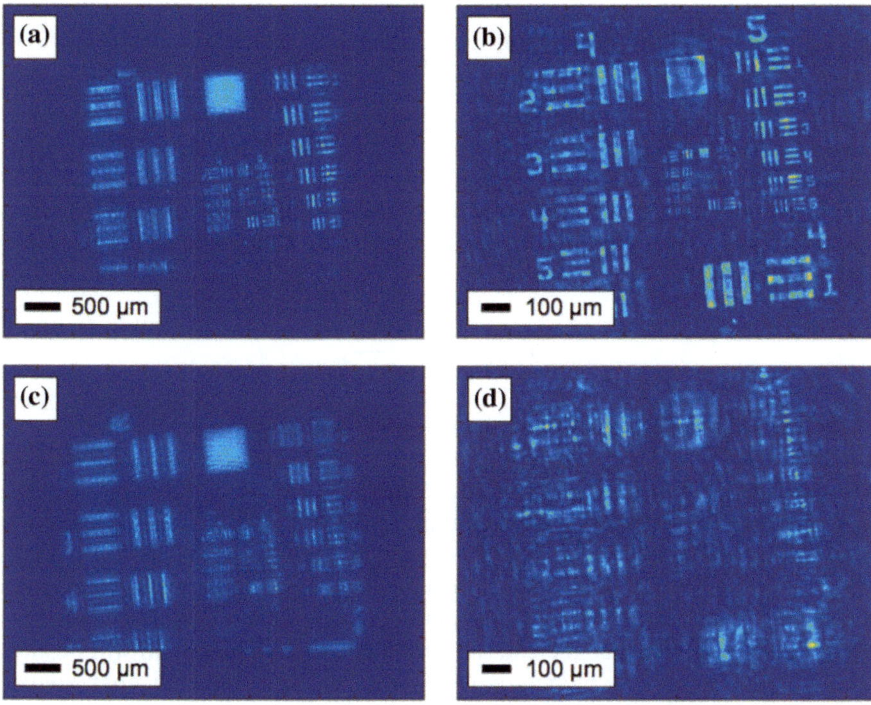

Fig. 4 Reconstruction of a USAF resolution test target. This is the same hologram that is shown in Fig. 3. **a** The reconstructed object at a reconstruction distance of $z = z_0 = 23.3$ mm. **b** Zoomed in region of the same, showing a high-resolution image recorded without any lenses. **c** The same hologram but with a reconstruction distance of $z = 20.0$ mm, i.e. not in the original object plane. **d** Zooming in as in (**b**) shows that the high-resolution region of the image is now quite blurred. In this way, the optical field can be reconstructed at any plane in 3D space

3.2 Resolution Considerations

Consider that Eq. (16) can also be written explicitly in terms of the wavelength as follows:

$$O(x, y; z_0) = \mathfrak{F}^{-1}\left[\mathfrak{F}[O(x, y; 0)] \exp\left[2\pi i \frac{z_0}{\lambda}\sqrt{1 - (\lambda u)^2 - (\lambda v)^2}\right]\right] \quad (17)$$

Here, the coordinates λu and λv are the direction cosines of the wavevector k and the components are given as $k_x = 2\pi u$ and $k_y = 2\pi v$. Therefore, the maximum value of $(\lambda u)^2_{\max} = \sin\theta_{\max} = 1$ corresponds to the maximum possible spatial frequency in the Fourier domain, i.e., the plane wave propagating from the object to the imager at the steepest angle $\theta_{\max}$ from the z axis, and hence representing the maximum image resolution. The Abbe resolution criterion fixes this at

$$\text{Lateral Resolution} = \frac{1}{2u_{\text{max}}} = \frac{\lambda}{2\sin\theta_{\text{max}}} = \frac{\lambda}{2\text{NA}}, \tag{18}$$

where $\text{NA} = \sin\theta_{\text{max}}$ is the numerical aperture of the setup. This resolution limit determines the minimum spacing at which two separated points can effectively be resolved. The resolution of digital holography is therefore also linked to the classical resolution limit where the NA of the imaging aperture, in this case the size of the imager array, determines the maximum resolution. This would be determined by the geometry of the recorded hologram

$$\theta_{\text{max}} = \tan^{-1}\frac{S/2}{z_0}, \tag{19}$$

where S is the size of the imager chip and z_0 is the distance from the object to the imager as shown in Fig. 2. Therefore, the maximum possible resolution obtainable is determined by both the wavelength used and the geometry of the recording setup, including the physical size of the recording medium and the object-to-imager distance. Using this relationship, it can be seen that resolution on the order of the wavelength is easily obtainable [9]. Of course, other experimental factors may also play a role, for example, the size of the pinhole. If the pinhole is small, on the order of only a few wavelengths, the emission cone of the reference beam will be large, thereby illuminating more of the sample and resulting in a higher quality hologram. If an incoherent light source such as an LED is used, the size of the pinhole will determine the spatial coherence of the reference beam and can be used to control coherent noise [23] by reducing background speckle, and simplifying the alignment tolerances and expense associated with laser sources [33]. The axial z resolution of the hologram is also given by the NA of the recording [11]:

$$\text{Axial Resolution} = \frac{\lambda}{2(\text{NA})^2} \tag{20}$$

This axial resolution is important for 3D imaging of particles and provides the minimum separation distance in which two particles could be sufficiently resolved in the z direction.

3.3 *Effect of the Pixels*

In digital holography, the hologram is sampled in space by the imager pixels. There are N_x pixels in the x direction and N_y pixels in the y direction. If the pixels have a size δx in the x direction and δy in the y direction, the coordinates (x, y) of the camera pixels centered on the z axis will become

$$(x,y)=\left(\left(n-\frac{(N_x-1)}{2}\right)\delta x,\left(m-\frac{(N_y-1)}{2}\right)\delta y\right) \tag{21}$$

with index integers $n=0,1,2,\ldots,N_x-1$ and $m=0,1,2,\ldots,N_y-1$ in the x and y directions, respectively. The pixel sizes also directly determine the wavevector components (k_x,k_y), e.g., the spatial frequencies that make up the recorded hologram. Since at least two pixels are needed to record a fringe, the largest spatial frequency k_x that the imager pixels can resolve is simply $k_{x(\max)}=\frac{\pi}{\delta x}$ with a spacing of $k_{x(\min)}=\frac{2\pi}{\delta x N_x}$. It is similar in the y direction. Therefore the values of (k_x,k_y) that form the physical image on the camera pixels are

$$(k_x,k_y)=\left(\left(n-\frac{(N_x-1)}{2}\right)\frac{2\pi}{\delta x N_x},\left(m-\frac{(N_y-1)}{2}\right)\frac{2\pi}{\delta y N_y}\right), \tag{22}$$

where k_x and k_y are centered on axis. Again, n and m are simply integers. Physically, as long as the pixel sizes are larger than the wavelength and the imager array is more than a few wavelengths from the object, then the criterion $k^2>k_x^2+k_y^2$ will be satisfied. Here we can see that the physical size of the imager array $\delta x N_x\times\delta y N_y=S_x\times S_y$ determines the maximum spatial frequency, or angle of the plane wave, that can be recorded. The terms S_x and S_y are the physical sizes of the recorded hologram in the x and y directions, respectively, as described above and in Fig. 2. According to Eq. (18), the lateral resolution was given by the physical NA of the imaging system. However, this will only be the case if the pixels can physically record these high spatial frequencies without aliasing. The maximum recordable spatial frequency will be $k_{x(\max)}=\frac{\pi}{\delta x}$. This corresponds to $u_{\max}=\frac{1}{2\delta x}$, again according to Eq. (18), and to a lateral resolution of

$$\text{Lateral Resolution}=\frac{1}{2u_{\max}}=\delta x \tag{23}$$

which is equal to the size of the pixels. Furthermore, the angular spectrum technique maintains the size of the pixels at any distance since plane waves do not diverge or converge [7]. With a typical pixel size of ~2 μm, it can seem that achieving sub-micron resolution with visible light, i.e., diffraction-limited imaging resolution, may not be possible. However, when using a point-source reference wave, the hologram has been geometrically magnified and higher resolution is indeed possible.

3.4 Magnification

The geometric magnification factor M in an in-line digital hologram using a point source reference wave as shown in Fig. 2 is given as

$$M = \frac{L}{L - z_0} \tag{24}$$

As the point source gets farther and farther away, i.e., $L \gg z_0$, the magnification factor approaches $M = 1$ which is the magnification that would be recorded with a plane wave reference beam. Revisiting the resolution of the hologram, then, we can see that the maximum lateral resolution of the hologram is now:

$$\text{Lateral Resolution} = \text{Max}\left(\frac{\delta x}{M}, \frac{\lambda}{2\text{NA}}\right), \tag{25}$$

where the Max() function shows explicitly that either the pixel size or the physical NA of the imaging setup will ultimately determine the resolution of the hologram. Therefore, different geometrical setups allow different magnifications and hence different effective pixel resolutions, down to a fundamental diffraction-limited resolution. This pixel resolution effect with magnification is shown in Fig. 5. For the hologram data shown, $L = 30.0$ mm and $z_0 = 21.6$ mm. This provides a magnification factor of $M = 3.6$. If the hologram is reconstructed as-is (shown in Fig. 5a) without "re-sizing" the effective pixel size, the resolution is rather poor, i.e., just the size of the camera pixels which are, in this case, $\delta x = \delta y = 4.65$ μm. While the object is reconstructed properly, i.e., the sizes of the resolution test target are correct and it appears in focus at the correct experimental location, the resulting resolution is limited by the pixel size. This comes from the constant field-of-view given by the angular spectrum method. However, since the holographic fringes have been magnified, the high-resolution information has indeed been stored in the hologram data.

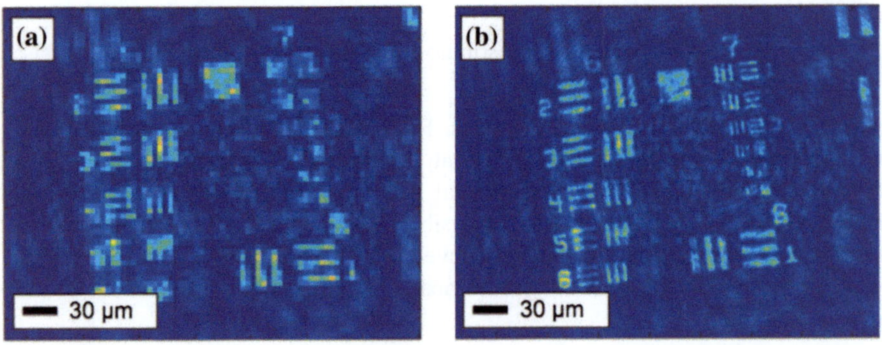

Fig. 5 Effect of magnification. **a** Reconstructing the hologram "as-is" without taking into account the magnification. While the object is reconstructed properly, i.e. the sizes of the resolution test target are correct and it appears in focus at the correct location, the resulting resolution is limited by the pixel size. **b** By reconstructing the hologram with a magnification factor, the effective pixel size is smaller and the high-resolution features are recovered. The smallest features visible on the test target in this hologram are roughly ~5.5 μm in size

Therefore, we can digitally change the effective size of the pixels by the magnification factor. In the digital reconstruction process, this magnification can be taken into account by setting the source-to-imager distance $L \to \infty$ as if it were a plane wave reference beam and then setting the reconstruction distance to $z = Mz_0$. Doing this during the numerical reconstruction process, the resulting hologram using the same exact data as in Fig. 5a is shown in Fig. 5b. The x and y axes have to also be scaled by the magnification factor accordingly. When this is done, the high-resolution regions of the object are resolved. Importantly, Fig. 5a and b show the exact same data but with a different reconstruction process. This modification allows the angular spectrum technique to achieve high resolution with point-source DHM [45].

Finally, as a general rule, it is easy to see that if the object is set directly on the pixels, so that $z_0 \to 0$, then the pixel size will determine the image resolution. However, the field of view will be the largest in this case. Furthermore, while not discussed here, by recording multiple holograms by either shifting the light source or the object, certain "sub-pixel" resolution holograms can be reconstructed [27]. On the other hand, if the object approaches the source where $z_0 \to L$, then the magnification increases and diffraction will start limiting the resolution. In that case, however, the field of view will become much smaller.

3.5 *Phase Imaging*

The phase of the reconstructed object beam can also be calculated from Eq. 16. Since $O(x, y) = |O(x, y)|\exp[i\phi(x, y)]$ is a complex number, the amplitude $|O(x, y)|$ and phase $\phi(x, y)$ can be calculated as

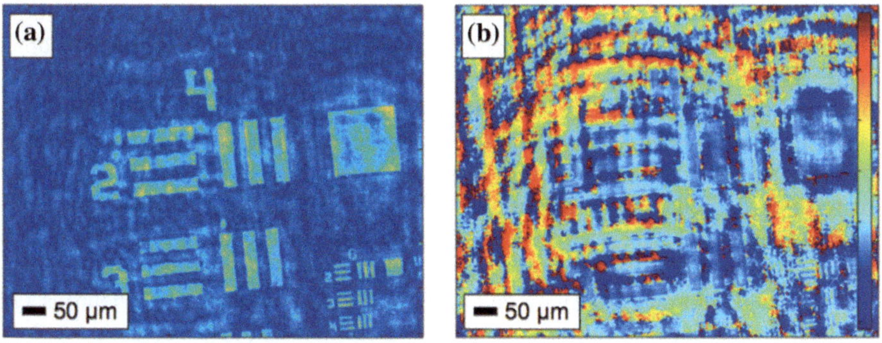

Fig. 6 Phase of the reconstructed object. **a** The amplitude of the reconstructed object beam. **b** The phase of the reconstructed object beam. The scale bar is $-\pi$ to $+\pi$ radians

$$|O(x,y)| = \sqrt{\text{Real}[O(x,y)]^2 + \text{Imag}[O(x,y)]^2} \tag{26}$$

$$\phi(x,y) = \text{Tan}^{-1}\left[\frac{\text{Imag}[O(x,y)]}{\text{Real}[O(x,y)]}\right] \tag{27}$$

The amplitude and phase of the reconstructed beam is plotted in Fig. 6. The phase curvature of the expanding reference beam should also be taken into account or removed.

4 Applications

In-line DHM has been used to image cells and other microorganisms [29], including bacteria [46], cancer cells [47], HeLa cells [46], yeast cells [24], and sperm cells [48]. By utilizing various tracking and focusing algorithms [13] and recording successive holograms, the positions and trajectories of particles [49], blood cells [50], algae [51], microfibers [52], bubbles [53], and microspheres [54] can be visualized. Due to its simplicity, the entire microscope setup can also be submerged for enhanced resolution or to image live cells in a natural environment [55]. Ultra-widefield digital holograms can be created with a commercial scanner [56]. Multiple colors can also be used to provide spectral information [22]. Cell identification and counting [22] as well as integration with antibody microarrays [57] is also feasible. In-line DHM has also been used to image microelectromechanical system [58], monitor microfabrication and materials deposition [59], replicate differential interference contrast microscopy [23, 60], segment and count cells of interest [47], analyze cell shape [61], provide cellphones with high-resolution microscopy capabilities [25], perform on-chip cytometry [24], and create full opto-fluidic systems-on-a-chip [62]. Some examples are shown in Fig. 7.

Finally, while a lensless in-line setup has been discussed at length in this chapter, many other arrangements exist [63]. They can use the reconstruction technique discussed here as well as other algorithms suited towards a particular application. Many techniques will typically split a single beam into the object beam and the reference using a beam-splitter [7] and increase the resolution by utilizing a microscope objective [64]. Indeed, all of these techniques, including the in-line methods discussed here, can be classified broadly as digital holographic microscopy. As a general method, therefore, DHM is extremely versatile and has been used to image and analyze living cells [65], red blood cells [66, 67], neurons [14, 68], cell death [15, 69], tumor cells [22, 70], fibroblasts [15], mechanics and 3D trajectories of transparent micro-particles [71], the refractive index of changes of cell cytoplasm [72], inflammation [73], circulating blood cells [67], cell dynamics [74], or the 3D distribution of the refractive index of living amoeba [75].

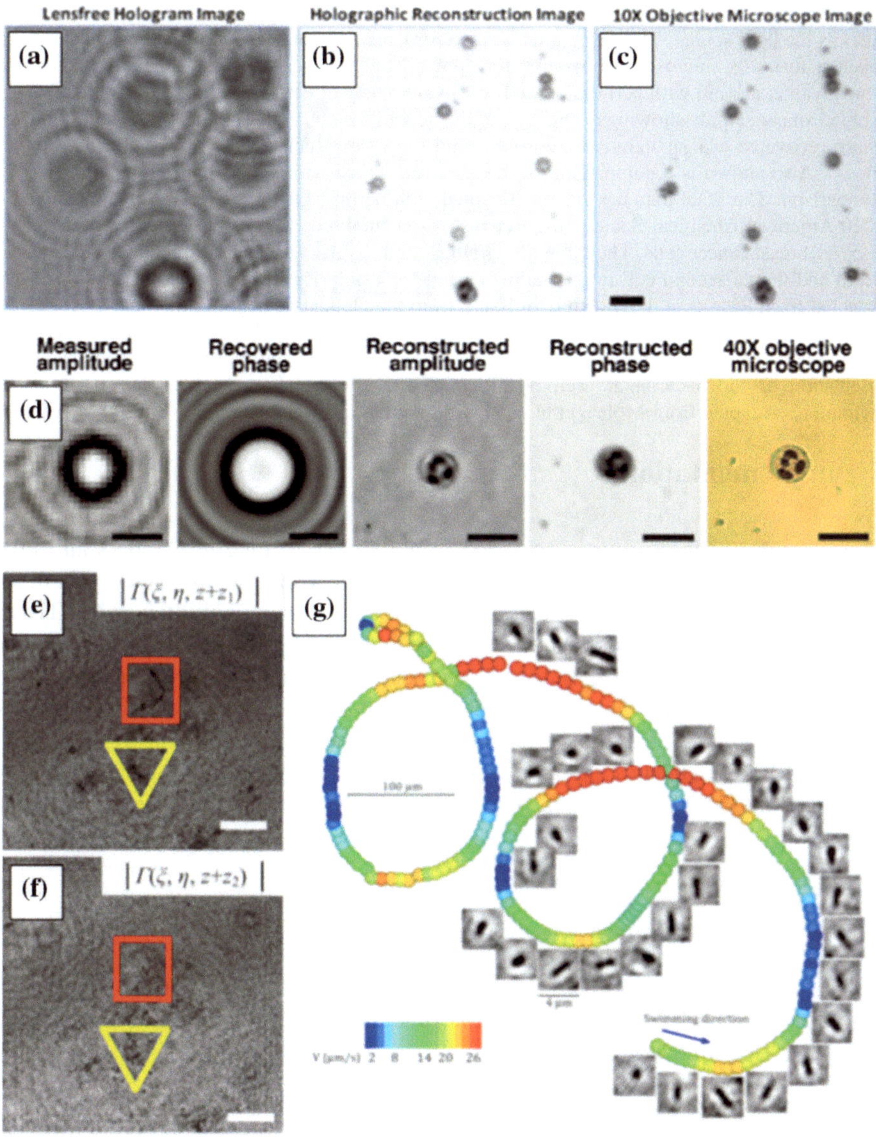
Lensfree Hologram Image
Holographic Reconstruction Image
10X Objective Microscope Image
(a)
(b)
(c)
Measured amplitude
Recovered phase
Reconstructed amplitude
Reconstructed phase
40X objective microscope
(d)
(e)
(f)
(g)
V (μm/s) 2 8 14 20 26

◀**Fig. 7** Various biological applications of in-line DHM. **a–c** Images of a blood smear using an LED as the light source. The hologram and reconstructed amplitude are shown alongside an image taken with a 10x microscope objective for comparison. The scale bar is 20 μm. Adapted from Mudanyali et al. [23] with permission of The Royal Society of Chemistry. **d** Holographic image of a blood smear sample showing a single granulocyte white blood cell. The measured amplitude, the phase recovered via an iterative algorithm, and the reconstructed amplitude and phase are all shown. Also shown is a comparison of the same cell taken with a 40x microscope objective for comparison. The scale bars are 20 μm. Adapted with permission from Seo et al. [77]. Copyright 2010 American Chemical Society. **e–f** Reconstructed amplitude images at two different depths of Hs578T breast cancer cells. The same hologram is used to image a first cell in the red square at one depth and then a second cell in the yellow triangle at a different depth. The scale bar is 100 μm. Adapted from Ryle et al. [78] with permission from the Optical Society of America. **g** Tracking a single E. coli bacteria swimming in a circle through 3D space. A single hologram records the scene in time and the 3D position of the bacteria can be localized. The individual images show the orientation of the bacteria at each position and the color scale represents the speed of the swimming. Adapted from Molaei et al. [79] with permission from the Optical Society of America

5 Implementation

This section of the chapter is provided as a resource to give researchers an initial "bare bones" place to start with an in-line DHM setup. A simplified schematic diagram is shown in Fig. 8. The computer code given here is written in MATLAB but should be generally portable to other scientific computing languages. It relies on a fast Fourier transform (FFT) to first compute the angular spectrum, propagate it, and then an inverse FFT to calculate the new field. The code is commented heavily using the % symbol. Example experimental parameters are given in SI units.

After recording the hologram on a CCD or CMOS imager chip, the images must be imported into MATLAB using the following code. It assumes a gray-scale bitmap file.

```
% Load the raw hologram image file
Hologram = double(imread('hologram.bmp'));

% Load the background reference image file
Background = double(imread('background.bmp'));
```

The contrast image is then calculated and the size of the hologram is extracted. An array of indices is computed to help generate the spatial coordinates (x, y) and the spectral coordinates (k_x, k_y).

```
% Calculate the contrast image
Uh = Hologram - Background;

% Get the size of the hologram image
[Ny,Nx] = size(Uh);

% Index arrays of the pixels
[II,JJ] = meshgrid(0:Nx-1,0:Ny-1);

% Compute the coordinates in real space
X = (II - (Nx - 1)/2)*dx;
Y = (JJ - (Ny - 1)/2)*dy;

% Compute the coordinates in Fourier space
kx = 2*pi/dx*(II - (Nx - 1)/2)/Nx;
ky = 2*pi/dy*(JJ - (Ny - 1)/2)/Ny;
```

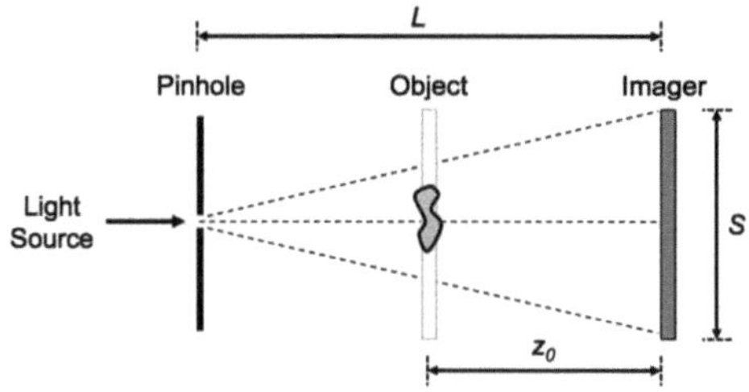

Fig. 8 Simplified schematic of an in-line DHM experimental setup. The light source can be either a laser or an LED, and the pinhole can be of various sizes—e.g. ~50–100 μm for LED illumination or ~1–10 μm for laser illumination. The "pinhole" can also be the exit end of a single-mode optical fiber. In some setups, the "pinhole" is simply a bare LED [20]. In other techniques, a laser is expanded to a plane wave and the pinhole is not necessary, or a microscope objective is used to simply provide a focused and then expanding reference beam. The object is mounted on a coverslip somewhere between the source and the imager. The imager is a bare CCD or CMOS pixel array connected to a computer. Typical total distances are $L = \sim 10–100$ mm providing an extremely compact yet high-resolution microscopy setup. The object to imager distance z_0 can be adjusted to change the magnification and resolution of the hologram. The reconstruction can be at any z location so 3D object information or different objects at multiple planes can be visualized

The physical parameters of the recorded hologram are programmed next.

```
% Set up the physical wave parameters
lambda = 633e-9;                   % Wavelength [m]
k = 2*pi/lambda;                   % Wavenumber [rad/m]

% Set up the physical CCD pixel parameters
dx = 4.65e-6;                      % Pixel size in X [m]
dy = dx;                           % Pixel size in Y [m]
```

The physical recording distance L and the reconstruction distance z are set.

```
% The reconstruction distance from the imager
z = 27.9e-3;                       % Distance in [m]

% The physical distance from the source to imager
L = 71.0e-3;                       % Source to imager [m]
```

The object wave is reconstructed with the angular plane wave spectrum technique.

```
% Compute the numerical reference beam
R = exp(-1i*k*sqrt(L.^2+X.^2+Y.^2))./sqrt(L.^2+X.^2+Y.^2);

% Compute the angular spectrum propagator
T = fftshift(exp(1i*z*sqrt(k.^2-kx.^2-ky.^2)));

% Compute the angular spectrum, propagate, and reconstruct
K = ifft2(fft2(Uh.*R).*T);
```

Finally, the amplitude and phase are plotted and analyzed.

```
% Plot the amplitude of the reconstructed object wave K
figure(1);
imagesc(X(1,:), Y(:,1),abs(K));

% Plot the phase of the reconstructed object wave K
figure(2);
imagesc(X(1,:), Y(:,1),angle(K));
```

This code provides a basic entry into numerical reconstruction of an in-line DHM hologram. It can also be modified easily to take into account the magnification M and enhancing the resolution by setting L to a very large number (e.g., 10^6 m) and multiplying the reconstruction distance z by M. The coordinates X and Y then need to be divided by M for correct plotting and image analysis. Other operations could also be performed, such as filtering in Fourier space [11], and more sophisticated reconstruction options are available through an open-source computational library [76].

6 Conclusion

Of the many imaging techniques available, lensless in-line DHM is likely the simplest arrangement, requiring no lenses, no mirrors, and typically only a light source, a sample to be imaged, and a digital imager chip such as a CCD or CMOS pixel array. Despite this simplicity, lensless in-line DHM is capable of producing high-resolution images over a wide field of view and has produced significant interest among many researchers around the world. Given the wide array of applications, digital holographic microscopy and the various lensless imaging techniques [20, 21] are poised to offer researchers with a range of new and novel capabilities for miniaturized and rapid biological detection, imaging, sensing, and analysis.

References

1. Gabor D (1948) A new microscopic principle. Nature 161:777–778
2. AB, NM (2014) The nobel prize in physics 1971. http://www.nobelprize.org/nobel_prizes/physics/laureates/1971/
3. Goodman JW (2005) Introduction to Fourier optics. Roberts and Company Publishers
4. Goodman JW, Lawrence R (1967) Digital image formation from electronically detected holograms. Appl Phys Lett 11:77–79
5. Lesem LB, Hirsch P, Jordan J Jr (1968) Scientific applications: computer synthesis of holograms for 3-D display. Commun ACM 11:661–674
6. Garcia-Sucerquia J, Xu W, Jericho SK, Klages P, Jericho MH, Kreuzer HJ (2006) Digital in-line holographic microscopy. Appl Opt 45:836–850
7. Kim MK (2010) Principles and techniques of digital holographic microscopy. J Photon Energy 018005–018005 (2010)
8. Kemper B, von Bally G (2008) Digital holographic microscopy for live cell applications and technical inspection. Appl Opt 47:A52–A61
9. Xu W, Jericho M, Meinertzhagen I, Kreuzer H (2001) Digital in-line holography for biological applications. Proc Natl Acad Sci 98:11301–11305
10. Schnars U, Jüptner WP (2002) Digital recording and numerical reconstruction of holograms. Meas Sci Technol 13:R85
11. Latychevskaia T, Fink H-W (2015) Practical algorithms for simulation and reconstruction of digital in-line holograms. Appl Opt 54:2424
12. Marquet P, Rappaz B, Pavillon N (2015) Quantitative phase-digital holographic microscopy: a new modality for live cell imaging. New Tech Dig Hologr 169–217
13. Yu X, Hong J, Liu C, Kim MK (2014) Review of digital holographic microscopy for three-dimensional profiling and tracking. Opt Eng 112306
14. Marquet P, Depeursinge C, Magistretti PJ (2014) Review of quantitative phase-digital holographic microscopy: promising novel imaging technique to resolve neuronal network activity and identify cellular biomarkers of psychiatric disorders. Neurophotonics 1:020901–020901
15. Alm K, El-Schich Z, Falck Miniotis M, Gjörloff Wingren A, Janicke B, Oredsson S (2013) Cells and holograms: holograms and digital holographic microscopy as a tool to study the morphology of living cells. https://doi.org/10.5772/54505
16. Kim MK (2011) Digital holographic microscopy 162
17. Kemper B, Langehanenberg P, Bredebusch I, Schnekenburger J, von Bally G (2007) Techniques and applications of digital holographic microscopy for life cell imaging. In: European conference on biomedical optics proceedings, pp 66330D–66330D
18. Schnars U, Jueptner W (2005) Digital holography: digital hologram recording. Numerical reconstruction, and related techniques. Springer
19. Ozcan A, McLeod E (2016) Lensless imaging and sensing. Annu Rev Biomed Eng 18:77
20. Greenbaum A, Luo W, Su T-W, Göröcs Z, Xue L, Isikman SO, Coskun AF, Mudanyali O, Ozcan A (2012) Imaging without lenses: achievements and remaining challenges of wide-field on-chip microscopy. Nat Methods 9:889–895
21. Roy M, Seo D, Oh S, Yang J-W, Seo S (2017) A review of recent progress in lens-free imaging and sensing. Biosens Bioelectron 88:130–143
22. Ryle JP, McDonnell S, Sheridan JT (2011) Lensless multispectral digital in-line holographic microscope. J Biomed Opt 16:126004–12600417
23. Mudanyali O, Tseng D, Oh C, Isikman SO, Sencan I, Bishara W, Oztoprak C, Seo S, Khademhosseini B, Ozcan A (2010) Compact, light-weight and cost-effective microscope based on lensless incoherent holography for telemedicine applications. Lab Chip 10:1417–1428
24. Seo S, Su T-W, Tseng DK, Erlinger A, Ozcan A (2009) Lensfree holographic imaging for on-chip cytometry and diagnostics. Lab Chip 9:777–787

25. Tseng D, Mudanyali O, Oztoprak C, Isikman SO, Sencan I, Yaglidere O, Ozcan A (2010) Lensfree microscopy on a cellphone. Lab Chip 10:1787–1792
26. Bishara W, Sikora U, Mudanyali O, Su T-W, Yaglidere O, Luckhart S, Ozcan A (2011) Holographic pixel super-resolution in portable lensless on-chip microscopy using a fiber-optic array. Lab Chip 11:1276–1279
27. Bishara W, Su T-W, Coskun AF, Ozcan A (2010) Lensfree on-chip microscopy over a wide field-of-view using pixel super-resolution. Opt Express 18:11181–11191
28. Ozcan A, Demirci U (2008) Ultra wide-field lens-free monitoring of cells on-chip. Lab Chip 8:98–106
29. Xu W, Jericho M, Meinertzhagen I, Kreuzer H (2001) Digital in-line holography for biological applications. Proc Natl Acad Sci 11301
30. Göröcs Z, Ozcan A (2013) On-chip biomedical imaging. IEEE Rev Biomed Eng 6:29–46
31. Zhang X, Khimji I, Gurkan UA, Safaee H, Catalano PN, Keles HO, Kayaalp E, Demirci U (2011) Lensless imaging for simultaneous microfluidic sperm monitoring and sorting. Lab Chip 11:2535–2540
32. Kim SB, Bae H, Cha JM, Moon SJ, Dokmeci MR, Cropek DM, Khademhosseini A (2011) A cell-based biosensor for real-time detection of cardiotoxicity using lensfree imaging. Lab Chip 11:1801–1807
33. Repetto L, Piano E, Pontiggia C (2004) Lensless digital holographic microscope with light-emitting diode illumination. Opt Lett 29:1132–1134
34. Garcia-Sucerquia J (2013) Noise reduction in digital lensless holographic microscopy by engineering the light from a light-emitting diode. Appl Opt 52:A232–A239
35. Isikman SO, Sencan I, Mudanyali O, Bishara W, Oztoprak C, Ozcan A (2010) Color and monochrome lensless on-chip imaging of Caenorhabditis elegans over a wide field-of-view. Lab Chip 10:1109–1112
36. Pedrini G, Fröning P, Fessler H, Tiziani HJ (1998) In-line digital holographic interferometry. Appl Opt 37:6262–6269
37. Barton J (1991) Removing multiple scattering and twin images from holographic images. Phys Rev Lett 67:3106
38. Yamaguchi I, Zhang T (1997) Phase-shifting digital holography. Opt Lett 22:1268–1270
39. Situ G, Ryle JP, Gopinathan U, Sheridan JT (2008) Generalized in-line digital holographic technique based on intensity measurements at two different planes. Appl Opt 47:711–717
40. Lu Y, Liu Y, Tian X, Fu Y, Zhao J (2015) An ultra-compact multiplexed holographic microscope using a multiple-pinhole aperture. Opt Express 23:26779–26793
41. Latychevskaia T, Fink H-W (2007) Solution to the twin image problem in holography. Phys Rev Lett 98:233901
42. Cuche E, Marquet P, Depeursinge C (2000) Spatial filtering for zero-order and twin-image elimination in digital off-axis holography. Appl Opt 39:4070–4075
43. Denis L, Fournier C, Fournel T, Ducottet C (2008) Numerical suppression of the twin image in in-line holography of a volume of micro-objects. Meas Sci Technol 19:074004
44. Ratcliffe J (1956) Some aspects of diffraction theory and their application to the ionosphere. Rep Prog Phys 19:188
45. Molony KM, Hennelly BM, Kelly DP, Naughton TJ (2010) Reconstruction algorithms applied to in-line Gabor digital holographic microscopy. Opt Commun 283:903–909
46. Jericho M, Kreuzer H, Kanka M, Riesenberg R (2012) Quantitative phase and refractive index measurements with point-source digital in-line holographic microscopy. Appl Opt 51:1503–1515
47. Ryle JP, Molony KM, McDonnell S, Naughton TJ, Sheridan JT (2009) Multispectral lensless digital holographic microscope: imaging MCF-7 and MDA-MB-231 cancer cell cultures. SPIE Opt Eng + Appl Proc 744206–744206
48. Su T-W, Xue L, Ozcan A (2012) High-throughput lensfree 3D tracking of human sperms reveals rare statistics of helical trajectories. Proc Natl Acad Sci 109:16018–16022
49. Sheng J, Malkiel E, Katz J (2006) Digital holographic microscope for measuring three-dimensional particle distributions and motions. Appl Opt 45:3893–3901

50. Choi Y-S, Lee S-J (2009) Three-dimensional volumetric measurement of red blood cell motion using digital holographic microscopy. Appl Opt 48:2983–2990
51. Xu W, Jericho M, Kreuzer H, Meinertzhagen I (2003) Tracking particles in four dimensions with in-line holographic microscopy. Opt Lett 28:164–166
52. Kempkes M, Darakis E, Khanam T, Rajendran A, Kariwala V, Mazzotti M, Naughton TJ, Asundi AK (2009) Three dimensional digital holographic profiling of micro-fibers. Opt Express 17:2938–2943
53. Tian L, Loomis N, Domínguez-Caballero JA, Barbastathis G (2010) Quantitative measurement of size and three-dimensional position of fast-moving bubbles in air-water mixture flows using digital holography. Appl Opt 49:1549–1554
54. Garcia-Sucerquia J, Xu W, Jericho S, Jericho M, Kreuzer HJ (2008) 4-D imaging of fluid flow with digital in-line holographic microscopy. Opt-Int J Light Electr Opt 119:419–423
55. Jericho S, Garcia-Sucerquia J, Xu W, Jericho M, Kreuzer H (2006) Submersible digital in-line holographic microscope. Rev Sci Instrum 77:043706
56. Shimobaba T, Yamanashi H, Kakue T, Oikawa M, Okada N, Endo Y, Hirayama R, Masuda N, Ito T (2013) In-line digital holographic microscopy using a consumer scanner. Sci Rep 3
57. Stybayeva G, Mudanyali O, Seo S, Silangcruz J, Macal M, Ramanculov E, Dandekar S, Erlinger A, Ozcan A, Revzin A (2010) Lensfree holographic imaging of antibody microarrays for high-throughput detection of leukocyte numbers and function. Anal Chem 82:3736–3744
58. Singh VR, Miao J, Wang Z, Hegde G, Asundi A (2007) Dynamic characterization of MEMS diaphragm using time averaged in-line digital holography. Opt Commun 280:285–290
59. Repetto L, Chittofrati R, Piano E, Pontiggia C (2005) Infrared lensless holographic microscope with a vidicon camera for inspection of metallic evaporations on silicon wafers. Opt Commun 251:44–50
60. Oh C, Isikman SO, Khademhosseinieh B, Ozcan A (2010) On-chip differential interference contrast microscopy using lensless digital holography. Opt Express 18:4717–4726
61. Zakrisson J, Schedin S, Andersson M (2015) Cell shape identification using digital holographic microscopy. Appl Opt 54:7442–7448
62. Bishara W, Zhu H, Ozcan A (2010) Holographic opto-fluidic microscopy. Opt Express 18:27499–27510
63. Kemper B, Langehanenberg P, Von Bally G (2007) Digital holographic microscopy. Opt Photon 2:41–44
64. Marquet P, Rappaz B, Magistretti PJ, Cuche E, Emery Y, Colomb T, Depeursinge C (2005) Digital holographic microscopy: a noninvasive contrast imaging technique allowing quantitative visualization of living cells with subwavelength axial accuracy. Opt Lett 30:468–470
65. Carl D, Kemper B, Wernicke G, von Bally G (2004) Parameter-optimized digital holographic microscope for high-resolution living-cell analysis. Appl Opt 43:6536–6544
66. Rappaz B, Barbul A, Emery Y, Korenstein R, Depeursinge C, Magistretti PJ, Marquet P (2008) Comparative study of human erythrocytes by digital holographic microscopy, confocal microscopy, and impedance volume analyzer. Cytometry Part A 73:895–903
67. Kim K, Choe K, Park I, Kim P, Park Y (2016) Holographic intravital microscopy for 2-D and 3-D imaging intact circulating blood cells in microcapillaries of live mice. Sci Rep 6
68. Jourdain P, Pavillon N, Moratal C, Boss D, Rappaz B, Depeursinge C, Marquet P, Magistretti PJ (2011) Determination of transmembrane water fluxes in neurons elicited by glutamate ionotropic receptors and by the cotransporters KCC2 and NKCC1: a digital holographic microscopy study. J Neurosci 31:11846–11854
69. Pavillon N, Kühn J, Moratal C, Jourdain P, Depeursinge C, Magistretti PJ, Marquet P (2012) Early cell death detection with digital holographic microscopy. PLoS ONE 7:e30912
70. Kemper B, Vollmer A, Rommel CE, Schnekenburger J, von Bally G (2011) Simplified approach for quantitative digital holographic phase contrast imaging of living cells. J Biomed Opt 16:026014–026014

71. Miccio L, Memmolo P, Merola F, Fusco S, Netti P, Ferraro P (2014) A new 3D tracking method for cell mechanics investigation exploiting the capabilities of digital holography in microscopy. SPIE BiOS Proc 89471L–89471L
72. Kemper B, Schubert R, Dartmann S, Vollmer A, Ketelhut S, von Bally G (2013) Improved quantitative phase contrast in self-interference digital holographic microscopy and sensing dynamic refractive index changes of the cytoplasm using internalized microspheres as probes. SPIE BiOS 85890M–85890M
73. Lenz P, Bettenworth D, Krausewitz P, Brückner M, Ketelhut S, von Bally G, Domagk D, Kemper B (2013) Digital holographic microscopy quantifies the degree of inflammation in experimental colitis. Integr Biol 5:624–630
74. Rappaz B, Marquet P, Cuche E, Emery Y, Depeursinge C, Magistretti P (2005) Measurement of the integral refractive index and dynamic cell morphometry of living cells with digital holographic microscopy. Opt Express 13:9361–9373
75. Charrière F, Pavillon N, Colomb T, Depeursinge C, Heger TJ, Mitchell EA, Marquet P, Rappaz B (2006) Living specimen tomography by digital holographic microscopy: morphometry of testate amoeba. Opt Express 14:7005–7013
76. Shimobaba T, Weng J, Sakurai T, Okada N, Nishitsuji T, Takada N, Shiraki A, Masuda N, Ito T (2012) Computational wave optics library for C++: CWO++ library. Comput Phys Commun 183:1124–1138
77. Seo S, Isikman SO, Sencan I, Mudanyali O, Su T-W, Bishara W, Erlinger A, Ozcan A (2010) High-throughput lens-free blood analysis on a chip. Anal Chem 82:4621–4627
78. Ryle JP, McDonnell S, Glennon B, Sheridan JT (2013) Calibration of a digital in-line holographic microscopy system: depth of focus and bioprocess analysis. Appl Opt 52:C78–C87
79. Molaei M, Sheng J (2014) Imaging bacterial 3D motion using digital in-line holographic microscopy and correlation-based de-noising algorithm. Opt Express 22:32119–32137

Biosensing on Digital Microfluidics: From Sample Preparation to Detection

Ehsan Samiei and Mina Hoorfar

1 Introduction

In recent years, the tendency to adopt digital microfluidics (DMF) for lab-on-a-chip (LOC) applications has increased extensively. This is due to the unique advantages that DMF offers comparing to other microfluidic systems. Many analytical and physical techniques for performing biochemical experiments have been reconfigured to be compatible with DMF systems, and numerous studies have shown the use of DMF technology for different biochemical applications. Along with the progress made towards the development of liquid handling and sample preparation aspects, biochemical detection techniques have continuously been reconfigured to be compatible with DMF chips, and nowadays, multiple DMF-based devices are developed with integrated detection systems for real-time- or on-line biosensing and diagnosis purposes.

This chapter briefly introduces DMF, along with detection techniques which have been integrated into these devices. Then, the development of the sample preparation techniques for DMF systems is discussed in detail. Finally, the applications of DMF platforms with integrated detection systems used in different biochemical applications are discussed.

E. Samiei · M. Hoorfar (✉)
University of British Columbia, 3333 University Way, Kelowna, BC V1V 1V7, Canada
e-mail: mina.hoorfar@ubc.ca

S.-H. Oh et al. (eds.), *Miniature Fluidic Devices for Rapid Biological Detection*, Integrated Analytical Systems, https://doi.org/10.1007/978-3-319-64747-0_7

2 Digital Microfluidics (DMF)

Miniaturized devices such as lab-on-a-chip (LOC), introduced for downscaling and automating biochemical processes, have attracted a significant attention in the past decades. LOC offers numerous advantages over the conventional in-laboratory methods used in performing biochemical experiments. These advantages include, but not limited to automation, low consumption of samples and reagents, high surface-to-volume ratio resulting high reaction rates, low manpower and human error, and high control on sample manipulation [1]. Microfluidics is the key to the development of LOC devices, as the manipulation of the samples is performed with microfluidic systems. The primary microfluidic systems introduced include permanently etched microchannels through which liquid sample flows using different pumping mechanisms. In the past two decades, digital microfluidic (DMF) platforms have been introduced which function based on the manipulation of liquid samples in the form of droplets in pico-liter to microliter sizes. The discrete nature of sample manipulation and control on individual droplets on DMF offers multiple advantages: (i) a single chip can be programed for different applications using one design of actuating electrodes; and (ii) the sample volume dispensed can be controlled with high accuracy; (iii) numerous droplets can be manipulated simultaneously, allowing for parallel and high throughput operations. Due to these features, DMF has been widely used for LOC applications. Many researchers have focused on the advancement of the operators on DMF for performing different processes; while other groups have implemented these operators for biochemical applications. Among different operators developed and used in many applications is biosensing. Different biosensing mechanisms have been integrated into DMF for off-chip and on-chip detection. This section briefly explains the DMF technology and the biosensing mechanisms integrated to these devices.

2.1 Droplet Actuation Mechanisms

Unlike the droplet microfluidics in which the sample droplets are formed and transported in a continuous flow of a carrier liquid, the droplets on DMF devices are manipulated individually on an array of electrodes. The array of electrodes could be designed in a 2D configuration, allowing for the transport of droplets in different directions on the planar chip surface. Such a configuration makes DMF versatile and reconfigurable for different applications, since the droplet path can be programed depending on the application on demand.

Droplet manipulation on DMF devices has been performed through several mechanisms, relying on surface and body forces applied on the liquid. These mechanisms include thermocapillary [2, 3], magnetic force [4, 5], surface acoustic wave [6, 7], dielectrophoresis (DEP) [8, 9], electrowetting-on-dielectric (EWOD) [10, 11], and optoelectrowetting [12, 13]. Although all of these mechanisms have shown to be functional in terms of droplet manipulation, EWOD is the most effective and widely used method in DMF devices. Thus, among the different methods, only the principal of EWOD is explained here.

2.1.1 Electrowetting-on-Dielectric

Electrowetting-on-dielectric was developed by Berge in early 1990s based on the electrocapillarity (electrowetting) effect which was introduced by Gabriel Lippmann in 1875 [14]. In electrowetting, when a conductive liquid is in contact with an electrode and a voltage is applied between them, the liquid spreads on the surface. This may result in a change in the apparent contact angle formed between the liquid and the substrate. Lippmann's electrowetting experiment was performed using mercury and an electrolyte. However, aqueous solutions are subject to electrolytic decomposition when the applied voltage is higher than a few hundred millivolts. Berge solved this problem by placing an insulating layer between the liquid and the electrode, introducing electrowetting-on-dielectric (EWOD) [14]. This phenomenon is schematically shown in Fig. 1 where the contact angle before and after applying voltage (U) is θ_0 and θ, respectively. Due to the application of the voltage, electrons and counter-ions will be accumulated in the interface of the insulating layer and the liquid, and therefore, a force will be applied to the three-phase contact line (TPCL), pulling TPCL and changing the apparent contact angle. The horizontal force and also the change in the apparent contact angle depends on the magnitude of the applied voltage, i.e., the higher the voltage, the higher the force. However, the change in the apparent contact angle is limited to a certain value (referred to as the saturation contact angle), above which the angle remains unchanged even with the application of a higher voltage.

Lippmann-Young equation (Eq. 1) is a simple correlation between the applied voltage and the change in the contact angle [15]

$$cos\theta = cos\theta_0 + \frac{1}{2}\frac{C}{\sigma_{lv}}U^2, \tag{1}$$

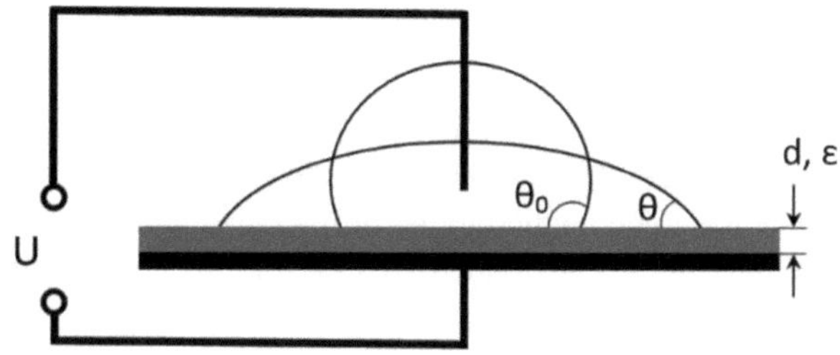

Fig. 1 Schematic of the electrowetting-on-dielectric (EWOD) phenomenon

where C and σ_{lv} are the capacitance of the insulating layer and the liquid–vapor interfacial tension, respectively. This equation, however, is valid only for a low range of the applied voltage. The results deviate from the experimental values when the contact angle is close to the saturation contact angle [15, 16]. Berthier et al. [15, 16] developed a more complicated correlation which was accurate even for higher applied voltages

$$\frac{cos\theta - cos\theta_0}{cos\theta_S - cos\theta_0} = coth\left(\frac{3CU^2}{2\sigma_{lv}(cos\theta_S - cos\theta_0)}\right) - \frac{1}{\frac{3CU^2}{2\sigma_{lv}(cos\theta_S - cos\theta_0)}} \tag{2}$$

In this equation θ_S is the saturation contact angle.

EWOD also exerts a force on the droplet. Several approaches have been developed to formulate the EWOD force. Here, electromechanical approach, which is the most accurate method in estimating the EWOD force, is briefly presented. With some simplifications [14], the stress acting on an infinitesimal element of liquid due to an electric field (generated by the applied voltage to the system (Fig. 1)) can be presented using Maxwell stress tensor [14]

$$T_{ik} = \varepsilon_0 \varepsilon \left(E_i E_k - \frac{1}{2} \delta_{ik} E^2 \right), \tag{3}$$

where ε_0, ε are the permittivity of vacuum and the insulating layer, respectively, δ_{ik} is the Kronecker delta function, E is the electric field and $i, k = x, y$ and z. The force acting on the liquid can be calculated using the following equation:

$$F_i = \ominus\ T_{ik} n_k dA, \tag{4}$$

where n is the unit normal vector to the liquid–vapor interface and dA is the area of an infinitesimal element at the interface. After integration and considering certain assumptions (the details can be found in [14, 17]) the horizontal component of the EWOD force per unit area can be calculated as

$$F_x = \frac{\varepsilon_0 \varepsilon}{2d} U^2, \tag{5}$$

where d is the thickness of the insulating layer.

2.1.2 EWOD Actuation on DMF

As explained earlier, EWOD is the common mechanism used for droplet manipulation on DMF devices. In order to transport a droplet on the chip, the EWOD force has to be applied to the droplet asymmetrically. For this purpose, the electrode on the surface is divided to multiple smaller electrodes, electrically addressable

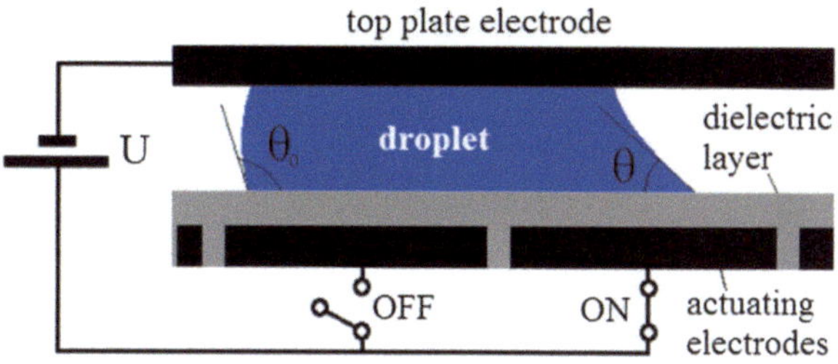

Fig. 2 Schematic of a typical DMF device

individually. In addition, the wire in Fig. 1 is replaced with a top plate coated with a metal layer. Figure 2 illustrates the schematic of a DMF device which consists of (i) a bottom layer containing the actuating electrodes and covered by the insulating layer, (ii) the top plate containing the ground electrode, and (iii) the liquid droplet sandwiched between the two plates. To transport a droplet on the chip, the chip surface has to be hydrophobic. Therefore, a thin hydrophobic layer is coated on both bottom and top plates.

2.1.3 Open and Closed DMF Systems

In the previous section, it was shown that in order to use EWOD for transporting a droplet on the chip, the top wire was replaced with a top plate containing an electrode. This configuration is called the "closed" or "sandwiched" system (Fig. 3a). However, that is not the only configuration that could be used for such a purpose. The ground electrode could also be patterned on the bottom plate, beside the actuating electrode. This configuration does not need a top plate and is referred to as the "open" system (Fig. 3b). The open system allows for simple liquid dispensing methods (such as pipetting) on to the chip, and has been used in multiple studies [18, 19]. However, it is not as common as the closed system since in the case of the open systems, the evaporation rate is high and droplet manipulation is not as straightforward as the closed system [1].

2.1.4 Basic Fluidic Operations

Performing biochemical experiments on the chip requires generating specific volumes of different samples and reagents, transporting the liquid volumes, and mixing them to reach a uniform mixture. On DMF devices, these operations are performed by applying EWOD using the patterned electrodes on the bottom plate. Transport of droplets on the chip is carried out by sequential actuation of the electrodes. As shown in Fig. 4a, when the adjacent electrode to the droplet is actuated, EWOD pulls the droplet towards the electrode; when it wets the entire electrode surface, the

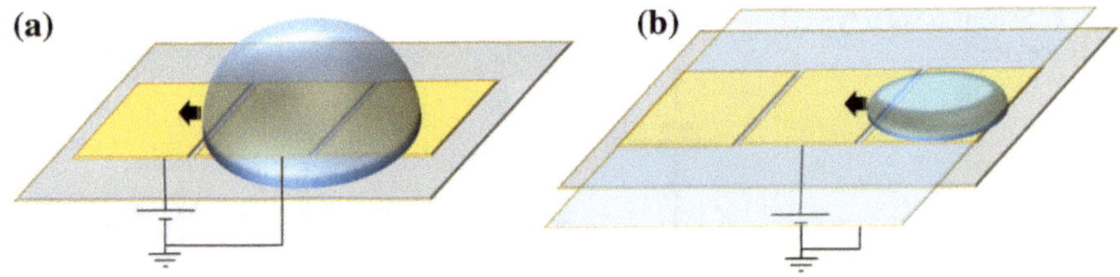

Fig. 3 Schematic of **a** open, and **b** closed (sandwiched) DMF systems

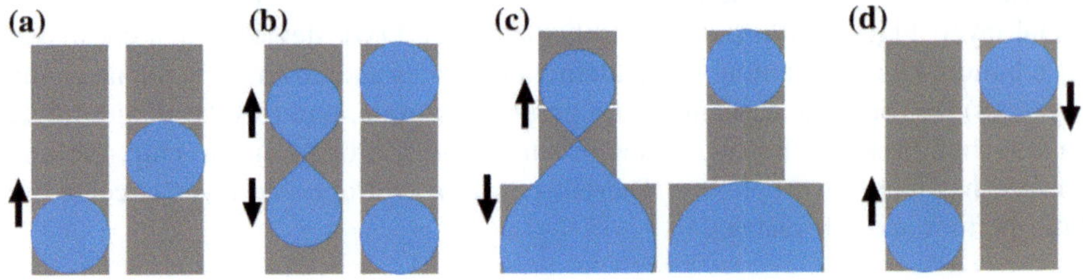

Fig. 4 Basic fluidic operations on DMF devices: **a** transport, **b** splitting, **c** dispensing, and **d** mixing

next electrode will be actuated to maintain the transport process. This is the basis of droplet transport on DMF which was introduced in [10].

Generating a specific volume of samples is performed by splitting a bigger droplet (referred to as mother droplet) into two smaller droplets (referred to as daughter droplets) or extracting a small droplet from a larger droplet in the reservoir. The former is called splitting and the latter is referred to as dispensing. To split a mother droplet into two daughter droplets, normally three electrodes are required. First the droplet is transported to the middle electrode, and then by actuating the two side electrodes (while the middle one is turned off) the droplet is split (Fig. 4b). Dispensing is also very similar to splitting. The droplet is pulled on the array of electrodes from the reservoir electrode to form a liquid finger, and then the reservoir and the second actuating electrode will be actuated to dispense a small droplet (Fig. 4c).

Mixing is also an important operation required for biochemical applications where more than one sample or component is involved. When two sample droplets are merged, due to the high aspect ratio of the droplet (diameter-to-height ratio) the content will not be mixed properly. Several methods have been developed in order to achieve proper mixing inside the droplet, which will be discussed in detail in Sect. 3. Here the basic mixing method is shown in Fig. 4d. In essence, it involves transport of the droplet back and forth on an array of three or four electrodes until the content is mixed properly.

2.2 Detection Systems Integrated into DMF Systems

Biosensing or detection is a key process and probably the final step in biochemical experiments, as after the sample preparation is carried out on the device, the final product is taken to the sensing component to detect a specific analyte [1]. Historically, right after the development of fluidic operations, DMF devices were used for performing biochemical assays which included detection. In early studies, the DMF devices were used only for a part of or the entire sample preparation step, and the product was taken off the chip and delivered to the detection device which was mainly fluorescence microscopy [20]. This mode of sensing is referred to as off-line sensing. The tendency to develop DMF-based LOC platforms for point-of-case applications (in which the entire assay process is required to be performed on the chip) resulted in the integration of the biosensors into DMF devices. Such a mode of sensing is referred to as online sensing, where the biosensing component is a part of the DMF chip. Several sensing mechanisms integrated into DMF devices will be discussed in this section.

2.2.1 Optical Systems

Optical detection systems widely coupled with DMF devices include absorbance, fluorescence, chemiluminescence, and surface plasmon resonance (SPR) detection systems [1, 20]. Absorbance techniques [21, 22] are based on UV/Vis absorption spectroscopy which works based on the measurement of the attenuation of the incident light in a range of the wavelengths. The part of the spectrum in which the attenuation peaks depends on the type of the analyte, which is used for the identification of the specific analytes that are present in the sample.

In fluorescence detection techniques [23, 24], the target analyte is labeled using a fluorescent dye. The fluorescence detection systems include a light source for excitation, and a fluorescent photodetector for detecting the signal. The fluorescent labels are excited using a high-intensity light source, and upon excitation, they emit low intensity, long wavelength light which is detected using a fluorescent detector (microscope, spectrometer, or well-plate reader).

Chemiluminescence [25, 26] techniques are conceptually similar to the fluorescence techniques in a way that the light emitted from the labeled analyte is detected. However, in chemiluminescence system, a chemical reaction occurs between the target analyte and an enzyme label instead of using a fluorescent dye and an exciting light source. This method is very sensitive and selective and its application is more straightforward as it is based on a chemical reaction and only requires the integration of a photodetector for detecting the signal [20].

Surface plasmon resonance (SPR) [27, 28] is a sensitive, label-free optical detection technique which is based on the measurement of the change in the

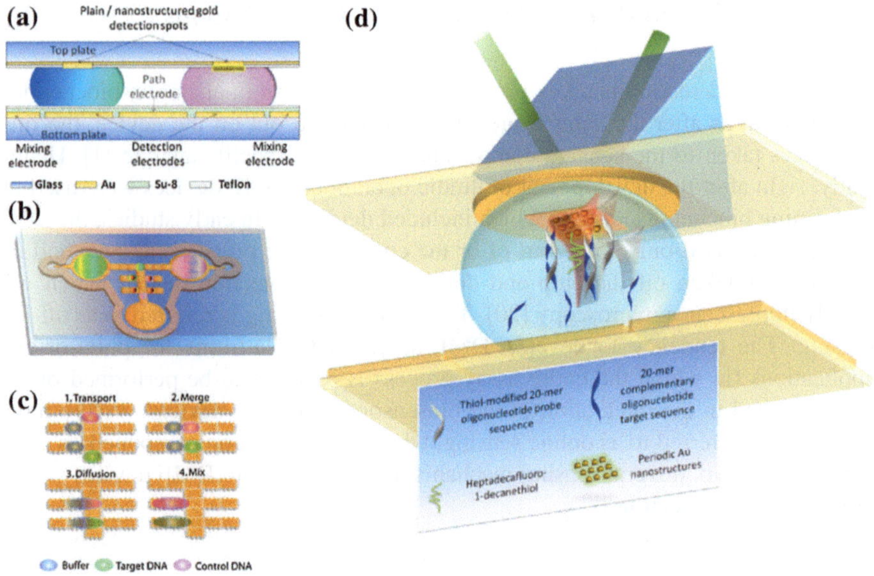

Fig. 5 Schematic of a DMF chip with integrated surface plasmon resonance (SPR) detection system. Reproduced from [28] with permission from Elsevier

refractive index of the reflected light from a light beam shined to the surface of the sensor through a prism. The SPR biosensor consists of a thin metal layer in contact with the liquid sample and a dielectric layer which is usually a prism. The frequency of the light source is adjusted with the natural frequency of the surface plasmon formed at the metal–dielectric interface. The refractive index of the reflected light is measured using a photodetector. As the target biochemical analytes bind to the receptors on the metal layer, the refractive index of the reflected light changes. A schematic of SPR biosensor integrated into a DMF platform is illustrated in Fig. 5.

2.2.2 Mass Spectrometry

Mass spectrometry is an analytical method which has been coupled with DMF for biosensing purposes. This method functions based on sorting the chemical species within the sample into a spectrum is dependent on the mass-to-charge ratio of the species [29]. The equipment used for mass spectrometry is large, and therefore, it is not feasible to miniaturize and integrate the device into the DMF chip [20]. Instead, these systems are interfaced with the DMF chips, normally using microscale tubes, to transfer the sample from the chip to the spectrometer [29]. Figure 6 shows

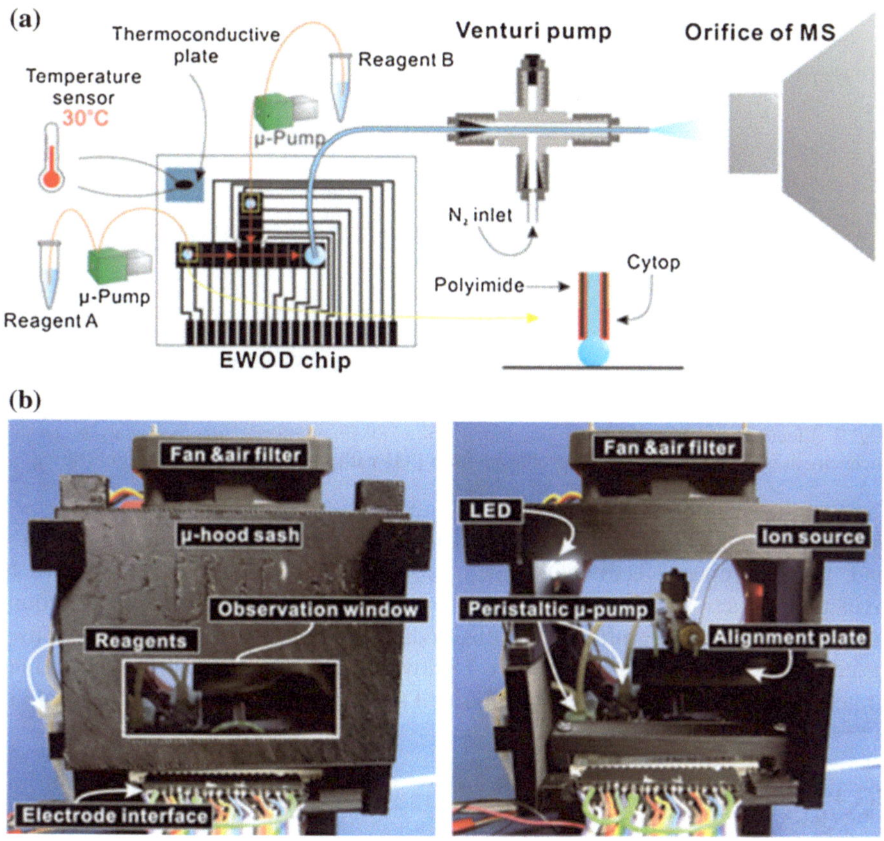

Fig. 6 A mass spectrometry system coupled with an open DMF device used for analytical analyses. Reproduced from [30] with permission from the Royal Society of Chemistry

coupling of a mass spectrometry system with a DMF device used for multiple analytical analyses [30].

2.2.3 Electrochemical Systems

Electrochemical detection is a sensing mechanism in which the change in an electrical signal (such as current, conductance, impedance, or potential) due to an enzymatic reaction on the sensing surface is measured [31]. The enzymatic reaction causes the chemical oxidation and reduction, which is associated with transfer of electrons, measured by the measurement equipment. Binding or the enzymatic reaction of the target analyte to the receptors immobilized on the sensing surface causes the change in the oxidation/reduction rate, which is used as a tool for

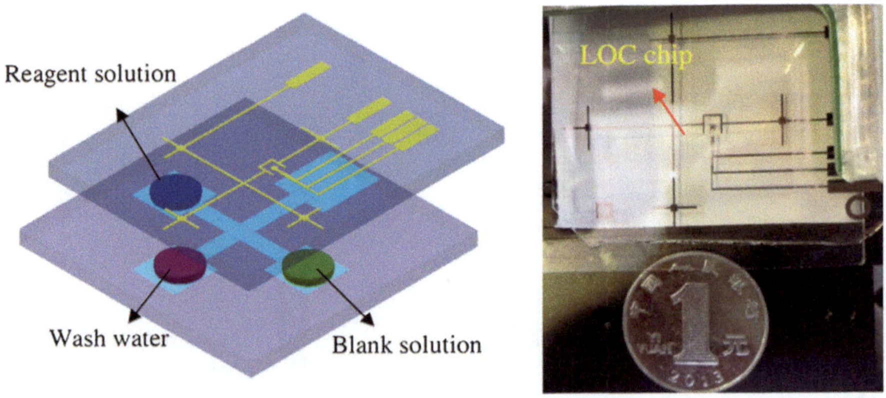

Fig. 7 Electrochemical biosensor integrated to a DMF platform for the analysis of ferrocenemethanol and dopamine. Reproduced from [33] with permission from IOP Publishing

detection [31]. These systems consist of three electrodes (the working, the auxiliary, and the reference electrodes) which are used for the measurement of the oxidation/reduction rates. These sensors are easy to fabricate, as they only require photolithographic patterning of the electrodes, making this method a widely used detection technique [20, 32]. Figure 7 shows an electrochemical biosensor integrated into a DMF device for the analysis of ferrocenemethanol and dopamine [33].

2.2.4 Other Systems

Optical, mass spectrometry, and electrochemical detection systems are the dominant mechanisms used for biochemical sensing on DMF devices. However, there are some other detection systems integrated into DMF devices which are usually electrical based. A field-effect transistor based biosensor has also been integrated into DMF systems for the detection of avian influenza virus [34]. These sensors are highly sensitive and function based on the change in the conductance of the semiconductor transducer due to the electric field around the biomolecules bound to their surface. Capacitive electrodes have also been used as transducers for biosensing purposes [35]. Other than these mechanisms, hydrodynamic manipulation of microbeads has also been used as a mean for the indication of the presence of low concentrations of DNA in liquid samples [36]: the presence of DNA in the sample changes viscoelasticity of the liquid which consequently changes the pattern of microbeads motion due to their hydrodynamic manipulation.

Regardless of the transduction mechanism used for biosensing and biochemical detection, biosensors are required to have high selectivity towards certain analytes in order to accurately detect them. The selectivity of the biosensors to a certain analyte is achieved by surface modifications. Different methods have been used for the modification of the sensing surface of the biosensors [1] and based on these methods, the biosensors can be categorized as antibody-based [34, 35], enzyme-based [37, 38] and aptamer-based [39–41] systems.

3 Advances in Digital Microfluidics (DMF)

This section explains the new developments and advances made in digital microfluidic (DMF) devices with regards to biochemical applications. The advances in the fabrication of the devices is explained first. Then fluid manipulation techniques and advanced operators developed for assay processes are discussed.

3.1 Device Fabrication

Within the past two decades, several enhancements have been made in the fabrication procedures and material selection for the DMF devices to increase durability and resolution, and decrease the cost of the devices. Here we explain the individual components forming the DMF devices separately.

3.1.1 Substrate

Depending on whether the resolution, cost, or throughput of the device is the main criterion, the substrate used for DMF chips may be different. Normally, for in-laboratory and proof-of-concept experiments glass substrates are used due to the availability and versatility of the microscope glass slides [42]. When the resolution of the chip is important, silicon wafers with a thermally grown oxide layer are used as the substrate as these wafers provide a very smooth surface, allowing for the fabrication of very small features [43, 44]. On the other hand, for the low cost and disposable chips paper substrates are used [45]. For high throughput and durable devices, printed circuit boards (PCB), allowing for the fabrication of a large number of electrodes [46], are used. PCB chips have also shown to be cost-effective and used for applications requiring disposable chips [47].

3.1.2 Electrodes

Electrodes are mostly fabricated using the photolithography technique using the thin metal layers deposited on the substrate. In such a method of fabrication, the material used as the electrode does not significantly affect the resolution of fabrication, and hence, multiple different materials have been used for this purpose (e.g., copper, silver, and carbon [42, 48]). However, for the smoothness of the surface, and avoiding the oxidation of the metal layer gold is the most widely used material [1]. One issue in the deposition of gold is its low adhesion to the substrates. In order to solve this issue, an adhesion layer such as a few nanometers of chromium or titanium is deposited first, and the gold layer is deposited on the top of this adhesion layer [21, 44]. In some applications, it is required to have optical access to the buffers on the chip for sensing purposes. For these cases, normally a transparent material such as indium-tin oxide (ITO) is used for the fabrication of electrodes [49]. The electrodes in disposable and flexible chips are fabricated using simpler methods such as screen or inkjet printing [50].

3.1.3 Insulating Layer

As explained in the Sect. 2.1.1, the insulating layer which separates the electrodes and the liquid (neglecting the hydrophobic layer), plays an important role in EWOD actuation as it prevents electrolysis of the liquid and allowing the application of higher voltages for manipulating the droplets. This layer has to be very uniform and free of pinholes to avoid the bridging of the electrical charges from the electrode to the liquid and the resultant electrolysis. Since the EWOD phenomena is a capacitive-based phenomenon, its magnitude is dependent to the dielectric constant and the thickness of the insulating layer, i.e., for higher dielectric constants and smaller thicknesses the EWOD effect is more significant [14, 51]. Figure 8 shows the contact angle of water as a function of the magnitude of the applied voltage in an EWOD setting with three different insulating layers of $(Ba_{0.7}Sr_{0.3})TiO_3$ (BST, $\varepsilon \approx 180$), silicon dioxide (SiO_2, $\varepsilon \approx 3.7$) and a fluoropolymer ($\varepsilon \approx 2-3$) [51]. The thicknesses of the three layers are 70 nm, 100 nm, and 120 nm, respectively. It could be observed that the change in the contact angle as a result of the applied voltage is significantly higher for BST than the other two materials. This is due to the higher dielectric constant and smaller thickness of BST as compared to the other layers.

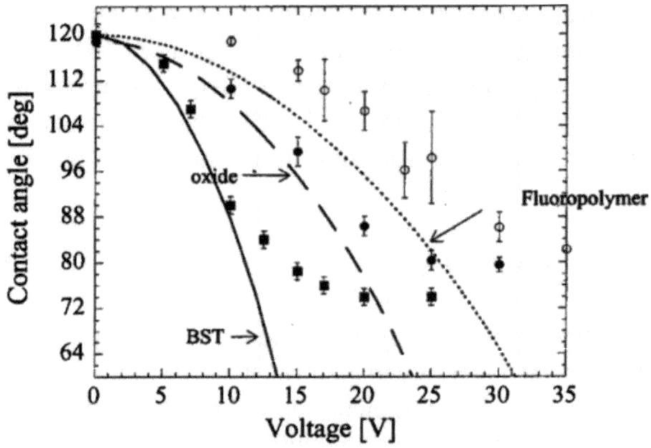

Fig. 8 Variation of the contact angle of water versus the applied voltage for three different materials as BST (70 nm), silicon dioxide (100 nm), and a fluoropolymer (120 n). Reproduced from [51] with permission from AIP Publishing

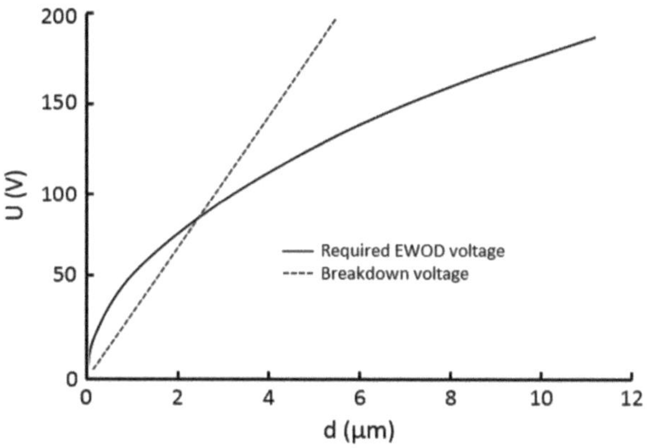

Fig. 9 Required electrowetting voltage (solid line) and the breakdown voltage (dashed line) versus the thickness of the insulating layer. The data belongs to water on a dielectric material with $\varepsilon = 2$ [14]

Similarly, the change in the contact angle for the SiO_2 layer is larger than that of the fluoropolymer layer.

Smaller thicknesses of the dielectric layer result in stronger EWOD effect for a constant applied voltage. However, for each material, there is a limit smaller than which breakdown of the insulating layer occurs upon applying high voltages. Figure 9 illustrates two curves, one representing the required electrowetting voltage for droplet movement, and the other showing the breakdown voltage versus the thickness of the dielectric layer (material with $\varepsilon = 2$) [14]. For this particular case, the two curves meet at the dielectric thickness of slightly higher than 2 μm. Below this value, the electrowetting voltage is higher, meaning that before the droplet moves the insulating layer breaks down. For larger thicknesses, the electrowetting voltage is smaller, and therefore, the droplet can be moved without dielectric breakdown.

For the fabrication of DMF devices a variety of materials have been used as the insulating layer. Silicon dioxide [52], parylene C [10], and polytetrafluoroethylene (PTFE) have been widely used due to their availability. Photoresists such as SU-8 [53] and S1813 [42] have also been used as the dielectric layer due to their ease of deposition. However, the photoresists have poor dielectric properties. Finally, materials with high dielectric constants (such as BST [51] and Ta_2O_5 [54]) have also been used to reduce the required voltage for droplet actuation [1].

3.1.4 Hydrophobic Layer

Hydrophobicity of the chip surfaces is crucial for the transport of liquid droplets on the DMF chips. For this purpose, low surface energy materials must be used as the hydrophobic layer. Fluoropolymer materials such as Teflon [52] and Cytop [43] have been widely used as the hydrophobic layer for DMF devices.

3.2 Fluid Manipulation Techniques

The basic fluidic operations explained in Sect. 2.1.4 are the basis of sample manipulation on DMF devices [52]. This section focuses on the advances made towards enhancing the precision and speed of these operations.

3.2.1 Droplet Transport

Droplets are transported on DMF devices by sequentially actuating the electrodes. To enhance the speed of transport, several attempts have been made including modulation of the applied voltage and using different surrounding media [14]. In early studies, DC voltages were used for droplet actuation. It was shown that applying AC voltages enhances sample manipulation and increases the chip lifetime [14]. Later on, it was shown that applying pulses of DC voltage rather than normal AC or DC voltages will further improve the chip lifetime and speed of transport [55]. The use of silicone oil [53, 56] or mineral oil [57] instead of air as the surrounding media also reduces the interfacial tension and contact angle hysteresis, enhancing the droplet motion. These oils were shown to be bio-compatible [57]. Figure 10 shows the transport of gelatin methacrylate hydrogel (GelMA) and growth media in mineral oil for cell patterning applications [57]. Hydrogels such as

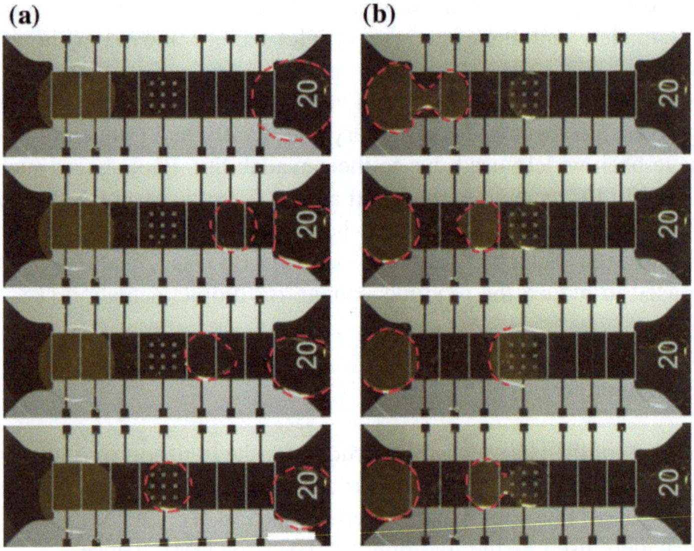

Fig. 10 Transporting gelatin methacrylate hydrogel (GelMA) and growth media in mineral oil for cell patterning applications. Reproduced from [57] with permission from the Royal Society of Chemistry

GelMA are highly viscous and sticky and cannot be transported on the chip when the surrounding media is air.

3.2.2 Droplet Splitting and Dispensing

The splitting and dispensing operations were explained in Sect. 2.1.4. Splitting is performed on an array of three electrodes by simultaneous actuation of the two side electrodes which results in pinching off and splitting the mother droplet into two daughter droplets. Dispensing from a reservoir droplet is also performed by actuating the reservoir electrode and the second adjacent electrode to the reservoir. This procedure without controlling the magnitude of the applied voltage will result in a deviation of 5% or higher in the volume of the split/dispensed droplets [58]. Several strategies have been proposed to enhance the accuracy of splitting and dispensing. For instance, the capacitance of the droplet is used as a mean to determine its volume. In this approach, a feedback control system is used to modulate the applied voltages to the two side electrodes [58] to accurately control the droplet volume. This strategy has improved the precision and reduced the deviation in the volume to 1%. In another study, to improve the precision of droplet dispensing, the middle electrode was turned on at the same time as the two side electrodes, and then the voltage applied to the middle electrode was ramped down until the droplet split [59]. Samiei et al. [42] showed that precise droplet splitting with an error less than 1% can be achieved using an applied voltage close to the threshold value for splitting. This method does not require any feedback control or voltage modulation system. Also, once the physical properties of the chip are determined this method is very easy to implement. They showed that the threshold splitting voltage is a function of the gap height when other parameters are kept the same. The relationship between the threshold voltage versus the gap height has been determined and these results are shown in Fig. 11 for three different electrode geometries. This

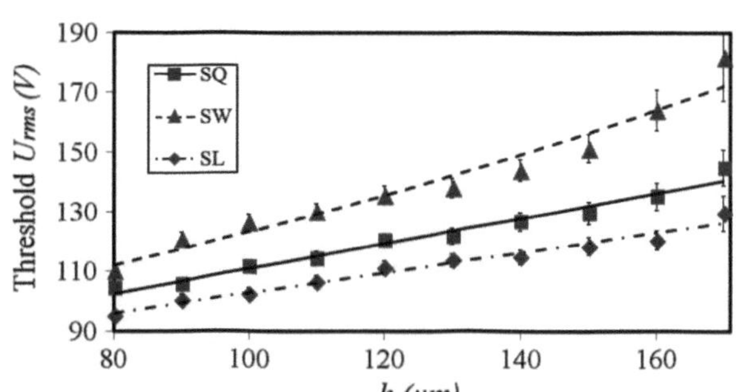

Fig. 11 Threshold voltage required for droplet splitting versus the gap height between the top and the bottom plates for three different electrode geometries as SQ (square), SW (stripped along the width of the electrode) and SL (stripped along the length of the electrode). Reproduced from [42] with permissions from IOP Publishing

figure shows that for a wide range of the gap height, this relationship is linear which facilitates the selection of a proper voltage based on the chosen gap height.

3.2.3 Droplet Mixing

It was explained that the basic mixing method on DMF devices is based on transporting the droplet over an array of three or four electrodes. However, this method is very inefficient and results in a relatively long mixing process. Later on, different 2D arrays of electrodes were studied for mixing and it was found that cyclic transport of the droplet over a 2 × 4 array of electrodes provides the best mixing performance [60]. This method also includes cyclic transport of the droplet over the chip surface which occupies a portion of the chip. The major problem with this method is that upon the presence of biological species (which have the tendency to adsorb to the surface) this method can increase the chance of chip breakdown. Most recently, an electrohydrodynamic (EHD) technique was implemented for rapid droplet mixing while the droplet is stationary [61]. The mixing using this method could be performed within less than 1 s, and the method was tested for different buffers including deionized water, phosphate-buffered saline (PBS), and sample containing DNA. The technique can be performed using the actuating electrodes on DMF devices and works for both open and closed systems. Figure 12 illustrates the EHD mixing method for different samples of polystyrene microbeads and safranin-o dye on both open and closed DMF devices.

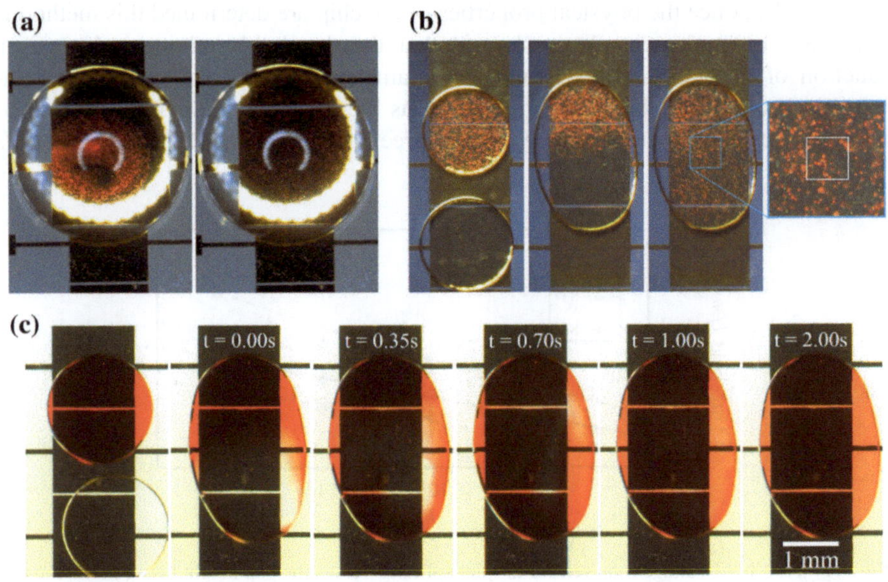

Fig. 12 Electrohydrodynamic mixing of **a** sample containing polystyrene microbeads (PS) on an open system DMF chip, **b** sample containing PS on a closed system DMF device, and **c** sample of safranin-o dye and DI water on a closed system DMF device. Reproduced from [61] with permission from the Royal Society of Chemistry

3.3 Advanced Operators

While the basic fluidic operations can be used for the manipulation of liquid samples, the entire sample preparation processes may not be feasible only using these operations. Some applications require variable volumes of different samples to be generated. In some applications which include microbeads, one needs to concentrate them within the droplet. In cell-based applications, patterning of the cells might be required. Also, heating is a part of multiple applications. For these purposes, specific strategies have to be adopted to perform the process. Here the techniques that have been developed for such processes are discussed briefly.

Droplet dispensing using the conventional method can produce constant volumes of droplets proportional to the size of the electrode. Therefore, if one needs to mix two samples with different volumes, either one of the samples has to have a large volume (and resultant contact area) which is difficult, if not impossible, to manipulate on the chip, or several dilution steps have to be used to reach the desired ratio of the samples in the mixed droplet. One way to avoid this issue is to have the ability to dispense variable volumes of liquid on the chip. The first strategy developed for dispensing variable volumes of droplets is based on applying different voltages to the reservoir and the dispensing electrode [58]. This study used capacitive sensing for determining the droplet size and a feedback control system to modulate the voltages applied to the reservoir and the splitting electrode. Lower voltages applied to the splitting electrode resulted in a lower size of the dispensed droplet. A different strategy, which was recently developed, is based on the geometrical modification of one of the electrodes by dividing it into several sub-electrodes [42]. It was shown that applying the threshold splitting voltage results in high-precision droplet dispensing. Also, it has been shown that the size of the droplet (measured based on its normalized area) has a linear relationship with the number of the sub-electrodes (Fig. 13), allowing for choosing the proper number of sub-electrodes based on the desired volume of the droplet.

Microbeads are used in several immunoassays for capturing the target analytes and labeling them for fluorescent microscopy [62]. Typically, magnetic microbeads

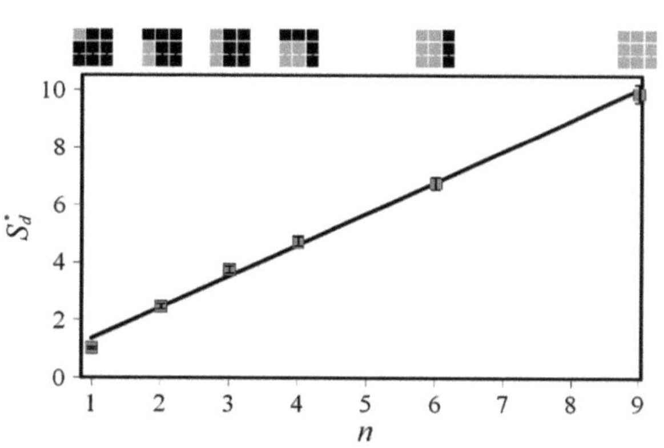

Fig. 13 The normalized area of the dispensed droplet versus the number of actuated sub-electrodes. Reprinted from [42] with permission from IOP Publishing

are used for this purpose where the functionalized beads are mixed in the sample to capture the target analyte which are then collected using an external magnet [63]. Recently, some techniques have been developed to eliminate the need for an external magnet for concentrating the beads inside the droplet. In a recent study, it has been shown that rotation of the droplet containing beads can focus the beads in one area. For this purpose, the chip includes a central electrode surrounded by several electrodes. One end of the droplet is fixed by continuous actuation of the central electrode, while the rest of the droplet spins around the central part by sequential actuation of the surrounding electrodes [36]. This rotational motion of the droplet causes the collection the beads in the center of the droplet due to the hydrodynamic-density based phenomenon. After collecting the beads, the droplet was then split to keep one side highly concentrated and the other side with a small concentration of the beads. In a different study, a dielectrophoretic-gravity driven (DGD) technique was developed for concentrating the microbeads [64]. By combining the effects of negative dielectrophoresis and gravity, the beads were collected in the center of one electrode. Then, the droplet was split into two parts one with majority of the beads and the other with a small concentration of the beads. The advantage of DGD technique is that it can be performed using the conventional electrodes and on any location of the chip. An example of the DGD method is shown in Fig. 14. The combination of the DGD technique and the methods proposed for dispensing variable volumes [42] of droplets can be a powerful method for pre-concentration of the samples for immunoassays.

Microheaters have also been integrated to the DMF devices for the applications requiring the control on the temperature of the assay [25, 65]. The heaters were either external modules added to the chip [25] or were fabricated on the chip using the similar fabrication method used for patterning the actuating electrodes [65].

3.4 Anti-biofouling Techniques

The hydrophobic layer on the DMF chips facilitates the motion of the droplets for both aqueous and non-aqueous buffers and prevents cross contamination. However, some biological species such as proteins and lipids have the tendency to stick to the surface through the hydrophobic adsorption or electrostatic interaction [1]. The biomolecular adsorption hinders the manipulation of liquid samples and increases the chance of cross contamination. The strategies used for preventing biofouling in conventional biological equipment [66] (test tubes) or continuous microfluidic devices [67] may not be practical for DMF chips as the addition of any coating may reduce hydrophobicity as well as the effect of EWOD. Thus, approaches specifically developed for preventing biofouling in DMF devices are discussed here.

The primary strategy used for reducing the tendency of the proteins to the surface is through controlling the pH of the buffer and modulating the applied voltage to the electrodes [68]. This technique reduces the biofouling effect, however, the permanent effects on the surface have been observed. The use of silicone

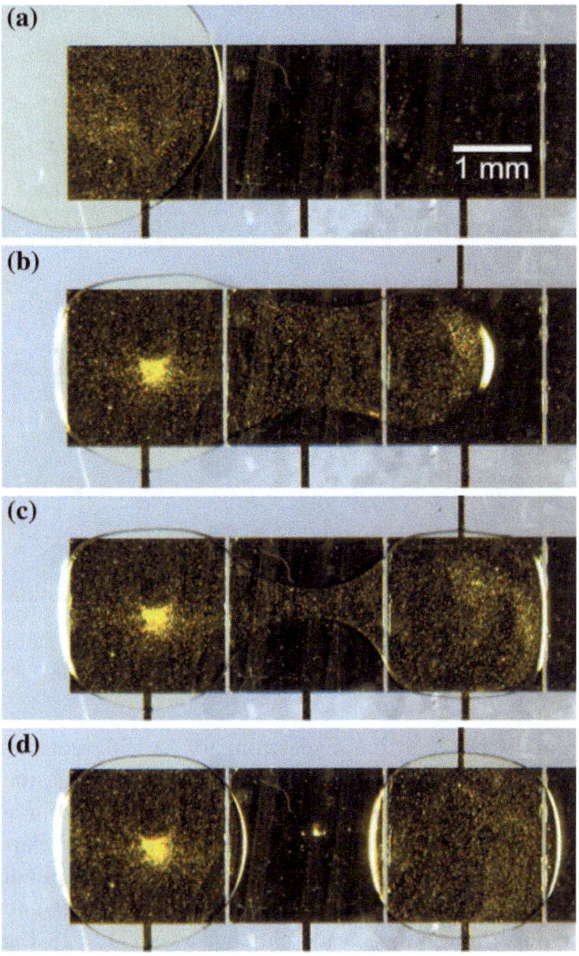

Fig. 14 The sequences of images from the concentration of silica beads using DGD technique. Reproduced from [64] with permission from AIP Publishing

oil as the surrounding media is another strategy which is very effective, preventing biofouling even for the whole blood for thousands of cycles of transport on the chip [22]. However, using oil might limit applications such as those with integrated biosensors on the chip surface. The use of silicone oil as a thin shell around the droplet rather than the entire surrounding media has also been proposed for anti-biofouling purposes [56]. In this method, before pipetting the samples in the reservoirs, the chip surface is touched with a pipette tip initially immersed in oil. This leaves a ~0.1 μL oil droplet on the surface. When the sample droplet is placed on the reservoir, the oil spontaneously forms a thin layer around the entire droplet [53]. A similar strategy was propped using Pluronic F127, which functions as a surfactant, forming a shell around the droplet to prevent biofouling [69]. Using Pluronic F127, samples containing up to 1 mg/mL proteins have been manipulated on the chip with negligible surface contamination, proved through confocal

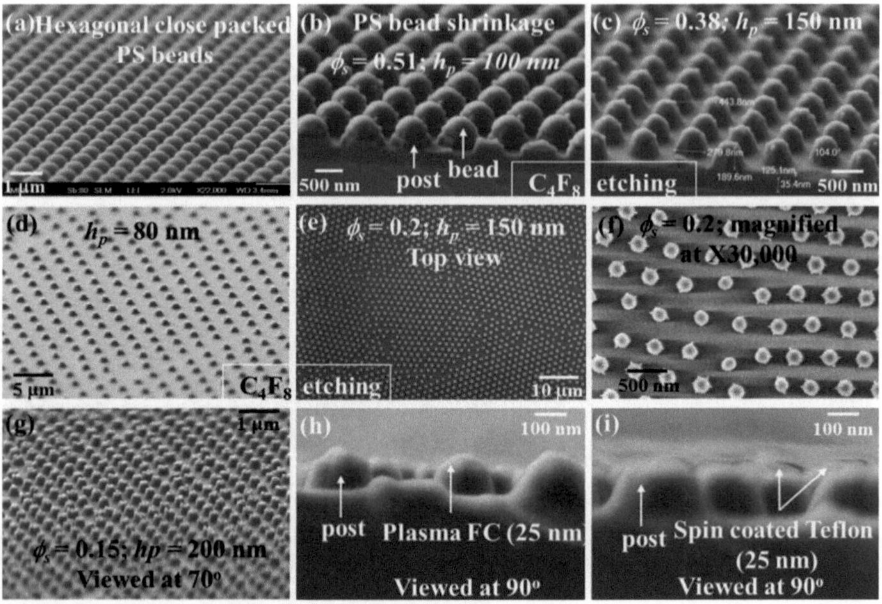

Fig. 15 Samples of different nano-structures made on the surface to create a superhydrophobic and anti-biofouling surface. Reprinted from [72] with permission from Elsevier

microscopy and mass spectrometry. A different method, simple but effective, has been the use of disposable polymer "skins" as the hydrophobic layer [70]. The polymer skins can significantly reduce the protein adsorption, and for every round of experiments, a new skin can be applied to the chip, allowing for reusability of the chip. Generation of a nanostructured superhydrophobic layer by the aid of methods such as plasma deposition of fluoropolymers is another technique to significantly reduce biofouling on the chip surfaces [71, 72]. Figure 15 shows different nanostructures for generating superhydrophobic surfaces and enhancing anti-biofouling effects [72].

4 Digital Microfluidic (DMF) Platforms for Biochemical Sensing Applications

In the previous sections, digital microfluidics (DMF) along with the biochemical sensing mechanisms integrated to or paired with DMF were introduced, and the advancements made in DMF for lab-on-a-chip (LOC) applications (specifically for biochemical detection purposes) were discussed. This section includes the applications of DMF platforms involving biochemical sensing processes.

As explained in Sect. 2.2, detection on DMF devices can be categorized into off-line and on-line modes. This classification is based on whether the detection is performed off the chip (off-line) or using a detection system integrated into the chip and as a part of the on-chip processes (on-line) [1, 20]. The off-line detection mode has widely been used since the introduction of DMF, as the equipment used for biochemical sensing such as microscopy and mass spectrometry devices) already existed in laboratories while DMF has been used for sample preparation, partially or entirely [1]. On the other hand, the tendency towards developing microfluidic packages for performing the entire assay processes, including biochemical sensing on the chip, has resulted in the miniaturization and integration of sensing techniques into the chip for on-line detection. Here we go over the implementation of DMF devices for applications involving biochemical sensing.

4.1 *Off-Line Mode*

As discussed earlier, in off-line detection modes either the sample or the chip with the sample therein is taken to the detection equipment, which is mostly a microscope or mass spectrometer [1]. While other sensing mechanisms might have also been used for off-line detection, here the optical and mass spectrometry mechanisms are discussed due to their popularity.

4.1.1 Optical Detection

Optical systems are probably the most widely used systems for off-line detection on DMF [73]. Optical systems are categorized into fluorescence, absorbance, chemiluminescence, and surface plasmon resonance (SPR) systems which were explained in Sect. 2.2.1.

Several studies have used fluorescence microscopy as the sensing mechanism for performing biochemical assays on DMF [74]. For instance, one study has shown the possibility of performing polymerase chain reactions (PCR) on DMF [75], in which the sample transport and mixing was performed by EWOD actuation of the sample droplets. To perform the amplification of the DNA sequences, a PCR chamber was created on the chip using the hydrophilic/hydrophilic surface and generating surface tension gradient. Microheaters and temperature control units were also integrated into the chip to control the reactions. The platform was used for the amplification and detection of Dengue II gene. Fluorescent microscopy was used for the detection of the amplified samples and their results showed significantly reduced amount of the consumed buffers and the operation time of the assay using the DMF platform in comparison with the macro-scale counterpart. Another study has shown the application of DMF for proteolytic digestion assays [70]. In this study, an open DMF system was used with the removable skins as the dielectric and hydrophilic layer. The dried spots of the digestive enzymes were formed on the chip surface. These so called "skin depots" were used for the protein digestion assay. Fluorescence measurements were performed for the evaluation of the shelf life of these skin depots when stored at −80 °C. The results showed that the activity of the dried enzymes can be retained with a high performance over a period of one month. Single cell analysis using DMF was shown in [76] in which a parallel plate DMF device was used for the manipulation of the cell droplets and isolating single cells, followed by cytotoxicity analysis. An array of hydrophilic microwells was formed on the chip surface where after transporting the droplet over the array, single yeast cells (*Saccharomyces cerevisiae*) were isolated. The effect of Amphotericin B agent was studied on the membrane integrity of the isolated cells by transporting the droplet of the agent over the array of isolated cells. This effect was monitored using time-lapse fluorescence microscopy. Figure 16 illustrates the schematic of the sample preparation, on-chip assay processes, and fluorescence microscopy for this cytotoxicity analysis.

The application of DMF coupled with off-line absorbance and chemiluminescence measurement techniques is shown for a magnetic bead based heterogeneous immunoassay on human insulin and interleukin-6 (IL-6) [77]. The steps required for the immunoassay including concentrating beads, washing steps, bead retention, and resuspension of beads were performed using a parallel plate DMF device. The collection of the magnetic beads and washing steps were carried out by the aid of an external permanent magnet. The efficiency of the washing protocol was monitored and optimized by absorbance measurement using a plate reader. Based on an 8 min enzymatic reaction and the results of the traditional washing method, it was found that five cycles of washing process are required to achieve similar absorbance signals. Chemiluminescence detection was carried out for detection of antibody binding using a photo-multiplier tube and a multi-well-plate reader, and the kinetic curves were formed for both insulin and IL-6.

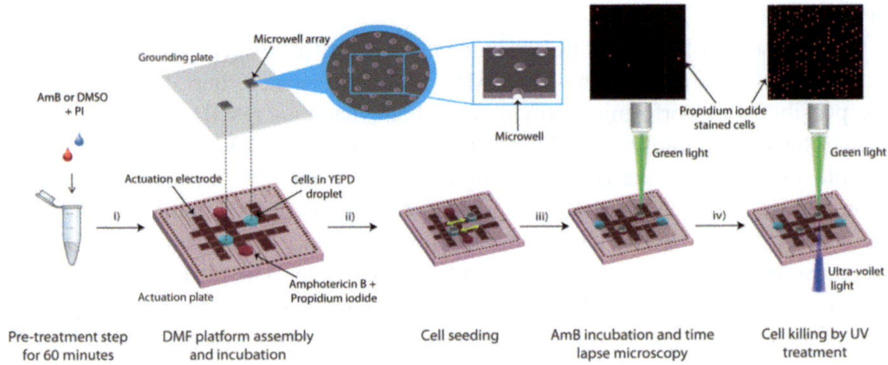

Fig. 16 Schematic of the sample preparation, on-chip assay processes, and fluorescence microscopy for the cytotoxicity analysis the yeast cells. Reproduced from [76] with permission from the Royal Society of Chemistry

4.1.2 Mass Spectrometry

As explained earlier, mass spectrometry is based on sorting the chemical species within the sample into a spectrum which depends on the mass-to-charge ratio of the species. This is carried out by ionizing the sample which results in the breakdown of the sample molecules into charged fragments. By passing these ions through an electric or magnetic field, the spectrum of the species within the sample is formed and the content is identified. Therefore, this method is used as an off-line method by coupling a bulky device with the DMF chip and transferring the sample from the chip to the spectrometer [78]. The study on proteolytic digestion assays, which was discussed earlier for fluorescence analysis of the shelf life of the dried enzyme spots, used mass spectrometry for the measurement of the enzymatic reaction products [70]. After sample preparation conducted by mixing several buffers and reagents, the final product was transported to the dried spot and incubated until the reaction was completed. Then, the sample was allowed to dry and finally the skin was peeled off of the device and processed using mass spectrometry.

Coupling the DMF sample processing unit with mass spectrometry has also been shown for the analysis of dried blood spots (DBS) from newborns [79]. The coupled platform is used for the study of succinylacetone (SA), which is a biomarker for hepatorenal tyrosinemia. The quantification of SA in DBS samples is performed using a nanoelectrospray ionization mass spectrometry (nESI-MS). The sample preparation is performed by actuating the droplets of buffers and reagents on the DBS. A glass capillary emitter which is sandwiched between the bottom and top plates of the DMF device is used for transferring the products from the DMF chip to nESI-MS device. Figure 17 shows the DMF device and the capillary emitter, as well as the schematic of the nESI-MS system coupled with the DMF device for the SA quantification assay.

4.2 On-Line Mode

Introduction of microfluidics (such as DMF systems) has been based on the desire to downscale biochemical assays, automate the processes, and develop point-of-care devices for the diagnosis purposes. As part of this miniaturization step, detection systems have been downscaled and integrated into the DMF devices to allow for the on-line detection of target analytes. Several challenges have been associated with the integration of these detection systems into DMF. These challenges are related to the geometrical configuration and the method of sample manipulation on the DMF devices. This section reports different detection technologies integrated into DMF devices for on-line detection purposes.

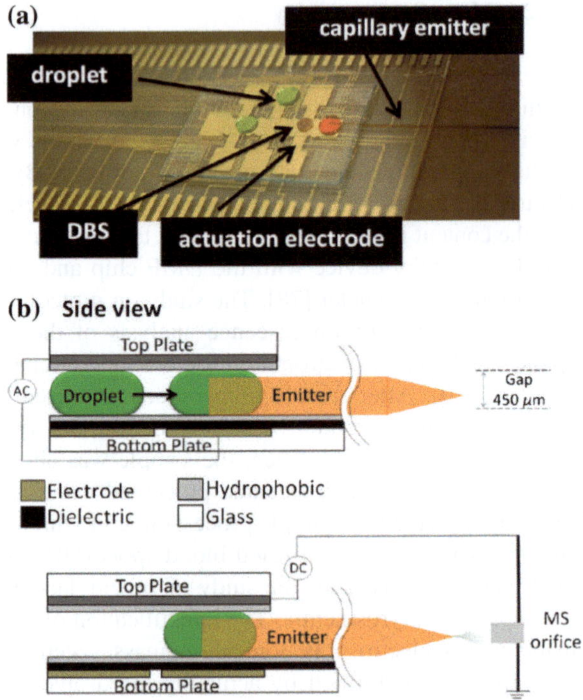

Fig. 17 **a** DMF device and the capillary emitter, and **b** the schematic of the nanoelectrospray ionization mass spectrometry system coupled with the DMF device for the analysis of DBS from newborns. Reproduced from [79] with permission from ACS Publications

4.2.1 Integrated Optical Detection Systems

Originally, the optical detection techniques were operated using bulky devices such as microscopes and plate readers. Development of miniaturized exciters and photodetectors has enabled the integration of the optical detection systems into microfluidic devices. In DMF devices, the integration of the majority of the optical-based detection systems requires at least one of the two plates to be transparent, which is achieved using ITO as the electrode [1].

On-line fluorescence detection has been shown for the magnetic bead based assays on a DMF platform [80]. The DMF device has been fabricated using a PCB which allows for easy assembly of the device. A magnet is used for collecting the beads for washing steps. A Peltie element and a thermal sensor were integrated into the chip for temperature control. The fluorescence detection is carried out using a CCD camera and two lasers, assembled to the device. The DMF platform and the integrated CCD is shown in Fig. 18. Another study shows the application of DMF with an integrated fluorescence measurement system for the enzymatic detection of r Fabry, Pompe, Gaucher, Hunter, and Hurler diseases [47]. Disposable cartridges were used as DMF chips, and the fluorometric enzymatic assay was performed on newborn DBSs punches on the cartridge. Lysosomal enzymatic activities were

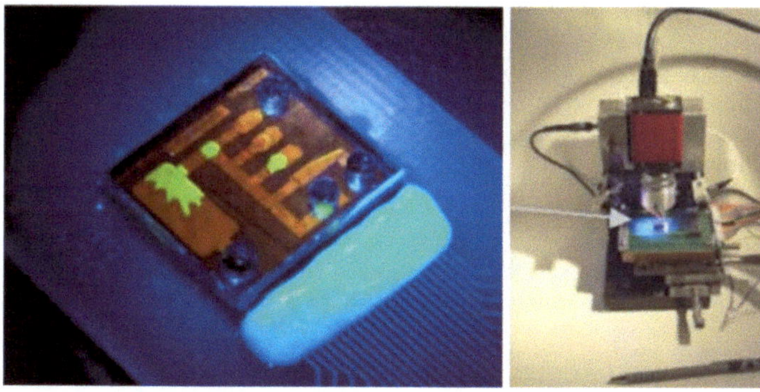

Fig. 18 The DMF device and the assembled CCD camera for an on-chip fluorescence measurement. Reproduced from [80] with permission from Springer

measured using the integrated fluorescence detection system. More than 98% linearity was shown for their range of measurements. In a different study, complementary-metal-oxide semiconductor (CMOS) integrated circuit was used for quantitative-polymerase-chain-reaction (qPCR) analyses with all required components integrated into the platform [23]. Temperature was controlled by the aid of resistive aluminum temperature sensors and polysilicon heaters. Geiger-mode single-photon avalanche diodes were integrated into the platform for the measurement of the fluorescence intensity. High voltage required for droplet actuation was also generated using an integrated circuit on the chip. Detection of *Staphylococcus aureus* was performed as a proof of concept. The integrated platform showed reliable and sensitive analysis of qPCR.

Integration of absorbance-based optical detection systems into DMF has been shown in multiple studies for online detection [21, 22]. Such a platform has been used for a colorimetric enzymatic glucose assay which was used for the analysis of body fluids (plasma, serum, urine, and saliva) [22]. The detection system consisted of an LED light source installed underneath the bottom plate, and a photodiode on the top plate to measure the intensity of the incident light. Thus, both bottom and top plates were fabricated using transparent electrodes (ITO) to allow for the optical connection between the light source and the detector.

Application of the on-line chemiluminescence detectors on DMF devices has been shown in several studies [49, 81, 82]. A portable compact DMF-based platform was developed with an open DMF system as the sample handling unit and a chemiluminescence detector for the analyte detection. It has been shown that ball-like droplet shape in the open DMF system focuses the fluorescence and improves the sensitivity of detection. It was also shown that such an improvement increases by increasing the contact angle of the droplet. Using this method, a detection limit of 10 μmol/L was found for H_2O_2-luminol (3-aminophthalhydrazide)-HRP. Figure 19 shows this open system DMF device with the integrated chemiluminescence detector. In another

Fig. 19 An open DMF platform integrated with a chemiluminescence detector. Reproduced from [49] with permission from the Royal Society of Chemistry

study, electrochemiluminescence (ECL) detectors were integrated onto the top plate of a parallel plate DMF device for microRNA analysis [82]. The light generated by electrically excited luminophores in the droplets of tris(phenanthroline)ruthenium(II) and tripropylamine (TPA) solutions was detected by the ECL detector. The integrated system was used for an oligonucleotide hybridization assay in which single nucleotide mismatches were detected. A detection limit of 1.5 femtomoles was found for the system.

Surface plasmon resonance (SPR) is highly sensitive, and due to its small size, integrated into DMF devices for on-line detection [28, 83, 84]. An example of such integration is shown in Fig. 20, where the sensing surface of the SPR detector is patterned on the top plate and in contact with the liquid sample. The measurement is performed using a laser source and the CCD camera, which emits and detects the laser beam through a prism placed or assembled on the top of the top plate [83]. As a proof of concept, such a platform has been used for the detection of different concentrations of NaCl in ultra-pure water. The same setup was used for the detection of DNA hybridization [84], where both DNA probe immobilization and the subsequent hybridization reactions were detected, as the immobilized probe density was actively controlled. It has been shown that a twofold increase in the SPR signal (as compared to passive probe immobilization) is achievable when immobilization of the probes is actively controlled using negative electrical potential. Using this integrated platform, a detection limit of 500 pM of DNA was obtained. In another study, the same device was further enhanced by patterning a

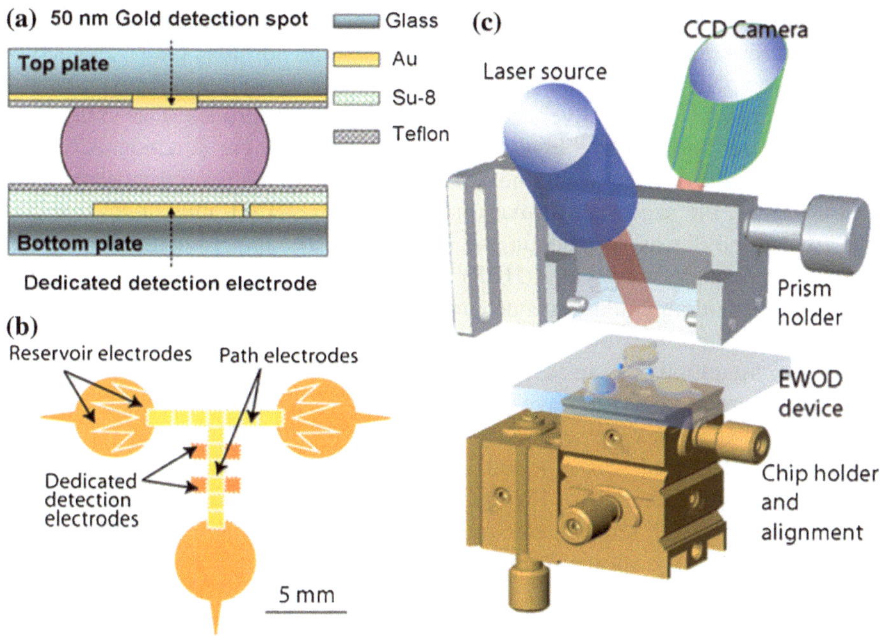

Fig. 20 Schematic of a DMF device integrated with a SPR detection system. **a** The components of the chip, **b** the design of the DMF platform, and **c** the assembly of the SPR with the DMF device. Reproduced from [83] with permission from the Royal Society of Chemistry

2D array of gold nanostructures. Colloidal-gold nanoparticles were used as the labels amplifying the SPR signal [28].

4.2.2 Integrated Electrochemical Detection Systems

Electrochemical detection systems are the most suitable detectors to be integration into DMF platforms, as they can easily be patterned and functionalized on the top plate of the DMF devices [1, 20]. Several studies have shown successful applications of these devices for biochemical sensing on DMF platforms [33, 85, 86]. The study which was previously shown in Fig. 7 illustrates an example of such an integration, in which the bottom plate contains the actuating electrodes for sample manipulation and the electrochemical sensor, along with the DMF ground electrode, is located on the top plate [33]. Through a series of cyclic voltammetry, an analysis of ferrocenemethanol (FcM) and dopamine (DA) was performed, illustrating the high compatibility and ease of integration of such a detector. A linear calibration curve was obtained for DA for a wide range of concentrations, Also, a detection limit of 420 nM was obtained for this detection system.

Recently, a DMF platform including an electrochemical sensor has been developed for selective detection of thyroid stimulating hormone (TSH) biomarkers [85]. The top plate includes the ITO ground electrode, as well as the patterned gold working electrode and the silver counter and reference electrodes. This assay was a magnetic bead based assay in which the microbeads were functionalized with primary antibodies for capturing the TSH antigens. Through several incubation and washing steps, the sandwich format (primary antibody-TSH antigen-enzyme conjugated antibody) was formed on the beads. For the amperometric sensing the horseradish peroxidase enzymes (HRP) were used which in reaction with 3,3′,5,5′-tetramethylbenzidine could generate proper oxidation-reduction. A limit of detection of 2.4 μIU/mL was found for the this integrated system. In a different study, a similar configuration was used for the detection of rubella virus (RV) IgG; the electrodes were modified to have a nanostructured surface. It was shown that the nanostructure increases the sensitivity of the electrochemical sensing significantly. The limit of detection was reported as 0.07 μIU/mL, which is a very low concentration for the detection of RV IgG. The schematic of the DMF platform, the three electrochemical sensing electrodes, and the nanostructured surface of the sensing electrodes are shown in Fig. 21.

4.2.3 Other Integrated Systems

Although majority of the detection systems integrated into DMF devices are optical and electrochemical based, some studies have used other detection methods for biosensing on DMF devices. For instance, a digital microfluidic impedance sensing platform was developed for the detection of Interleukin-6 through an ELISA assay [87]. A detection limit of 50 pM was reported for the integrated system. Another study has shown the integration of a capacitive biosensor into a DMF platform for the detection of *Cryptosporidium* oocysts in water samples [35]. Samples with different concentrations of cells were tested, and a calibration curve was obtained for the detection system. Integration of field-effect-transistor based biosensors into DMF devices have also been shown for the detection of avian influenza antibodies [34]. This study is based on the integration of the biosensor within the actuating electrodes on an open DMF device, as shown in Fig. 22. Although the fabrication of the device is very complicated (involving multi-layer patterning of the electrodes, dielectric separators, and semi-conductive material) the developed device showed a very high sensitivity with a limit of detection of 6.67 fM.

4.3 Post-sensing Sample Removal

Integration of different types of biochemical detection systems into DMF devices was shown in the previous section. Significant improvements in detection of proteins and cells were shown by the combination of the DMF sample preparation unit

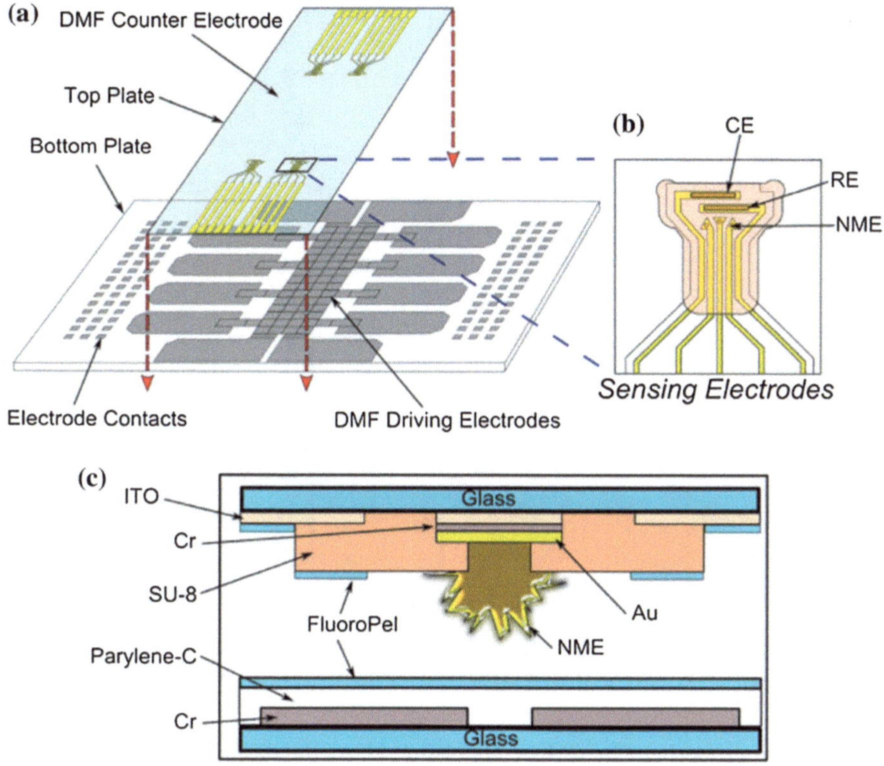

Fig. 21 The schematic of the DMF platform, the three electrochemical sensing electrodes, and the nano-structured surface of the sensing electrodes used for the detection of rubella virus IgG. Reproduced from [86] with permission from the Royal Society of Chemistry

and the integrated sensing system. However, one important issue is yet to be addressed, which is the complete removal of the sample from the sensing zone. Detection in many optical-based systems (e.g., fluorescence and absorbance) occurs through the liquid and such systems do not include a sensing surface. On the other hand, many other systems such as SPR, electrochemical, impedance, and FET systems include a sensing surface which consists of the transducer surface, and the immobilized receptors in many cases. The sensing surfaces are exposed to the sample through a window in which the hydrophobic layer is removed. In the case of the detection systems without the immobilized receptors, such as electrochemical systems, the sensor can be reused multiple times. The reusability of the systems with immobilized receptors is limited to biochemical regeneration of the sensing surface. However, incubation of the target analyte, washing cycles and sensing might require multiple steps of transportation of the buffer to and removing from the detector. The sensing surfaces have hydrophilic properties which hinder the transport of droplets on the DMF devices [35]. Therefore, if the sensing surface is

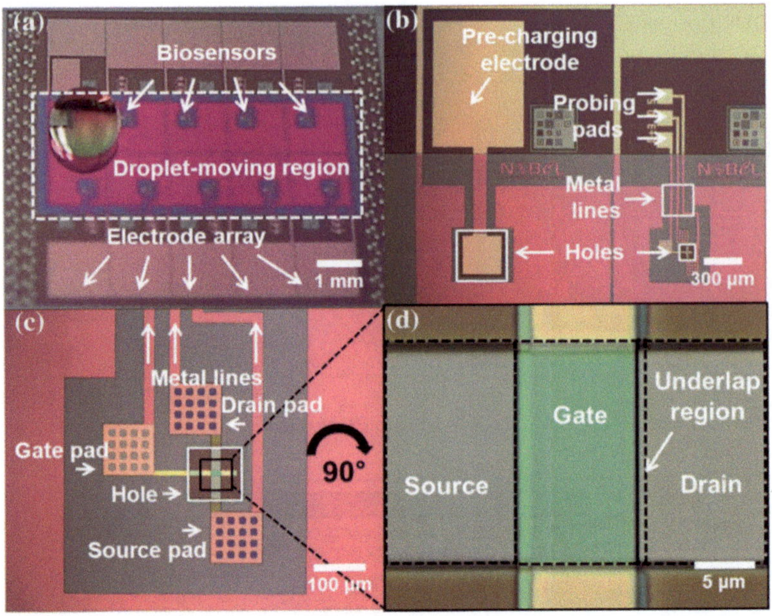

Fig. 22 An open DMF device with integrated FET biosensors for the detection of avian influenza antibodies. Reproduced from [34] with permission from the Royal Society of Chemistry

not designed properly or its surface area is too large, the droplet cannot be removed from the sensing unit after the detection process. In some studies the sensing window is designed very small to allow for the removal of the droplet [33, 84]. Although these studies have shown successful detection with the designed sensing area, it is desired to find the largest possible sensing area for the detection system without the limitation on droplet removal. A systematic study performed in [35] shows the optimized configuration and dimension of the rectangular hydrophilic and super-hydrophilic surfaces (HS) sensing area on which EWOD-based droplet removal from the detector is feasible. In this work, the minimum required voltage for removal of the droplet was studied for different surface wettability conditions, aspect ratios of the HS, and gap height of the parallel plate DMF device. Figure 23a shows examples of successful droplet removal from a gold HS surface. This study suggests the smaller the HS and the higher the aspect ratios (longer and lower width) the smaller the required applied voltage for droplet removal (Fig. 23b). Also, it has been shown that for super-hydrophilic surfaces, if the gap height is less than a certain threshold (e.g., 11% of the length of the electrode), the droplet will split on the HS, hindering complete droplet removal. They reported successful droplet removal from super-hydrophilic HS (e.g., gold with immobilized antibodies) as large as 13.3% and normal hydrophilic surfaces (e.g., sputtered gold) over 22% of the area of the actuating electrodes.

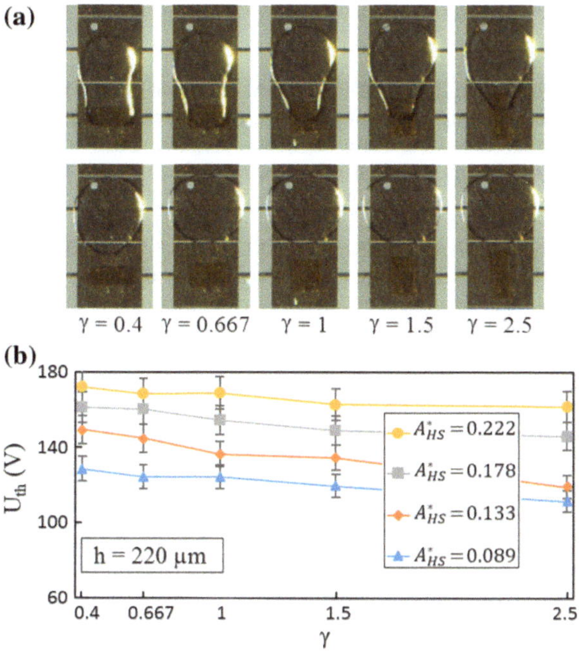

Fig. 23 **a** EWOD droplet removal from gold HS with different aspect ratios, and **b** the minimum required voltage for droplet removal for different sizes and aspect ratios of HS. Reproduced from [35] with permission from Elsevier

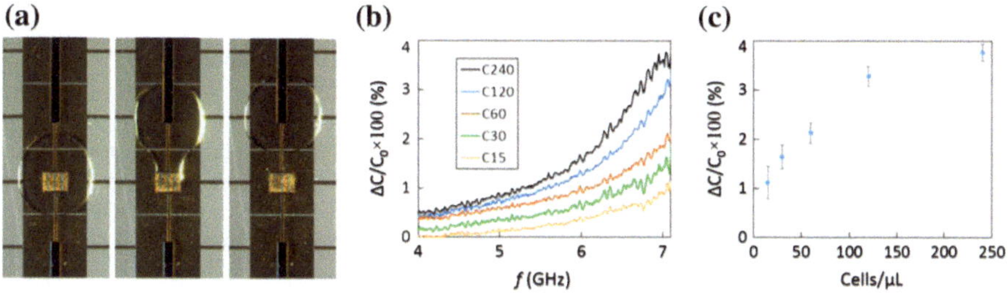

Fig. 24 A capacitive biosensor integrated into a DMF platform for the detection of different concentrations of *Cryptosporidium* oocysts. Reproduced from [35] with permission from Elsevier

Based on the optimum geometry found in the above study [35], a capacitive biosensor was developed and integrated into a DMF platform for detecting different concentrations of *Cryptosporidium* (as explained in the previous section). Such a configuration allowed for the detection of *Cryptosporidium* and complete droplet removal, as shown in Fig. 24.

References

1. Samiei E, Tabrizian M, Hoorfar M (2016) A review of digital microfluidics as portable platforms for lab-on a-chip applications. Lab Chip 16(13):2376–2396
2. J. Z. Chen, S. M. Troian, A. A. Darhuber, and S. Wagner, "Effect of contact angle hysteresis on thermocapillary droplet actuation," *J. Appl. Phys.*, vol. 97, no. 1, 2005
3. Darhuber AA, Valentino JP, Troian SM, Wagner S (2003) Thermocapillary actuation of droplets on chemically patterned surfaces by programmable microheater arrays. J. Microelectromechanical Syst. 12(6):873–879
4. Lehmann U, Hadjidj S, Parashar VK, Vandevyver C, Rida A, Gijs MAM (2006) Two-dimensional magnetic manipulation of microdroplets on a chip as a platform for bioanalytical applications. Sensors Actuators, B Chem. 117(2):457–463
5. Guo ZG, Zhou F, Hao JC, Liang YM, Liu WM, Huck WTS (2006) 'Stick and slide' ferrofluidic droplets on superhydrophobic surfaces. Appl Phys Lett 89(8):1–4
6. Renaudin A, Tabourier P, Camart JC, Druon C (2006) Surface acoustic wave two-dimensional transport and location of microdroplets using echo signal. J Appl Phys 100(11):1–4
7. Guttenberg Z et al (2005) Planar chip device for PCR and hybridization with surface acoustic wave pump. Lab Chip 5(3):308–317
8. Hunt TP, Issadore D, Westervelt RM (2008) Integrated circuit/microfluidic chip to programmably trap and move cells and droplets with dielectrophoresis. Lab Chip 8(1):81–87
9. Velev OD, Prevo BG, Bhatt KH (2003) On-chip manipulation of free droplets. Nature 426 (6966):515–516
10. Pollack MG, Fair RB, Shenderov AD (2000) Electrowetting-based actuation of liquid droplets for microfluidic applications. Appl Phys Lett 77(82):1725–1726
11. Fair RB (2007) Digital microfluidics: Is a true lab-on-a-chip possible? Microfluid. Nanofluidics 3(3):245–281
12. Pei SN, Valley JK, Neale SL, Jamshidi A, Hsu HL, Wu MC (2010) Light-actuated digital microfluidics for large-scale, parallel manipulation of arbitrarily sized droplets. In: Proceedings of IEEE international conference on mMicro electro mechanical systems, pp 252–255
13. Chiou PY, Moon H, Toshiyoshi H, Kim CJ, Wu MC (2003) Light actuation of liquid by optoelectrowetting. Sensors Actuators, A Phys. 104(3):222–228
14. Mugele F, Baret J-C (2005) Electrowetting: from basics to applications. J Phys: Condens Matter 17(28):R705–R774
15. Berthier J (2008) Microdrops and digital microfluidics
16. Berthier J et al (2005) An analytical model for the prediction of microdrop extraction and splitting in digital microfluidics systems. In: 2005 NSTI Nanotechnology conference trade show—NSTI nanotechnology 2005 technical proceedings, vol 1, no c, pp 664–667
17. Kang KH (2002) How electrostatic fields change contact angle in electrowetting. Langmuir 18 (26):10318–10322
18. Jebrail MJ et al (2014) A solvent replenishment solution for managing evaporation of biochemical reactions in air-matrix digital microfluidics devices. Lab Chip 15(1):151–158
19. Yi U-C, Kim C-J (2006) Characterization of electrowetting actuation on addressable single-side coplanar electrodes. J Micromech Microeng 16 (16):2053–2059
20. Malic L, Brassard D, Veres T, Tabrizian M (2010) Integration and detection of biochemical assays in digital microfluidic LOC devices. Lab Chip 10(4):418–431
21. Srinivasan V, Pamula VK, Fair RB (2004) Droplet-based microfluidic lab-on-a-chip for glucose detection. Anal Chim Acta 507(1):145–150
22. Srinivasan V, Pamula VK, Fair RB (2004) An integrated digital microfluidic lab-on-a-chip for clinical diagnostics on human physiological fluids. Lab Chip 4(4):310–315
23. Norian H, Field RM, Kymissis I, Shepard KL (2014) An integrated CMOS quantitative-polymerase-chain-reaction lab-on-chip for point-of-care diagnostics. Lab Chip 14(20):4076–4084

24. Kühnemund M, Witters D, Nilsson M, Lammertyn J (2014) Circle-to-circle amplification on a digital microfluidic chip for amplified single molecule detection. Lab Chip 14(16):2983–2992
25. Sista R et al (2008) Development of a digital microfluidic platform for point of care testing. Lab Chip 8(12):2091
26. Luan L, Evans RD, Jokerst NM, Fair RB (2008) Integrated optical sensor in a digital microfluidic platform. IEEE Sens J 8(5):628–635
27. Miller EM, Ng AHC, Uddayasankar U, Wheeler AR (2011) A digital microfluidic approach to heterogeneous immunoassays. Anal Bioanal Chem 399(1):337–345
28. Malic L, Veres T, Tabrizian M (2011) Nanostructured digital microfluidics for enhanced surface plasmon resonance imaging. Biosens Bioelectron 26(5):2053–2059
29. Sarkar PK, Prajapati PK, Shukla VJ, Ravishankar B, Choudhary AK (2015) Advances in coupling microfluidic chips to mass spectrometry. Mass Spectrom Rev 34:535–557
30. Hu J-B, Chen T-R, Chang C-H, Cheng J-Y, Chen Y-C, Urban PL (2015) A compact 3D-printed interface for coupling open digital microchips with Venturi easy ambient sonic-spray ionization mass spectrometry. Analyst 140(5):1495–1501
31. Han KN, Li CA, Seong GH (2013) Microfluidic chips for immunoassays. Annu. Rev. Anal. Chem. 6:119–141
32. Jebrail MJ, Bartsch MS, Patel KD (2012) Digital microfluidics: a versatile tool for applications in chemistry, biology and medicine. Lab Chip 12(14):2452–2463
33. Yu Y, Chen J, Zhou J (2014) Parallel-plate lab-on-a-chip based on digital microfluidics for on-chip electrochemical analysis. J Micromech Microeng 24(1):15020
34. Choi K, Kim J-Y, Ahn J-H, Choi J-M, Im M, Choi Y-K (2012) Integration of field effect transistor-based biosensors with a digital microfluidic device for a lab-on-a-chip application. Lab Chip 12(8):1533–1539
35. Samiei E, Luka GS, Najjaran H, Hoorfar M (2016) Integration of biosensors into digital microfluidics: impact of hydrophilic surface of biosensors on droplet manipulation. Biosens Bioelectron 81:480–486
36. Nejad HR, Samiei E, Ahmadi A, Hoorfar M (2015) Gravity-driven hydrodynamic particle separation in digital microfluidic systems. RSC Adv 5(45):35966–35975
37. Zhou Y, Li B, Wang M, Yang Z, Yin H, Ai S (2014) Enzyme-based electrochemical biosensor for sensitive detection of DNA demethylation and the activity of DNA demethylase. Anal Chim Acta 840:28–32
38. Amine A, Arduini F, Moscone D, Palleschi G (2016) Recent advances in biosensors based on enzyme inhibition. Biosens Bioelectron 76:180–194
39. Lin H, Zhang W, Jia S, Guan Z, Yang CJ, Zhu Z (2014) Microfluidic approaches to rapid and efficient aptamer selection. Biomicrofluidics 8(4):41501
40. Jia X, Dong S, Wang E (2016) Engineering the bioelectrochemical interface using functional nanomaterials and microchip technique toward sensitive and portable electrochemical biosensors. Biosens Bioelectron 76:80–90
41. Vergauwe N et al (2011) A versatile electrowetting-based digital microfluidic platform for quantitative homogeneous and heterogeneous bio-assays. J Micromech Microeng 21:54026
42. Samiei E, Hoorfar M (2015) Systematic analysis of geometrical based unequal droplet splitting in digital microfluidics. J Micromech Microeng 25:55008
43. Lin Y, Welch ERF, Fair RB (2012) Low voltage picoliter droplet manipulation utilizing electrowetting-on-dielectric platforms. Sens Actuators B Chem 173:338–345
44. Lin Y, Evans RD, Welch E, Hsu B, Madison AC, Fair RB (2010) Low voltage electrowetting-on-dielectric platform using multi-layer insulators. Sens Actuators B Chem 150(1):465–470
45. Fobel R, Kirby AE, Ng AHC, Farnood RR, Wheeler AR (2014) Paper microfluidics goes digital 2838–2843
46. Gong J, Kim C (2008) Direct-referencing two-dimensional-array digital microfluidics using multilayer printed circuit board. J Microelectromech Syst 17(2):257–264
47. Sista RS et al (2013) Multiplex newborn screening for Pompe, Fabry, Hunter, Gaucher, and Hurler diseases using a digital microfluidic platform. Clin Chim Acta 424:12–18

48. Mohamed Y, Saurabh S, Homayoun N (2015) Fabrication of digital microfluidic devices on flexible paper-based and rigid substrates via screen printing. J Micromech Microeng 25(5):57001
49. Zeng X et al (2013) Chemiluminescence detector based on a single planar transparent digital microfluidic device. Lab Chip 13(14):2714–2720
50. Ko H et al (2014) Active digital microfluidic paper chips with inkjet-printed patterned electrodes. Adv Mater 26(15):2335–2340
51. Moon H, Cho SK, Garrell RL, Kim CJ (2002) Low voltage electrowetting-on-dielectric. J Appl Phys 92(7):4080–4087
52. Cho SK, Moon H, Kim CJ (2003) Creating, transporting, cutting, and merging liquid droplets by electrowetting-based actuation for digital microfluidic circuits. J. Microelectromech Syst 12(1):70–80
53. Foudeh AM, Brassard D, Tabrizian M, Veres T (2015) Rapid and multiplex detection of Legionella's RNA using digital microfluidics. Lab Chip 15(6):1609–1618
54. Li Y et al (2008) Anodic Ta2O5 for CMOS compatible low voltage electrowetting-on-dielectric device fabrication. Solid State Electron 52(9):1382–1387
55. Dong C et al (2015) On the droplet velocity and electrode lifetime of digital microfluidics: voltage actuation techniques and comparison. Microfluid Nanofluid 18(4):673–683
56. Brassard D, Malic L, Normandin F, Veres T, Tabrizian M (2008) Water-oil core-shell droplets for electrowetting-based digital microfluidic devices. Lab Chip 8(8):1342–1349
57. Nestor BA et al (2016) Digital microfluidic platform for dielectrophoretic patterning of cells encapsulated in hydrogel droplets. RSC Adv 6(62):57409–57416
58. Gong J, Kim C-JCJ (2008) All-electronic droplet generation on-chip with real-time feedback control for EWOD digital microfluidics. Lab Chip 8(6):898–906
59. Liu Y, Banerjee A, Papautsky I (2014) Precise droplet volume measurement and electrode-based volume metering in digital microfluidics. Microfluid. Nanofluid 17(2): 295–303
60. Paik P, Pamula VK, Fair RB (2003) Rapid droplet mixers for digital microfluidic systems. Lab Chip 3(4):253–259
61. Samiei E, Diaz de Leon Derby M, van den Berg A, Hoorfar M (2016) An electrohydrodynamic technique for rapid mixing in stationary droplets on digital microfluidic platforms. Lab Chip
62. Tsaloglou MN, Jacobs A, Morgan H (2014) A fluorogenic heterogeneous immunoassay for cardiac muscle troponin cTnI on a digital microfluidic device. Anal Bioanal Chem 406(24):5967–5976
63. Ng AHC, Lee M, Choi K, Fischer AT, Robinson JM, Wheeler AR (2015) Digital microfluidic platform for the detection of rubella infection and immunity: A proof of concept. Clin Chem 61(2):420–429
64. Samiei E, Rezaei Nejad H, Hoorfar M (2015) A dielectrophoretic-gravity driven particle focusing technique for digital microfluidic systems. Appl Phys Lett 106(20):204101
65. Li Z, Ho TY, Chakrabarty K (2014) Optimization of heaters in a digital microfluidic biochip for the polymerase chain reaction. In: 2014—20th International Workshop on Thermal. Investigation of ICs and Systems Proceedings (THERMINIC), vol. 2014, pp. 1–5
66. Krishnan S, Weinman CJ, Ober CK (2008) Advances in polymers for anti-biofouling surfaces. J Mater Chem 18:3405
67. Patel P, Choi CK, Meng DD (2010) Superhydrophilic Surfaces for Antifogging and Antifouling Microfluidic Devices. JALA J Assoc Lab Autom 15(2):114–119
68. Yoon JY, Garrell RL (2003) Preventing biomolecular adsorption in electrowetting-based biofluidic chips. Anal Chem 75(19):5097–5102
69. Luk VN, Mo GC, Wheeler AR (2008) Pluronic additives: a solution to sticky problems in digital microfluidics. Langmuir 24(12):6382–6389
70. Yang H, Luk VN, Abelgawad M, Barbulovic-nad I, Wheeler AR (2009) A world-to-chip interface for digital microfluidics. Anal Chem 81(December 2008):1061–1067

71. P. Bayiati, A. Tserepi, P. S. Petrou, S. E. Kakabakos, K. Misiakos, and E. Gogolides, "Electrowetting on plasma-deposited fluorocarbon hydrophobic films for biofluid transport in microfluidics," *J. Appl. Phys.*, vol. 101, no. 10, 2007
72. Prakash R, Papageorgiou DP, Papathanasiou AG, Kaler KVIS (2013) Dielectrophoretic liquid actuation on nano-textured super hydrophobic surfaces. Sens Actuators B Chem 182:351–361
73. Myers FB, Lee LP (2008) Innovations in optical microfluidic technologies for point-of-care diagnostics. Lab Chip 8(12):2015–2031
74. Bender BF, Aijian AP, Garrell RL (2016) Digital microfluidics for spheroid-based invasion assays. Lab Chip 16(8):1505–1513
75. Chang YH, Bin Lee G, Huang FC, Chen YY, Lin JL (2006) Integrated polymerase chain reaction chips utilizing digital microfluidics. Biomed Microdevices 8(3):215–225
76. Kumar PT et al (2015) Digital microfluidics for time-resolved cytotoxicity studies on single non-adherent yeast cells. Lab Chip 15(8):1852–1860
77. Sista RRS, Eckhardt AEA, Srinivasan V, Pollack MG, Palanki S, Pamula VK (2008) Heterogeneous immunoassays using magnetic beads on a digital microfluidic platform. Lab Chip 8(12):2188–2196
78. Kirby AE, Lafrenière NM, Seale B, Hendricks PI, Cooks RG, Wheeler AR (2014) Analysis on the go: quantitation of drugs of abuse in dried urine with digital microfluidics and miniature mass spectrometry. Anal Chem 86(12):6121–6129
79. Shih SCC et al (2012) Dried blood spot analysis by digital microfluidics coupled to nanoelectrospray ionization mass spectrometry. Anal Chem 84(8):3731–3738
80. Fouillet Y, Jary D, Chabrol C, Claustre P, Peponnet C (2008) Digital microfluidic design and optimization of classic and new fluidic functions for lab on a chip systems. Microfluid Nanofluid 4(3):159–165
81. Zeng Z, Zhang K, Wang W, Xu W, Zhou J (2016) Portable Electrowetting Digital Microfluidics Analysis Platform For Chemiluminescence Sensing. IEEE Sens J 16(11): 4531–4536
82. Shamsi MH, Choi K, Ng AHC, Dean Chamberlain M, Wheeler AR (2016) Electrochemiluminescence on digital microfluidics for microRNA analysis. Biosens Bioelectron 77: 845–852
83. Malic L, Veres T, Tabrizian M (2009) Two-dimensional droplet-based surface plasmon resonance imaging using electrowetting-on-dielectric microfluidics. Lab Chip 9(3):473–475
84. Malic L, Veres T, Tabrizian M (2009) Biochip functionalization using electrowetting-on-dielectric digital microfluidics for surface plasmon resonance imaging detection of DNA hybridization. Biosens Bioelectron 24(7):2218–2224
85. Shamsi MH, Choi K, Ng AHC, Wheeler AR (2014) A digital microfluidic electrochemical immunoassay. Lab Chip 14(3):547–554
86. Rackus DG et al (2015) A digital microfluidic device with integrated nanostructured microelectrodes for electrochemical immunoassays. Lab Chip 15(18):3776–3784
87. Mok J, Mindrinos MN, Davis RW, Javanmard M (2014) Digital microfluidic assay for protein detection. Proc Natl Acad Sci 111(6):2110–2115

Miniature Fluidic Microtissue Culturing Device for Rapid Biological Detection

Patrick M. Misun, Andreas Hierlemann and Olivier Frey

Abstract Microfluidics is becoming a technology of growing interest to build miniature culturing systems, capable of mimicking tissue functions and multi-tissue interactions in so-called "body-on-a-chip" applications while featuring integrated readout functionalities. This chapter presents a highly versatile, modular and scalable analytical platform technology, which combines microfluidic hanging-drop networks with multi-analyte biosensors for in situ monitoring of the metabolism of 3D microtissues. The microfluidic platform is based on the hanging-drop network technology, which has been designed for formation, cultivation, and analysis of fluidically interconnected organotypic spherical 3D microtissues that can be obtained from various different cell types. The sensor modules were designed as small glass plug-ins, which allow for convenient functionalization and calibration of the sensors and do not interfere with the microfluidic functions. They were placed in the ceiling substrate, from which the hanging drops that host the spheroid cultures were suspended. The detection of secreted lactate of single microtissue spheroids will be presented. Further, we will demonstrate that it is possible to monitor microtissue lactate secretion and glucose consumption in parallel.

1 Combining Microfluidic Systems, 3D Microtissue Culturing and Biosensor Systems

Microfluidics technology offers engineering of a variety of new methods and platforms for culturing and analyzing human cells and tissue structures [14, 32]. Tissue models that reproduce in vivo conditions as closely as possible are important for

P.M. Misun · A. Hierlemann · O. Frey
Bio Engineering Laboratory, Department of Biosystems Science and Engineering, ETH Zurich, Mattenstrasse 26, 4058 Basel, Switzerland

O. Frey (✉)
InSphero AG, Wagistrasse 27, 8952 Schlieren, Switzerland
e-mail: olivier.frey@insphero.com

S.-H. Oh et al. (eds.), *Miniature Fluidic Devices for Rapid Biological Detection*, Integrated Analytical Systems, https://doi.org/10.1007/978-3-319-64747-0_8

understanding organ-specific cell behavior for investigating diseases and for finding new compounds and therapies in the drug discovery process [16, 28, 42, 45].

Testing new compounds, and studying cellular responses and metabolism in preclinical trials by using monolayer cell systems in standard well plates has the advantage that one can rely on established techniques and laboratory equipment for cell handling and imaging at high throughput. Conventional 2D cell cultures, however, have limitations in mimicking functional living tissue, as the native microenvironment including cell-to-cell interaction, mechanical cues, and spatiotemporal biochemical gradients is not present [2, 4, 31]. Advancing cell culturing systems that potentially overcome these limitations is of great interest and gains more and more momentum in pharmaceutical industry and basic research [8].

Major efforts are made in developing 3D cell culture models that better reproduce the physiological tissue morphology and properties. A wide spectrum of 3D cell culture methods and materials are currently being developed [23, 33, 54]. Spheroids, for example, are a popular choice as scaffold-free 3D microtissues. Different types of spheroids can be formed through self-aggregation in hanging drops to mimic functional tissue of several organs, while they retain their specific characteristics and functions during an extended period of time [6, 25, 29, 38, 47, 56].

A second aspect of advanced culture systems is the integration of continuous perfusion capabilities and the possibility to integrate fluid-dynamic and mechano-dynamic cues to even better mimic the in vivo environment [60]. Perfusion, for example, can be modulated to provide tissue-specific environmental conditions [15, 46] and to simulate shear forces that are present at the interface of tissues in their fluidic environments [5, 35, 57]. Flexible substrates can be implemented to mimic the oscillatory mechanical stress in lung alveoli [26, 51]. Interconnected fluidic networks, finally, offer the possibility to combine different cell cultures or tissue types to realize "body-on-a-chip" configurations [27, 30, 36, 50, 52, 58, 61].

The third aspect relates to established analysis methods, which predominantly have been designed to meet the requirements of 2D cell cultures. Analysis methods and assays have to be adapted for 3D cell cultures and the corresponding platforms in order to fully capitalize on the advantages of 3D cell culture formats. Most metabolic processes are dynamic and occur within a few minutes. Thus, studying time-resolved responses of cell cultures upon environmental changes or upon defined compound dosage often requires continuous readout in order not to miss the occurrence of important events [34]. In well-based approaches, liquid handling is discrete, which entails a large risk that one may miss events. Moreover, conventional cell assays usually include relatively large sample volumes. Frequent sampling of the cell culture medium interrupts and disturbs the overall culturing process, increases the risk of contaminations, and may entail extensive dilution of important markers or metabolites in the process through the accompanying addition of new medium. As cell-to-medium volume ratios are comparably low, frequent sampling also entails the risk that metabolite concentrations fall below the limit of detection.

Microfluidics, in contrast, offer precise flow handling of small liquid volumes and accurate control over microenvironmental parameters that are important in analytical cell culture devices [14]. Additionally, microfabrication techniques offer multiplexing capabilities, and allow for straightforward integration of novel modules, such as pumping systems, actuators, and microsensors [12, 41]. Microsensor technologies have been developed since decades for a variety of applications. Their miniature size and versatile features, their high sensitivity and low detection limits enable monitoring of various analytes in cell and tissue culture setups at high temporal and spatial resolution.

Many multi-sensor systems were designed as probes, mostly comprising of electrochemical biosensors, and have been applied to conventional cell culture assays to monitor microphysiological conditions over time [13, 22, 24, 37]. Miniaturized versions have been used as scanning probes to monitor the glucose and lactate metabolism of single cells [10, 44]. Flow-through-type sensor systems have been devised to analyze cell culture conditions and metabolites in liquid samples downstream of cell cultures [1, 7, 11, 18, 40, 43, 49, 53]. Silicon-based sensor chips for multi-parameter online monitoring have been mounted in a perfused cell culture unit [9], and multiple sensing electrodes were directly incorporated into transparent microfluidic systems to measure glucose consumption of single cardiac cells [20], lactate production of various cell types [21] and for monitoring cancer cell metabolism in combination with an optical readout [59].

The combination and integration of microfluidics and microsensors with cell culturing units poses several challenges. Fabrication methods of the different components need to be compatible, and all target features and functions need to be preserved over the duration of an experiment; cell cultures need to stay viable and functional, microfluidic handling needs to be robust and precise, sensors need to meet the detection specifications including sensitivity and selectivity. The use of 3D cell cultures in microfluidic setups and the realization of multi-tissue configurations render this integration even more complex.

This chapter describes a technological approach to combine 3D microtissue spheroids, used as a 3D cell culture model, with a microfluidic perfusion system and the integration of biosensor-based online monitoring. For further details also refer to [39].

2 Microfluidic Hanging-Drop Networks

Scaffold-free 3D spherical microtissues can be formed by seeding a defined number of cells into hanging drops of a specific culture medium. Cells sediment by gravity force, self-aggregate and form a spherical microtissue at the liquid–air interface [29]. The liquid–air interface is an important component, as it prevents cell adhesion and promotes subsequent scaffold-free cell culturing and tissue maturation. A next step is to connect the hanging drops to each other by using miniature channels, which the yields hanging-drop networks (HDNs). HDNs combine isolated hanging drops into fully interconnected networks of hanging drops and enable controlled liquid flow between the hanging drops [17]. Adding perfusion substantially expands the experimental options for 3D spherical microtissues by adding the features of continuous medium exchange, application of compound-dosage protocols, and the biochemical interaction of different microtissues through the liquid medium in multi-tissue or so-called "body-on-a-chip" setups.

2.1 Design

Hanging-drop networks are designed as completely open microfluidic systems at the bottom of an inverted, surface-patterned substrate. Figure 1a shows a basic five-drop-row configuration with a single fluidic inlet and a single fluidic outlet. The surface patterns guide the liquid by surface tension and capillary forces. Rim structures are used to distinguish wetted regions from dry regions and prevent the liquid from flowing over the whole surface in an uncontrolled way. The design of the rim structures defines, where drops are formed (Fig. 1b). Circular patterns induce the formation of hanging drops (cross-section 1), whereas narrow regions yield channel-like structures (cross-section 2). The dimensions have been optimized with regard to cell loading, drop and network stability, and perfusion characteristics [17]. Many different arrangements and configurations of hanging-drop networks can be realized through the variation of feature dimensions. An example is given in Fig. 1c depicting an array of four by four drops connected to a preceding

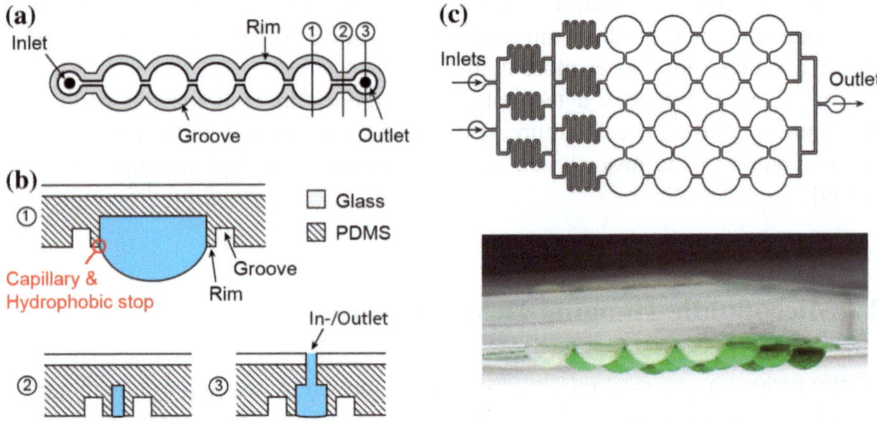

Fig. 1 Design of microfluidic hanging-drop networks. **a** A row of five interconnected hanging drops. Small rim structures confine the fluidic channels and drop structures. Hanging drops are formed at the circular regions interconnected through narrow channels. Inlet and outlet enable liquid perfusion. **b** Cross-sectional view through (1) hanging-drop structure, (2) channels, and (3) inlet and outlet position. The rim and groove act as phase guides. **c** Layout of a 4-by-4 hanging-drop network with integrated gradient generator. The photograph shows a side view of the chip during operation. Hanging drops of identical shapes are formed underneath the substrate. They were perfused with green food dye and a concentration gradient was formed across the array. Adapted from [17]

microfluidic gradient generator, which produces four different concentrations of a target compound by making use of integrated flow-splitting and -mixing structures.

Hanging-drop networks inherently fully exploit the benefits of the liquid–air interface, which include little or no cell adhesion and reduced compound adsorption on channel walls. Further, their open nature ensures gas exchange, prevents bubble formation and gives access to the liquid phase and the microtissues at every position in the network. The design is very versatile, so that a broad range of configurations can be designed.

2.2 *Fabrication*

The hanging-drop network structures are made of poly(dimethylsiloxane) (PDMS), casted from a microfabricated SU-8 mold (Fig. 2). The mold comprises of two layers of SU-8 that are successively spin-coated onto a 4 inch silicon wafer and processed using a standard photolithographic protocol (Fig. 2a–d). The two layers define the rim and the drop structures.

PDMS is prepared by mixing of elastomer and curing agent and pouring it onto the SU-8 mold (Fig. 2e). The cured ~3 mm thick PDMS replica is then peeled off the mold and cut into individual microfluidic chips (Fig. 2f). Holes are punched at predefined in- and outlet sites to enable fluidic connection via tubing. Microscopy

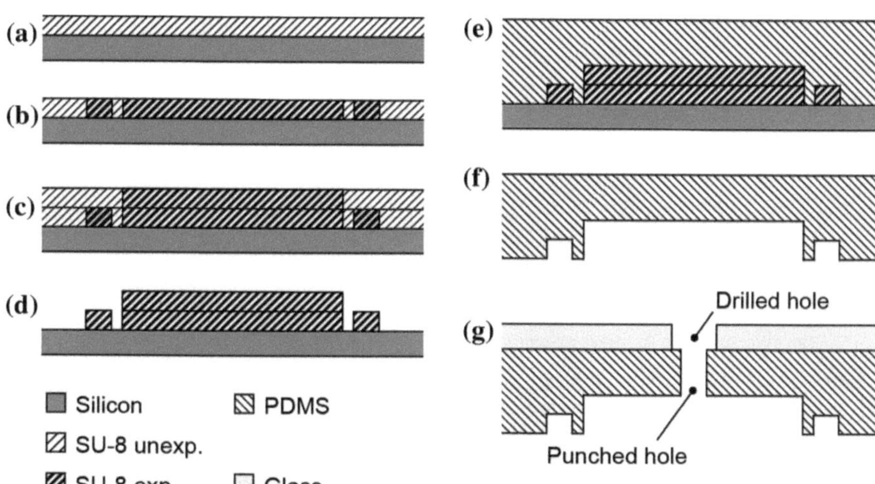

Fig. 2 Fabrication of the SU-8 mold and the microfluidic PDMS chip (schematic side view). **a** Silicon wafer with first layer of spin-coated SU-8 photoresist. **b** UV-exposed and cross-linked first SU-8 layer using a transparency mask. **c** Spin-coated and exposed second SU-8 layer. **d** Development of the non-cross-linked SU-8. **e** Casted and cured PDMS layer. **f** Cured PDMS layer removed from the SU-8 mold. **g** PDMS chip bonded to a glass substrate to make the chip more stable

slides are prepared with liquid access holes, and the PDMS chips are bonded to these slides after oxygen plasma surface activation. The glass slide improves planarity and mechanical stability of the system (Fig. 2g). The chips are placed in a custom-made chip holder and connected to conventional pumps via tubing.

3 Biosensor Integration

The biosensors are integrated into the hanging-drop networks in a modular approach. The final device consists of two parts: (i) the microfluidic PDMS chip and (ii) the small glass plug-ins featuring the biosensor electrodes (Fig. 3a). In this version, the hanging-drop network consisted of eight drops, arranged in two rows of four drops with three inlets and one outlet. The medium can be guided in parallel through the microfluidic structures allowing, for example, two different experimental conditions to be tested simultaneously.

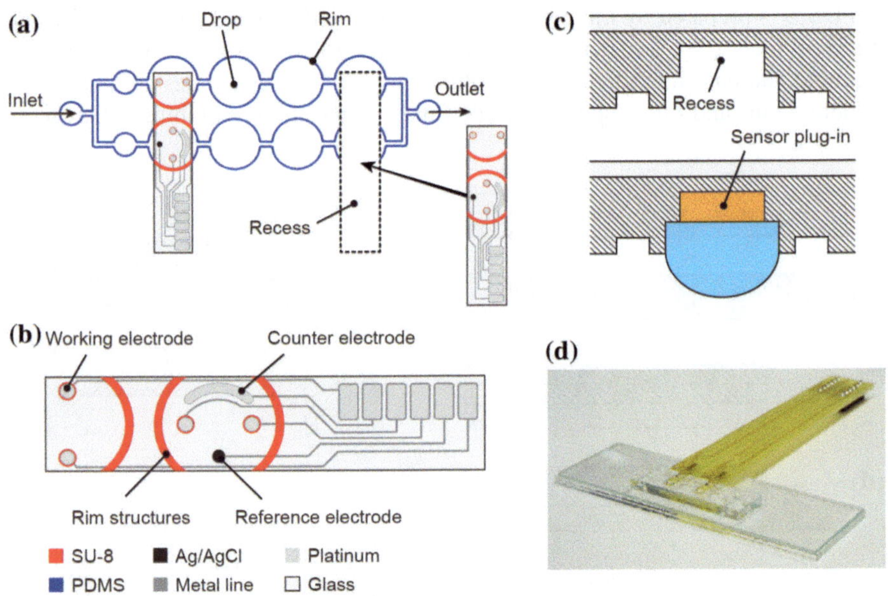

Fig. 3 Design and assembly of the hanging-drop-biosensor device. **a** Schematics of a 2-by-4 hanging-drop network showing the position and layout of the sensor plug-ins. The device consists of a microfluidic PDMS chip and two sensor glass plug-ins, which are inserted into a small recess. The PDMS rim (blue) and the SU-8 rim in the sensor plug-in (red) are complementary and close the fluidic network. **b** Layout of the sensor glass plug-in featuring four platinum working electrodes, one Ag/AgCl reference and a counter electrode. **c** Schematic cross-sectional view through a drop structure. The sensor is located at the ceiling of the hanging-drop structure. **d** Final assembly of the device. The sensor glass plug-in is connected to a PCB that facilitates the handling and provides an electric interface. Adapted from [39]

The sensor plug-ins extend over two drops and comprise six platinum electrodes and connection pads located on a glass substrate (Fig. 3b). For a seamless integration, the sensor units are inserted into a small rectangular recess in the PDMS chip, so that the electrodes are located at the ceiling of the hanging drop once the device is flipped upside down in the final assembly (Fig. 3c). The recess is produced with a third layer of the SU-8 mold.

Two circular working electrodes are integrated in each drop. Further, a counter electrode and an Ag/AgCl pseudo-reference electrode are integrated in one of the drops. SU-8 structures on the glass plug-in complete the rim patterns on the PDMS to obtain a leakage-free operation of the fluidic network. A completely assembled device is shown in Fig. 3d.

3.1 Fabrication

The sensor glass plug-ins are fabricated on a 4 inch glass wafer by using standard photolithography processes (Fig. 4). Thin-film platinum is sputtered on the glass and patterned in a lift-off process. A 500 nm thick Si_3N_4 passivation layer is then deposited onto the wafer using a plasma-enhanced chemical vapor deposition (PECVD) process. This passivation layer is re-opened at specified electrode and pad sites. Complementary rim structures and rings surrounding the electrodes on the sensor glass plug-in are fabricated by using SU-8 resist. The 20 μm SU-8 and the rim structures are fabricated with three layers of SU-8 according to the same procedure as described for the PDMS mold (Fig. 4f). At the end, the glass wafer is diced into single sensor glass plug-ins using a precision saw.

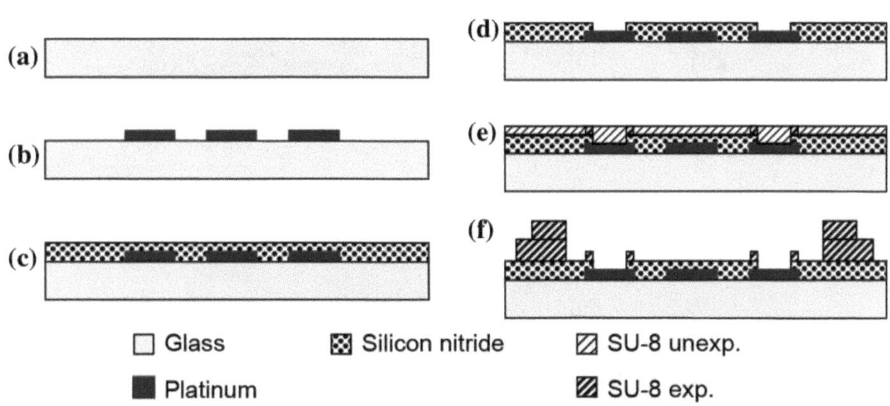

Fig. 4 Fabrication process of the sensor glass plug-in. **a, b** Platinum is deposited and patterned by using a lift-off process on a glass wafer by a photolithographic microfabrication processes. **c** A silicon nitride passivation layer is deposited onto the wafer surface for insulation. **d** The passivation layer is selectively removed at the electrode sites. **e** SU-8 is repeatedly spin-coated and UV exposed through a transparency mask in a three-layer process to structure the rim and liquid-phase guiding structures directly on the glass surface. **f** Final cross-section with developed SU-8

The individual glass plug-ins are glued onto a custom-made PCB, wire-bonded and packaged using epoxy glue. Connector pins were soldered to the PCB for electrical connections.

3.2 Assembly

For the final assembly of the device, sensor modules are inserted into the designated recesses on the microfluidic PDMS chip (Fig. 5a). Epoxy can be poured between the microscopy slide carrying the PDMS structures and the PCB to fix and stabilize the device.

Prior to each experiment, the assembled device is activated by an oxygen plasma treatment, while a thin PDMS mask with openings at the inlet, outlet, and all drop sites is used to cover all rim structures of the microfluidic hanging-drop device. This selective plasma treatment activates the inherently hydrophobic PDMS inside the circular drop structures and channel locations into a hydrophilic state. The covered rim structures remain hydrophobic and ensure that the liquid is confined.

The assembled device is clamped upside down onto a custom-made chip holder and placed into an OmniTray cultivation box, covered with a lid; wet cotton pads

(a)

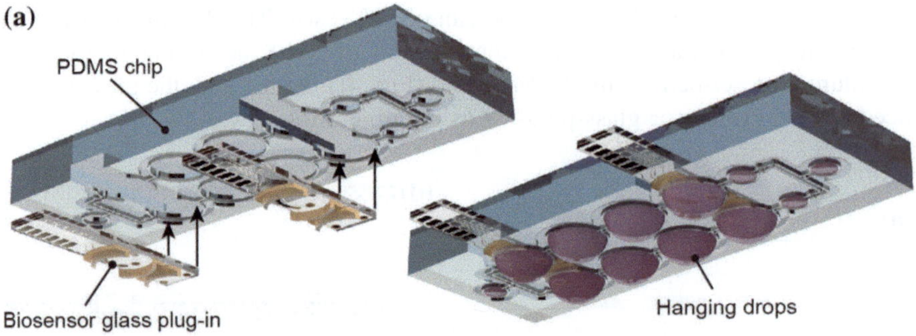

(b)

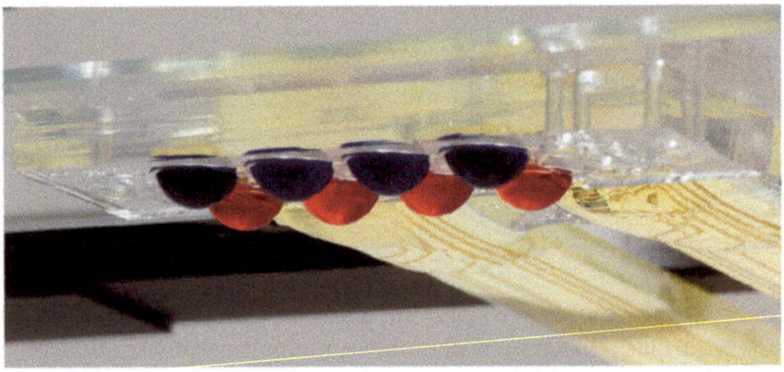

Fig. 5 Device assembly and operation. **a** Three-dimensional exploded view of the biosensor device. The sensor glass plug-ins are inserted into the recesses of the microfluidic PDMS chip. **b** Equally shaped hanging drops are formed directly underneath the patterned surface of the device. Two drop rows can be perfused with different media (indicated through different colors). Adapted from [39]

are included to provide additional humidity and to minimize evaporation. The sensor unit is electronically connected to an in-house fabricated CMOS-based multi-potentiostat for parallel current readout [48]. Precision syringe pumps are used for controlling the flow. The device is placed into an incubation box at controlled 37 °C, 5% CO_2, and 95% humidity for optimal culture conditions. Inlet tubing is connected to a heatable perfusion cannula so as to ensure constant temperature.

The device is loaded by pipetting 120 µl of liquid in one of the access holes, and eight hanging drops are formed underneath the circular regions (Fig. 5b). Each hanging drop comprises of a volume of ~10 µl, which includes the liquid in the cylindrical recess in the glass/PDMS substrate and the spherical drop fraction. The remaining liquid volume is in the channel and connection structures. The completely open PDMS chip architecture allows for easy access to the fluidic system and facilitates loading and harvesting of samples and spheroids during the experiments.

3.3 Electrode Preparation

An important advantage of the modular approach is that functionalization and calibration of the sensor can be carried out independently and before plugging the sensor unit into the microfluidic PDMS substrate.

All sensor units are initially sterilized with 70% ethanol, isopropanol and cleaned with oxygen plasma. The electrodes are tested by recording cyclic voltammograms in sulfuric acid solution. The reference electrode (RE) is coated with silver (Ag) through galvanostatic electrodeposition; the silver on the RE is subsequently partially transformed into AgCl. All four working electrodes (WEs) are coated with an m-polyphenylenediamine layer (mPPD) for rejection of potential electro-active interferant molecules. To improve the adhesion of the sensor hydrogel on the working electrodes, an additional (3-aminopropyl)triethoxysilane (APTES) layer is deposited through a vapor deposition process.

3.4 Biosensor Functionalization

A cross-section through the assembled microfluidic hanging-drop device and sensor unit is shown in Fig. 6a. The sensing electrodes allow for monitoring of the medium around the microtissues in the corresponding hanging drops. A simple drop-coating process is used to coat and functionalize the platinum working electrodes with a hydrogel containing glucose oxidase (GOx) or lactate oxidase (LOx). A schematic close-up of the hydrogel layer located directly on top of the sensing electrodes is shown in Fig. 6b. The enzymes catalyze an oxidation reaction of glucose or lactate that produces hydrogen peroxide, which, in turn, is

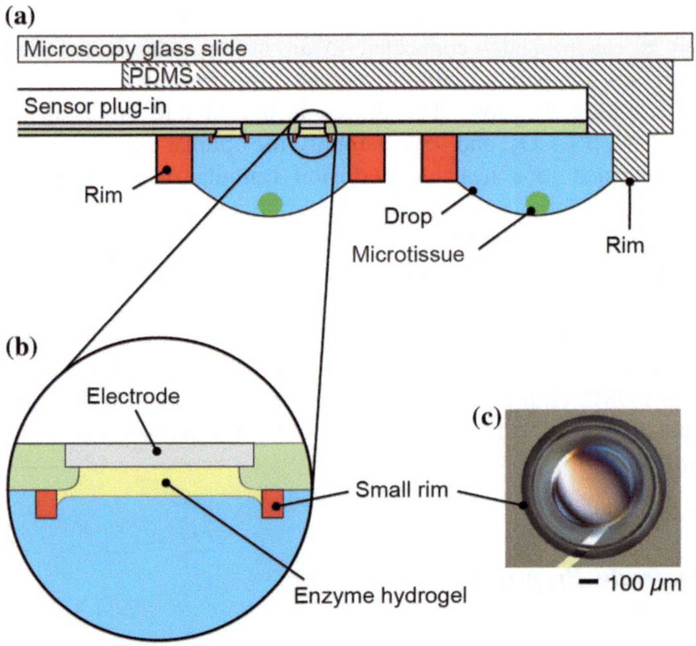

Fig. 6 Configuration and functionalization of the device. **a** Schematic cross-sectional view of the microfluidic biosensor device. Hanging drops form between the SU-8 rims on the glass plug-in and the microfluidic PDMS substrate. The biosensors are located at the ceilings of the hanging drops. **b** Close-up view showing the coating and functional layer of the electrodes. Enzymes (e.g. glucose oxidase or lactate oxidase) are immobilized in a hydrogel layer located directly on the electrodes. Analytes can be electrochemically detected on the electrode surface. **c** Photograph of a hydrogel-coated electrode showing the small SU-8 ring facilitating precise hydrogel deposition. Adapted from [39]

amperometrically detected on the Pt electrodes at a potential of 0.65 V versus Ag/AgCl. The current signal directly correlates with the analyte concentration of interest. A coated working electrode is shown in Fig. 6c. The enzyme-hydrogel is precisely located on the electrode area within the SU-8 ring structure. The ring structure facilitates the coating procedure and helps for achieving uniform hydrogel depositions on all electrodes.

The electrode coating is based on a co-cross-linking of the enzymes with bovine serum albumin (BSA) and glutaraldehyde (GA) [18]. Sub-microliter volumes of an aqueous solution containing one of the enzymes, BSA and GA are manually transferred onto the electrodes by touching the small rim structure surrounding the electrodes with a 2 μl pipette using appropriate tips. The hydrogel is formed directly on the electrodes by curing the solution at room temperature for at least 3 h before use.

3.5 Characterization of the Glucose and Lactate Biosensors

The biosensors were separately characterized through multiple calibration experiments “off-chip” and then after mounting into the microfluidic network “on-chip”. Raw data of an off-chip calibration of three functionalized LOx working electrodes and one blank BSA electrode in a conventional 50 ml beaker is shown in Fig. 7a. After a 20 min settling at the beginning of the recording, lactate was successively added to the RPMI 1640 medium to increase the concentration. A specific current response to lactate was recorded. The limit of detection (LoD) was calculated to be 7.07 ± 2.73 μM (three times the background noise at the end of the settling time) for the lactate biosensor during an off-chip calibration at 37 °C under continuous stirring.

The biosensors showed reproducible characteristics with regard to linearity, sensitivity, and reproducibility for both, glucose and lactate (Fig. 7b). A linear relationship was observed for up to 2 mM glucose after coating of the electrode with a single GOx membrane ($R^2 > 0.99$). This sensor features a relatively high sensitivity of 322 ± 41 nA mM^{-1} mm^{-2}. The biosensor characteristics could be tuned by implementing additional layers on top of the enzyme membrane. Adding, for example, a diffusion-limiting BSA layer on top of the enzyme-hydrogel membrane increased the linear range up to 7 mM while decreasing the sensitivity to 88 nA mM^{-1} mm^{-2}. Implementing a catalase membrane on top of the enzyme membrane further increased the linear range to 11 mM. The respective sensitivity was determined to 30 nA mM^{-1} mm^{-2} [39]. The lactate sensor, based on a simple

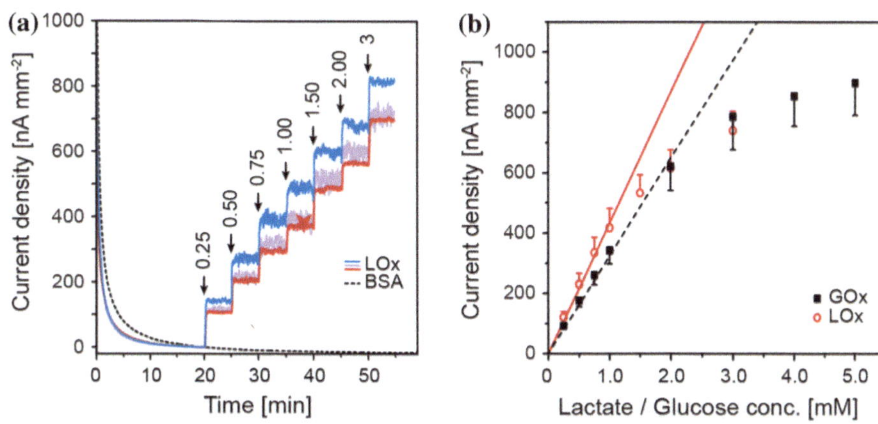

Fig. 7 Calibration and characterization of the glucose and lactate biosensors in RPMI 1640 medium at 37 °C. **a** Transient lactate current signals of three LOx- functionalized electrodes and one blank bovine serum albumin (BSA) electrode as a control. **b** Glucose and lactate calibration curves showing linear range and sensitivity of the biosensors. Adapted from [39]

LOx membrane coating on the electrodes, shows a linear response up to 1 mM and a sensitivity of 435 ± 66 nA mM^{-1} mm^{-2} in RPMI 1640 cell culture medium.

The biosensors were tested in the chip system in flow-through mode. Figure 8a displays simultaneously recorded raw signal traces of lactate biosensors and blank sensors during consecutive on-chip perfusion of 0, 0.25 and 0.50 mM lactate in PBS at a flow rate of 20 µl min^{-1} (10 µl min^{-1} in each row). The response of the biosensors was selective and reversible. The two lactate sensors showed slightly different sensitivities, which can be attributed to variations in the manual membrane deposition process. These variations demonstrate the importance of calibrations to be executed before any biological measurement. For this characterization experiment, several syringe pumps supplying the respective calibration solutions were connected to the device inlet via tubing. An artifact, which is caused by the change of the infusing syringe pump, indicates the change of the calibration solution. Tubing between syringes and microfluidic device produced a delayed sensor signal response. Based on the background noise, an LoD of 1.84 ± 0.34 µM was calculated for this calibration.

Figure 8b presents raw signal traces of a lactate and a glucose biosensor, combined in the same drop, in response to consecutive perfusion of PBS containing 1 mM glucose and 1 mM lactate at a constant flow rate of 5 µl min^{-1}. No sensor cross talk was observed in the continuous-flow mode.

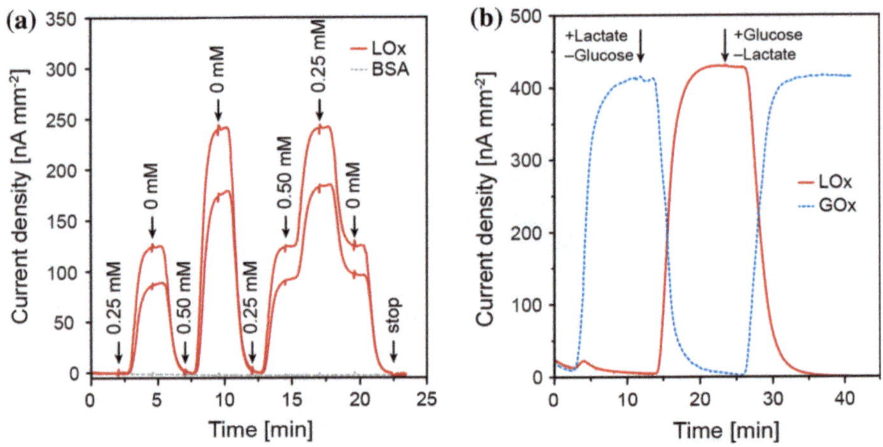

Fig. 8 Sensor on-chip response curves for lactate and glucose in PBS at 37 °C. **a** Lactate current signals recorded from two hanging drops in parallel. Electrodes were functionalized with lactate oxidase and BSA (blank electrode) in each of the two drops. Different concentrations of lactate were sequentially perfused. **b** On- chip lactate and glucose signals simultaneously recorded in PBS under an alternating analyte flow of glucose and lactate. Adapted from [39]

4 Real-Time Online Analyte Recording

4.1 Cell Culture and Microtissue Production

All cell-based experiments were carried out using the fluorescent human colon carcinoma cell line HCT116 eGFP. The cells were cultured in RPMI 1640 growth medium. Prior to an experiment, cells were harvested from the culture flasks for microtissue fabrication. The cell suspension was diluted with fresh medium containing all supplements. Cell concentration was adjusted with due regard to the experimental requirements and desired microtissue sizes.

GravityPLUS plates from InSphero AG, Switzerland, were used to form and grow microtissues from HCT116 eGFP cell suspensions inside hanging drops. After 3 days of incubation in the incubator, the microtissues were transferred to the GravityTRAP plate (InSphero) for maintenance and media exchange every 3 days.

4.2 Measurement of Analyte Secretion

Glucose and lactate biosensors were calibrated to determine the specific sensitivity per area (nA $mmol^{-1}$ L mm^{-2}) of each electrode. During the experiments, the current density (nA mm^{-2}) was constantly measured at a 10 Hz sampling rate. The corresponding analyte concentration was then calculated according to the previously determined sensitivity of the respective electrodes. This procedure enabled to continuously monitor the analyte concentration (mmol L^{-1}) in the hanging drops over time.

Continuous measurements, while perfusion was periodically switched on or off, were carried out, and the concentration change in the 10 μl hanging-drop compartment was recorded in real time (mmol L^{-1} h^{-1}). Secretion rates (mmol h^{-1}) were finally obtained by taking the volume of 10 μl into account.

A typical characteristic of cancerous tissue is an enhanced glucose uptake rate due to an increased glycolysis metabolism, through which pyruvate is fermented to lactate under aerobic conditions [19]. The performance of the device was assessed by measuring the metabolism of HCT116 eGFP cancer microtissues. Glucose uptake and lactate secretion were recorded in real time in the microfluidic hanging-drop network using the two developed biosensors.

Figure 9a presents measurements of lactate that was secreted from an HCT116 eGFP microtissue (480 μm in diameter) inserted in one of the sensor drops. The measurement was carried out in PBS, supplemented with 10 mM glucose, while perfusion was stopped. After the perfusion had been stopped, the lactate signal increased reproducibly to 73.8 ± 4.4 μmol l^{-1} as a result of lactate accumulation in the drop and then decreased again during the washing phase, when the drop liquid volume was exchanged with fresh PBS at a flow rate of 5 μl min^{-1}. A lactate secretion rate of 2.21 nmol h^{-1} was calculated. This value is comparable to lactate secretion rates of growing HCT116 microtissues found in the literature [55]. As a negative control, the microtissue was removed from the hanging drop after four perfusion cycles. No lactate was detected after removal, which confirmed the microtissue being the source of the lactate. Further, no signal that may originate from other electro-active species was recorded on the blank electrode. A three-point calibration was performed prior to microtissue loading and after the experiment to calculate lactate concentrations.

Figure 9b shows in situ real-time measurements of glucose consumption and simultaneous lactate production of four HCT116 eGFP microtissues of similar size (274 ± 9 μm), which were inserted in pairs into two sensing drops. One drop comprised a glucose sensor and a blank BSA electrode for glucose detection. In the other drop, the electrodes were functionalized as lactate sensor and blank BSA. The measurement was carried out by switching between PBS and glucose-supplemented PBS solution (0.5 mM) using an on-off-perfusion protocol and a flow rate of 5 μl min^{-1}. After microtissue insertion and the infusion of the glucose solution (clearly visible through the steep glucose signal increase) the microtissues were cultured for 60 min without flow. The glucose signal decreased from its initial value of 500 μM. The consumption rate was calculated to 2.57 nmol h^{-1} for two microtissues. At the same time, lactate was secreted at a rate of 2.40 nmol h^{-1}. Comparable values for lactate secretion [55] and glucose consumption [3, 55] were found in the literature. The blank BSA working electrodes showed little cross talk. This cross-talk signal originates from hydrogen peroxide diffusion in the same

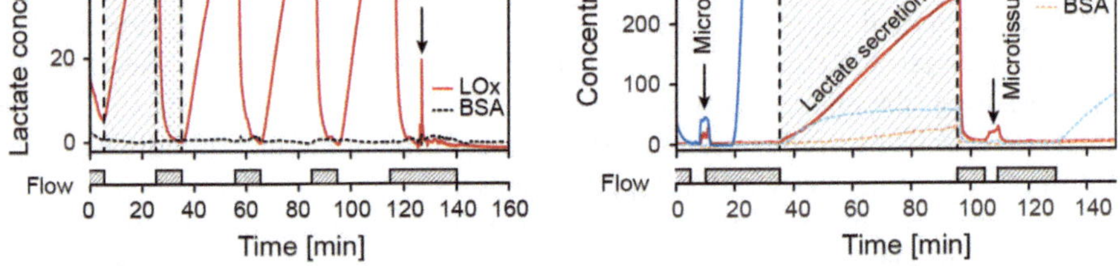

Fig. 9 Measuring the metabolism of human colon cancer microtissues (HCT116 eGFP) in the hanging-drop chip. The microtissues are cultured in glucose-enriched PBS at 37 °C. The measurements were carried out during an on–off medium perfusion protocol. **a** Lactate secretion was directly measured from one microtissue. **b** Real-time and simultaneous measurement of the glucose consumption and lactate secretion of cancer microtissues. Adapted from [39]

hanging-drop. The liquid was exchanged after 1 h restoring the initial sensor signal values for both sensor types. Microtissues were then removed from both sensing-drop sites. Perfusion was stopped again, and the conducted measurements confirmed that the recorded analyte transients indeed originated from the metabolism of the microtissues.

5 Conclusion

The modular approach of the analytical platform preserves all advantages of the hanging-drop network technology, including the adhesion-free liquid–air interface for spheroid cultures, a precise control of the liquid flow, bubble-free operation, inherent gas exchange, and full continuous access to the fluidic network, which enables loading and harvesting of liquid samples and spheroids for downstream analysis. Decoupling of the sensor unit from the microfluidic chip substantially reduces fabrication, assembly, and operation complexity of the integrated multi-functional platform. Critical procedures, such as biosensor functionalization and calibration, can be performed independently off-chip. Biosensors can, for example, be prepared, stored and plugged in just before use, when the microtissue structures are ready for the experiments. The developed functionalization procedure is simple, and its reproducibility has been improved through implementation of SU-8 electrode rims that support drop deposition.

The technological approach is highly versatile. Different versions of the microfluidic chip were fabricated, in which the biosensor modules could be plugged in at different locations in the hanging-drop network. The approach can be applied to more complex and larger hanging-drop networks including different spheroid types in multi-tissue configurations. Further, the readout capability of the system can be easily expanded through integration of additional plug-ins that comprise more and/or tissue-specific sensor types. The platform is very flexible in how sensors and hanging drops hosting the microtissues can be arranged.

Despite the fact that the presented approach allows for continuous sensor operation at sufficient sensitivity for approximately 1 day, biosensors always face the challenge of extended lifetime and good measurement reproducibility. We tried to address this challenge by using the plug-in approach: Sensors with satisfactory characteristics can be selected, pre- and post-calibration can be performed off-chip or by using an automated on-chip perfusion protocol during longer experiments.

Lactate and glucose metabolism of individual GFP-induced human colon carcinoma microtissues could be measured in situ inside a small, 10 µl volume by an automated perfusion setup so that any disturbance and influence of external effects could be effectively excluded. The obtained data provided real-time information in the time range of minutes on the metabolic state of the microtissues under different culture conditions.

Acknowledgements This work was financially supported by FP7 of the EU through the projects "Body on a chip", ICT-FET-296257 and the ERC Advanced Grant "NeuroCMOS" (contract 267351) as well as by an individual Ambizione Grant 142440 of the Swiss National Science Foundation for Olivier Frey.

References

1. Arquint P, Koudelka-Hep M, van der Schoot BH et al (1994) Micromachined analyzers on a silicon chip. Clin Chem 40:1805–1809
2. Asthana A, Kisaalita WS (2012) Microtissue size and hypoxia in HTS with 3D cultures. Drug Discov Today 17:810–817. https://doi.org/10.1016/j.drudis.2012.03.004
3. Aykin-Burns N, Ahmad IM, Zhu Y et al (2009) Increased levels of superoxide and H2O2 mediate the differential susceptibility of cancer cells versus normal cells to glucose deprivation. Biochem J 418:29–37. https://doi.org/10.1042/BJ20081258
4. Baharvand H, Hashemi SM, Kazemi Ashtiani S, Farrokhi A (2006) Differentiation of human embryonic stem cells into hepatocytes in 2D and 3D culture systems in vitro. Int J Dev Biol 50:645–652. https://doi.org/10.1387/ijdb.052072hb
5. Balcells M, Fernández Suárez M, Vázquez M, Edelman ER (2005) Cells in fluidic environments are sensitive to flow frequency. J Cell Physiol 204:329–335. https://doi.org/10.1002/jcp.20281
6. Beauchamp P, Moritz W, Kelm JM et al (2015) Development and characterization of a scaffold-free 3D spheroid model of induced pluripotent stem cell-derived human cardiomyocytes. Tissue Eng Part C Methods 21:852–861. https://doi.org/10.1089/ten.TEC.2014.0376
7. Boero C, Olivo J, De Micheli G, Carrara S (2012) New approaches for carbon nanotubes-based biosensors and their application to cell culture monitoring. IEEE Trans Biomed Circuits Syst 6:479–485. https://doi.org/10.1109/TBCAS.2012.2220137
8. Breslin S, O'Driscoll L (2013) Three-dimensional cell culture: the missing link in drug discovery. Drug Discov Today 18:240–249. https://doi.org/10.1016/j.drudis.2012.10.003
9. Brischwein M, Motrescu ER, Cabala E et al (2003) Functional cellular assays with multiparametric silicon sensor chips. Lab Chip 3:234–240. https://doi.org/10.1039/b308888j
10. Ciobanu M, Taylor DE, Wilburn JP, Cliffel DE (2008) Glucose and lactate biosensors for scanning electrochemical microscopy imaging of single live cells. Anal Chem 80:2717–2727. https://doi.org/10.1021/ac7021184
11. Dempsey E, Diamond D, Smyth MR et al (1997) Design and development of a miniaturised total chemical analysis system for on-line lactate and glucose monitoring in biological samples. Anal Chim Acta 346:341–349. https://doi.org/10.1016/S0003-2670(97)90075-1
12. Dittrich PS, Manz A (2006) Lab-on-a-chip: microfluidics in drug discovery. Nat Rev Drug Discov 5:210–218. https://doi.org/10.1038/nrd1985
13. Eklund SE, Taylor D, Kozlov E et al (2004) A microphysiometer for simultaneous measurement of changes in extracellular glucose, lactate, oxygen, and acidification rate. Anal Chem 76:519–527. https://doi.org/10.1021/ac034641z
14. El-Ali J, Sorger PK, Jensen KF (2006) Cells on chips. Nature 442:403–411. https://doi.org/10.1038/nature05063
15. Esch MB, Prot J-M, Wang YI et al (2015) Multi-cellular 3D human primary liver cell culture elevates metabolic activity under fluidic flow. Lab Chip 15:2269–2277. https://doi.org/10.1039/c5lc00237k

16. Esch MB, Smith AST, Prot J-M et al (2014) How multi-organ microdevices can help foster drug development. Adv Drug Deliv Rev 69–70:158–169. https://doi.org/10.1016/j.addr.2013.12.003
17. Frey O, Misun PM, Fluri DA et al (2014) Reconfigurable microfluidic hanging drop network for multi-tissue interaction and analysis. Nat Commun 5:4250. https://doi.org/10.1038/ncomms5250
18. Frey O, Talaei S, van der Wal PD et al (2010) Continuous-flow multi-analyte biosensor cartridge with controllable linear response range. Lab Chip 10:2226–2234. https://doi.org/10.1039/c004851h
19. Gatenby RA, Gillies RJ (2004) Why do cancers have high aerobic glycolysis? Nat Rev Cancer 4:891–899. https://doi.org/10.1038/nrc1478
20. Ges IA, Baudenbacher F (2010) Enzyme electrodes to monitor glucose consumption of single cardiac myocytes in sub-nanoliter volumes. Biosens Bioelectron 25:1019–1024. https://doi.org/10.1016/j.bios.2009.09.018
21. Ges IA, Baudenbacher F (2010) Enzyme-coated microelectrodes to monitor lactate production in a nanoliter microfluidic cell culture device. Biosens Bioelectron 26:828–833. https://doi.org/10.1016/j.bios.2010.05.030
22. Grieshaber D, MacKenzie R, Vörös J, Reimhult E (2008) Electrochemical biosensors—sensor principles and architectures. Sensors 8:1400–1458. https://doi.org/10.3390/s8031400
23. Griffith LG, Swartz MA (2006) Capturing complex 3D tissue physiology in vitro. Nat Rev Mol Cell Biol 7:211–224. https://doi.org/10.1038/nrm1858
24. Hafner F (2000) Cytosensor Microphysiometer: technology and recent applications. Biosens Bioelectron 15:149–158. https://doi.org/10.1016/S0956-5663(00)00069-5
25. Hirschhaeuser F, Menne H, Dittfeld C et al (2010) Multicellular tumor spheroids: an underestimated tool is catching up again. J Biotechnol 148:3–15. https://doi.org/10.1016/j.jbiotec.2010.01.012
26. Huh D, Matthews BD, Mammoto A et al (2010) Reconstituting organ-level lung functions on a chip. Science 328:1662–1668. https://doi.org/10.1126/science.1188302
27. Huh D, Torisawa Y, Hamilton GA et al (2012) Microengineered physiological biomimicry: organs-on-chips. Lab Chip 12:2156–2164. https://doi.org/10.1039/c2lc40089h
28. Justice BA, Badr NA, Felder RA (2009) 3D cell culture opens new dimensions in cell-based assays. Drug Discov Today 14:102–107. https://doi.org/10.1016/j.drudis.2008.11.006
29. Kelm JM, Timmins NE, Brown CJ et al (2003) Method for generation of homogeneous multicellular tumor spheroids applicable to a wide variety of cell types. Biotechnol Bioeng 83:173–180. https://doi.org/10.1002/bit.10655
30. Kim J-Y, Fluri DA, Marchan R et al (2015) 3D spherical microtissues and microfluidic technology for multi-tissue experiments and analysis. J Biotechnol 205:24–35. https://doi.org/10.1016/j.jbiotec.2015.01.003
31. Knight E, Przyborski S (2014) Advances in 3D cell culture technologies enabling tissue-like structures to be created in vitro. J Anat. https://doi.org/10.1111/joa.12257
32. Kovarik ML, Gach PC, Ornoff DM et al (2012) Micro total analysis systems for cell biology and biochemical assays. Anal Chem 84:516–540. https://doi.org/10.1021/ac202611x
33. Lee J, Cuddihy MJ, Kotov NA (2008) Three-dimensional cell culture matrices: state of the art. Tissue Eng Part B Rev 14:61–86. https://doi.org/10.1089/teb.2007.0150
34. Link H, Fuhrer T, Gerosa L et al (2015) Real-time metabolome profiling of the metabolic switch between starvation and growth. Nat Methods 12:1091–1097. https://doi.org/10.1038/nmeth.3584
35. Lu H, Koo LY, Wang WM et al (2004) Microfluidic shear devices for quantitative analysis of cell adhesion. Anal Chem 76:5257–5264. https://doi.org/10.1021/ac049837t

36. Maschmeyer I, Lorenz AK, Schimek K et al (2015) A four-organ-chip for interconnected long-term co-culture of human intestine, liver, skin and kidney equivalents. Lab Chip 15:2688–2699. https://doi.org/10.1039/c5lc00392j
37. McKenzie JR, Palubinsky AM, Brown JE et al (2012) Metabolic multianalyte microphysiometry reveals extracellular acidosis is an essential mediator of neuronal preconditioning. ACS Chem Neurosci 3:510–518. https://doi.org/10.1021/cn300003r
38. Messner S, Agarkova I, Moritz W, Kelm JM (2013) Multi-cell type human liver microtissues for hepatotoxicity testing. Arch Toxicol 87:209–213. https://doi.org/10.1007/s00204-012-0968-2
39. Misun PM, Rothe J, Schmid YRF et al (2016) Multi-analyte biosensor interface for real-time monitoring of 3D microtissue spheroids in hanging-drop networks. Microsystems Nanoeng 2:16022. https://doi.org/10.1038/micronano.2016.22
40. Moser I, Jobst G, Urban GA (2002) Biosensor arrays for simultaneous measurement of glucose, lactate, glutamate, and glutamine. Biosens Bioelectron 17:297–302
41. Nge PN, Rogers CI, Woolley AT (2013) Advances in microfluidic materials, functions, integration, and applications. Chem Rev 113:2550–2583. https://doi.org/10.1021/cr300337x
42. Pampaloni F, Reynaud EG, Stelzer EHK (2007) The third dimension bridges the gap between cell culture and live tissue. Nat Rev Mol Cell Biol 8:839–845. https://doi.org/10.1038/nrm2236
43. Perdomo J, Hinkers H, Sundermeier C et al (2000) Miniaturized real-time monitoring system for L-lactate and glucose using microfabricated multi-enzyme sensors. Biosens Bioelectron 15:515–522
44. Pitta Bauermann L, Schuhmann W, Schulte A (2004) An advanced biological scanning electrochemical microscope (Bio-SECM) for studying individual living cells. Phys Chem Chem Phys 6:4003. https://doi.org/10.1039/b405233a
45. Polini A, Prodanov L, Bhise NS et al (2014) Organs-on-a-chip: a new tool for drug discovery. Expert Opin Drug Discov 9:335–352. https://doi.org/10.1517/17460441.2014.886562
46. Powers MJ, Janigian DM, Wack KE et al (2002) Functional behavior of primary rat liver cells in a three-dimensional perfused microarray bioreactor. Tissue Eng 8:499–513. https://doi.org/10.1089/107632702760184745
47. Rimann M, Laternser S, Gvozdenovic A et al (2014) An in vitro osteosarcoma 3D microtissue model for drug development. J Biotechnol 189:129–135. https://doi.org/10.1016/j.jbiotec.2014.09.005
48. Rothe J, Frey O, Stettler A et al (2014) Fully integrated CMOS microsystem for electrochemical measurements on 32 × 32 working electrodes at 90 frames per second. Anal Chem 86:6425–6432. https://doi.org/10.1021/ac500862v
49. Satoh W, Hosono H, Yokomaku H et al (2008) Integrated electrochemical analysis system with microfluidic and sensing functions. Sensors 8:1111–1127. https://doi.org/10.3390/s8021111
50. Sonntag F, Schilling N, Mader K et al (2010) Design and prototyping of a chip-based multi-micro-organoid culture system for substance testing, predictive to human (substance) exposure. J Biotechnol 148:70–75. https://doi.org/10.1016/j.jbiotec.2010.02.001
51. Stucki AO, Stucki JD, Hall SRR et al (2015) A lung-on-a-chip array with an integrated bio-inspired respiration mechanism. Lab Chip 15:1302–1310. https://doi.org/10.1039/c4lc01252f
52. Sung JH, Kam C, Shuler ML (2010) A microfluidic device for a pharmacokinetic-pharmacodynamic (PK-PD) model on a chip. Lab Chip 10:446–455. https://doi.org/10.1039/b917763a

53. Talaei S, van der Wal PD, Ahmed S et al (2015) Enzyme SU-8 microreactors: simple tools for cell-culture monitoring. Microfluid Nanofluidics 19:351–361. https://doi.org/10.1007/s10404-015-1562-8
54. Tibbitt MW, Anseth KS (2009) Hydrogels as extracellular matrix mimics for 3D cell culture. Biotechnol Bioeng 103:655–663. https://doi.org/10.1002/bit.22361
55. Timmins NE, Dietmair S, Nielsen LK (2004) Hanging-drop multicellular spheroids as a model of tumour angiogenesis. Angiogenesis 7:97–103. https://doi.org/10.1007/s10456-004-8911-7
56. Tung Y-C, Hsiao AY, Allen SG et al (2011) High-throughput 3D spheroid culture and drug testing using a 384 hanging drop array. Analyst 136:473–478. https://doi.org/10.1039/c0an00609b
57. Van der Meer AD, Poot AA, Feijen J, Vermes I (2010) Analyzing shear stress-induced alignment of actin filaments in endothelial cells with a microfluidic assay. Biomicrofluidics 4:11103. https://doi.org/10.1063/1.3366720
58. Van Midwoud PM, Merema MT, Verpoorte E, Groothuis GMM (2010) A microfluidic approach for in vitro assessment of interorgan interactions in drug metabolism using intestinal and liver slices. Lab Chip 10:2778–2786. https://doi.org/10.1039/c0lc00043d
59. Weltin A, Slotwinski K, Kieninger J et al (2014) Cell culture monitoring for drug screening and cancer research: a transparent, microfluidic, multi-sensor microsystem. Lab Chip 14: 138–146. https://doi.org/10.1039/c3lc50759a
60. Young EWK, Beebe DJ (2010) Fundamentals of microfluidic cell culture in controlled microenvironments. Chem Soc Rev 39:1036–1048. https://doi.org/10.1039/b909900j
61. Zhang C, Zhao Z, Abdul Rahim NA et al (2009) Towards a human-on-chip: culturing multiple cell types on a chip with compartmentalized microenvironments. Lab Chip 9: 3185–3192. https://doi.org/10.1039/b915147h

Index

A
Active mixing, 72
Alloy, 27–29, 31, 32, 34

B
Biochemical gradient, 208
Biomolecule separation, 11
Biosensing, 62, 63, 70, 72, 75, 76, 80, 81, 87, 90, 93, 98, 99, 107, 112, 117, 119, 120, 123, 142, 143, 171, 172, 177, 178, 180, 198
Biosensor, 69–73, 75–81, 86–88, 90–93, 98, 99
Body-on-a-chip, 207, 208, 210

C
Cancer biomarker, 61
CCD, 49, 50, 56, 128, 147–150, 152, 156, 164–166, 194–196
Cell metabolism, 209
Cell-to-cell interaction, 208
Chaotic mixing, 81, 82, 86
Chemiluminescence detector, 195, 196
Cholesterol, 134
Circulating tumor cell, 69, 72, 73
CMOS, 147–150, 156, 164–166, 195, 215
COMSOL, 78, 81, 82, 89, 117, 127, 132, 133, 138
Conductivity, 14, 15, 106, 125, 128, 129, 140, 141
Contact angle, 115, 173, 174, 182–184, 195
Convolution, 155

D
3D cell culturing, 208, 209
Dielectrophoresis DEP, 106, 124, 173, 188
Diffusion, 2, 4, 5, 11, 13, 15, 16, 33, 46, 48, 71, 73, 74, 76–79, 87, 97, 106, 108, 113, 114, 123, 124, 126, 130, 131, 138, 139, 141, 217, 220
Digital Holographic Microscopy (DHM), 148, 150, 153, 154, 162, 166
Digital microfluidics, 171, 172, 181, 190, 198
3D microtissue, 207, 208
DNA hybridization, 16, 27, 46–48, 51, 59, 63, 196
Dopamine, 42, 45, 180, 197
Droplet, 52, 110, 111, 113, 114, 120, 136, 172–176, 182–189, 191–193, 195, 196, 199–201
Droplet actuation, 172, 183, 184, 195
3D tissue culturing, 207, 209, 210

E
Electrochemical sensing, 198, 199, 209
Electrode, 15, 120, 125–143, 172– 176, 180–182, 185–188, 194, 195, 198, 200, 209, 212, 213, 215–221
Electrohydrodynamics, 126, 186
Electrokinetics, 70, 124, 131
Electrophoresis, 13
Electrowetting, 173, 183
Environmental monitoring, 63, 70
Evaporation, 52, 106–108, 110, 111, 113–115, 119, 124, 128, 133, 175, 215
Evaporation-driven flow, 107, 109
Extinction, 27, 36–38, 41
Extraordinary optical transmission, 126

F
Fast Fourier Transform (FFT), 164
Field enhancement, 38, 39, 61
Field gradient, 106, 115–117, 127, 128, 132
Filtering, 148, 152, 154, 166
Filtration, 1
Finite Difference Time Domain (FDTD), 38, 128
Flow-through nanohole array, 126, 131, 132

S.-H. Oh et al. (eds.), *Miniature Fluidic Devices for Rapid Biological Detection*, Integrated Analytical Systems, https://doi.org/10.1007/978-3-319-64747-0

Fluorescence, 14, 26, 35, 38, 40, 46, 58, 60–62, 90, 91, 94, 105, 107, 109, 134–136, 140, 177, 192–194
Fourier transform, 155, 156
Fresnel, 155

G
Gold disks, 28, 31, 33, 60, 61
Gradient force, 123, 132
Guanine quadruplex (G4), 55

H
Hairpin probe, 47, 60–62
Hanging microfluidics, 98, 99, 106, 111, 207, 209, 214–216, 219
Hindered transport of molecules, 2
Hologram, 147–162, 164–166
Holography, 147–151, 153, 154, 158
Huygens-Fresnel, 156
Hydrodynamics, 3, 180

I
In-plane resonance, 36–38, 41
Integrated detection system, 171
Interface, 25, 107, 110, 113, 123, 125, 135, 173, 174, 178, 208, 209, 211, 212
Interference, 8, 9, 148–153, 162
Ionic transport, 11

L
Label-free, 15, 16, 27, 33, 42, 46, 47, 52, 55, 58, 59, 63, 108, 114, 137, 140, 177
Lab-on-a-chip, 2, 18, 149, 171, 172, 190
Laminar flow, 33, 72, 82
Localized Surface Plasmon Resonance (LSPR), 25, 36
LOD, 42, 45, 62, 131

M
Malachite Green (MG), 28, 55, 63
Mass spectrometry, 178–180, 190, 191, 193, 194
MATLAB, 128, 149, 164
Membrane fabrication, 5, 7, 8
Membrane functionalization, 10, 12, 17, 63, 119, 215, 221
Metallic nanoparticle, 25
Microelectrode, 137, 212
Microfluidic actuation, 172, 175, 182, 185
Microfluidic device, 18, 27, 32, 34, 42, 63, 188, 194, 218
Microfluidic hanging drop networks, 207, 209, 210–212, 221
Microfluidics, 33, 69, 113, 120, 147, 149, 171, 172, 193, 207–209
Micromixer, 69, 72, 81, 90, 91, 94
Microscopy, 29, 60, 94, 132, 134–136, 147, 148, 153, 162, 165, 177, 187, 190–192, 211, 214
Microsensor, 209
Mixing, 33, 34, 53, 69, 71, 72, 77, 79, 81–83, 85–87, 91–93, 98, 99, 211
Molecule separation, 1, 2, 13, 16
Multiplexed sensing, 33, 53
Multiplexing, 63, 209

N
Nanochannel, 18, 106
Nanofabrication, 5, 8
Nanogap, 124, 137–142
Nanohole, 107–114, 124, 126–128, 132, 142
Nanoimprinting, 112
Nanophotonics, 142
Nanoplasmonics, 62, 63, 110, 112, 124, 132, 136, 137, 142
Nanopore, 1, 2, 5, 11, 16–18, 35, 107, 110, 113, 124
Nanopore array, 16, 110
Nanopore membrane, 1, 2, 11, 13, 16–18
Nanoporous gold, 26, 29, 31, 33–35, 40, 42, 63
Nanosphere lithography, 28, 34
Numerical Aperture (NA), 158

O
Optical imaging, 149
Out-of-plane resonance, 36, 37, 41

P
Passive mixing, 72
Phase, 25, 48, 49, 78, 79, 83, 147–150, 156, 161, 162, 164, 210, 211, 219
Plasmonics, 16, 17, 25–28, 33, 36–38, 41, 63, 105, 107, 108, 112–119, 124, 130, 132, 134, 136, 138, 142
Point-of-care detection, 32, 63, 120, 193
Poly-dimethylsiloxane (PDMS), 6, 33–35, 37, 52, 62, 94, 211–216
Porosity, 2, 4, 6, 7, 13, 31
Preconcentration, 124
Pyramid, 108, 116, 117, 132

R
Raman, 26, 33, 34, 38–42, 45, 48, 49, 53, 57, 58, 105, 117– 119, 132, 134–137, 143
Real-time detection, 16, 107, 129, 135, 148, 171

Reconstruction, 148, 149, 153–156, 161, 162, 166
Reference beam, 148–151, 153–155, 158, 160–162
Refractive index, 16, 26, 36, 41, 54, 105, 108, 115, 123, 138, 149, 162
Rhodamine 6G (R6G), 35, 42, 60

S
Self-assembly, 61
Shannon-Nyquist sampling criterion, 150
Spatial frequency, 148, 150, 157, 159
Spectrum, 26, 27, 40, 49, 54, 62, 114, 127–130, 134, 148, 155, 156, 160, 164, 177, 193, 208
Spheroid tissue culture, 209, 215, 221
Surface-Enhanced Infrared Absorption (SEIRA), 26
Surface-Enhanced Near-Infrared Absorption (SENIRA), 26
Surface functionalization, 2, 10, 63, 119
Surface plasmon, 14, 16, 25, 26, 35, 38, 90, 110, 116, 123, 132, 177, 178, 191, 196
Surface Plasmon Resonance (SPR), 16, 26, 36, 110, 123, 177, 191, 196
Surface tension, 106, 113, 114, 119, 191, 210
Synthesis, 25

T
Tip, 90, 116, 117, 132–136, 189
Tissue engineering, 207
Tissue perfusion, 208, 209, 220
Transport, 2–4, 10, 13, 14, 69, 127, 130, 131, 172, 175, 176, 183, 184, 186, 189, 191
Transport phenomena, 84, 98
Trapping, 33, 114, 128, 130, 132, 134, 137–139, 142

U
Urea, 45, 46, 52–54

V
Vesicle, 124, 134, 135, 137, 143

W
Wavefront, 149, 150, 153–156
Wavelet, 156

MIX
Papier aus verantwortungsvollen Quellen
Paper from responsible sources
FSC® C105338

If you have any concerns about our products,
you can contact us on
ProductSafety@springernature.com

In case Publisher is established outside the EU,
the EU authorized representative is:
Springer Nature Customer Service Center GmbH
Europaplatz 3, 69115 Heidelberg, Germany

Printed by Libri Plureos GmbH
in Hamburg, Germany